Progress in Hybrid RANS-LES Modelling

Papers Contributed to the 3rd Symposium on Hybrid RANS-LES Methods, Gdansk, Poland, June 2009

Shia-Hui Peng
Piotr Doerffer
Werner Haase
(Editors)

Springer

Prof. Shia-Hui Peng
Department of Aeronautics and
Systems Technology
Swedish Defence Research Agency, FOI
SE-164 90 Stockholm
Sweden
E-mail: peng@foi.se

and

Department of Applied Mechanics
Chalmers University of Technology
SE-412 96 Göteborg
Sweden
E-mail: peng@chalmers.se

Prof. Piotr Doerffer
Transonic Flows and Numerical Methods
Institute of Fluid Flow Machinery
Polish Academy of Sciences
Fiszera 14
80-952 Gdansk
Poland
E-mail: doerffer@imp.gda.pl

Dr. Werner Haase
Höhenkirchener Str. 19 d
85662 Hohenbrunn
Germany
E-mail: office@haa.se

ISBN 978-3-642-14167-6 e-ISBN 978-3-642-14168-3

DOI 10.1007/978-3-642-14168-3

Notes on Numerical Fluid Mechanics
and Multidisciplinary Design ISSN 1612-2909

Library of Congress Control Number: 2010932017

Typeset & Cover Design: Scientific Publishing Services Pvt. Ltd., Chennai, India.

Printed on acid-free paper

5 4 3 2 1 0

springer.com

NNFM Editor Addresses

Preface

Hybrid modelling of turbulent flows, combining RANS and LES techniques, has received increasing attention over the past decade to fill the gap between (U)RANS and LES computations in aerodynamic applications at industrially relevant Reynolds numbers. With the advantage of hybrid RANS-LES modelling approaches, being considerably more computationally efficient than full LES and more accurate than (U)RANS, particularly for unsteady aerodynamic flows, has motivated numerous research and development activities. These activities have been increasingly stimulated by the provision of modern computing facilities.

The present book contains the contributions presented at the Third Symposium on Hybrid RANS-LES Methods, held in Gdansk, Poland, 10-12 June 2009. To a certain extent, this conference was a continuation of the first symposium taking place in Stockholm (Sweden, 2005) and the second in Corfu (Greece, 2007). Motivated by the extensive interest in the research community, the papers presented at the Corfu symposium were published by Springer in the book entitled "*Advances in Hybrid RANS-LES Modelling*" (in Notes on Numerical Fluid Mechanics and Multidisciplinary Design, Vol. 97).

At the Gdansk symposium, along with four invited keynotes, given respectively by S. Fu, U. Michel, M. Sillen and P. Spalart, another 28 papers were presented on the following topics: *Unsteady RANS, LES, Improved DES Methods, Hybrid RANS-LES Methods, DES versus URANS and other Hybrid Methods, Modelling-related Numerical Issues* and *Industrial Applications*. After the symposium all full papers have been further reviewed and revised for publication in the present book. In addition, 2 presentations have been adapted as short communications and included also in the book. To highlight different views on "how CFD will proceed", in particular on how hybrid RANS-LES methods will make their way in both scientific and industrial environments, a lively round-table discussion took place during the symposium. A conclusive summary of this discussion, being of interest to researchers and engineers working on CFD and turbulence modelling, is preceding the full papers – reflecting views from different experts on the development and application of hybrid RANS-LES methods.

The present book communicates recent activities and new progress on hybrid RANS-LES methods, contributed by a number of leading academic researchers and industrial experts in this field. The book is a useful reference for researchers, university graduate students, industrial engineers, and intended to facilitate further development and applications of advanced RANS-LES modelling approaches for both fundamental and industrial flow problems.

The publication of the book is a result of a collective effort, primarily by the participants in the Gdansk symposium, the paper contributors and the reviewers.

The editors are grateful for the excellent organization of the conference and we thank all colleagues of the Institute of Fluid Flow Machinery of the Polish Academy of Sciences in Gdansk for making the conference a real success.

Finally, we wish to express our sincere gratefulness for the valuable sponsorship to the symposium, provided kindly by ANSYS, ERCOFTAC, EUROCOPTER, INTEL and NUMECA, as well as by the two European Union projects ATAAC and KATnet-II.

April 2010

Shia-Hui Peng
Piotr Doerffer
Werner Haase

Contents

Short Communications

Hybrid RANS-LES Modelling: How to Proceed? The Round Table Forum Discussion

Minuted and compiled by C. Mockett

1 Introduction

The second day of the 3rd Symposium on Hybrid RANS-LES Methods closed with a panel discussion on the state of affairs in hybrid RANS-LES methods and future perspectives. The discussion was lively, informative, wide in scope and, perhaps necessarily, unstructured. The approach taken for this written summary has therefore been to identify some key recurring themes rather than to provide a chronological account. As such, most comments are paraphrased rather than direct citations. All participants were given the opportunity to contribute comments and revisions to the draft of this document.

2 Panel Members and Contributions from the Audience

The discussion was hosted and guided by Werner Haase [WH] and Shia-Hui Peng [SHP], and the invited panel members were:

- S. Fu [SF], Tsinghua University, Beijing, China
- T. Gatski [TG], Université de Poitiers, France
- F. Menter [FM], ANSYS Germany, Otterfing, Germany
- C. Mockett [CM], CFD Software GmbH & Technische Universität Berlin, Germany
- M. Sillén [MSi], SAAB AB, Linköping, Sweden
- P. Spalart [PS], Boeing Commercial Airplanes, Seattle, USA
- M. Strelets [MSt], New Technologies and Services (NTS) & St. Petersburg Technical University, Russia

For brevity, the panel members will be referred to by the initials given in square brackets above. In addition, several contributions were made from the audience:

- G. Barakos, The University of Liverpool, UK
- M. Braza, Institut de Mécanique des Fluides de Toulouse, France
- B. Chaouat, ONERA, Palaiseau, France
- L. Davidson, Chalmers University, Gothenburg, Sweden

Charles Mockett
CFD Software GmbH & Technische Universität Berlin, Institut für Strömungsmechanik und Technische Akustik, TU-Berlin, Sekr. MB1, Müller-Breslau Str. 8, D-10623 Berlin, Germany
charles.mockett@cfd.tu-berlin.de

S.-H. Peng et al. (Eds.): Progress in Hybrid RANS-LES Modelling, NNFM 111, pp. 1–6.
springerlink.com

- P. Doerffer, Polish Academy of Sciences, Gdańsk, Poland
- I. Mary, ONERA, Chatillon, France
- M. Morzyński, Poznan University of Technology, Poland
- B.R. Noack, Technische Universität Berlin, Germany
- J. Rokicki, Warsaw University of Technology, Poland
- S. Rolfo, The University of Manchester, UK

M. Leschziner of Imperial College, London, UK, who was not present in person, also contributed some slides of a technical nature to the discussion.

3 Status of Hybrid RANS-LES Methods around the World

One key outcome of the discussion was a snapshot of the general state-of-affairs in hybrid RANS-LES research and funding around the world. Concerning funding, it was reported that after decades of exaggerated promises in turbulence modelling proposals, those in charge of funding decisions are increasingly demanding results [WH, M. Leschziner]. This underlying frustration was expressed in terms of repeatedly "seeing light at the end of the tunnel, before the roof caves in" [WH]. Encouragement was given to separate technical and political/funding-related issues [P. Doerffer].

A generally encouraging picture of the level of **proliferation of hybrid methods** in industry was given, drawing on the example of the long-running cooperation between Boeing and NTS [PS, MSt]. In particular, LES was successfully applied in the design of a quieter jet engine nozzle and DES is already being applied to smaller industrial problems (the example of an aeroacoustic simulation of a flap edge was given) [PS]. Hybrid methods are now routinely available in commercial CFD software [FM, MSt] and chapters on hybrid modelling are appearing in textbooks [MSt]. The series of EU projects pursuing improvements in hybrid modelling (FLOMANIA, DESider, ATAAC) were widely assessed to have played an important role in the industrialisation and propagation of hybrid methods [PS, MSt]. However, alongside these positive examples, some central problems were described concerning a perceived increase in the level of expertise demanded of industrial users (relative to RANS technology), expertise- and cost-based barriers to the widespread adoption of hybrid methods by industry and a resulting gap between research and industry [MSi]. These topics are dealt with in the subsequent sections.

Concerning progress in **method development**, generally positive opinions were expressed [TG, FM, MSt], for example the acknowledgement that "we now know much more than we did in 1997" [MSt]. A general consensus was evident that the trend towards unsteady methods is considered essential in tackling a wide range of problems. Although many developments in the interfacing of RANS and LES have been proposed, it is often hard to identify the influence that these have on the quality of results. This is primarily due to the difficulties in separating multiple sources of error [I. Mary]. The significant efforts of collaborative research (such as the EU initiatives) on this front was emphasized [WH, L. Davidson], although the remaining importance of this problem was acknowledged. A listing of various hybridisation strategies and concrete conceptual issues concerning the formulation of hybrid RANS-LES methods in a wall-modelled LES (WMLES) setting were identified in the slides of M. Leschziner. Among these, a contrast between "top-down" and "bottom-up" modelling strategies (in terms of the importance of unsteady effects near the wall) was outlined.

The levels of **activity in hybrid methods around the world** could be approximately gauged on the basis of the international panel and audience. The high levels of activity in the USA and Europe (in particular the European collaborative projects) have already been touched upon. The activity in Russia was also estimated at a high level (MSt). Concerning China, the general situation of an existing expertise lag and a determined effort to reduce this was portrayed [SF].

Although the majority of discussion concerned hybrid methods in particular, the discussion inevitably widened to **the status of CFD in general**. The fundamental importance of CFD was expressed, particularly with regard to the uncertainties and expense inherent in experimental techniques (in the sense of an acknowledgement that both CFD and experiments are imperfect) [PS, MSt]. The general reliance on experiments was assessed as higher in China, although the trend towards more simulation is also underway there [SF]. The closing gap between CFD and experiment was also noted, drawing on examples of CFD application in the definition of experiments and on the increasing trend to redefine the purpose of many experiments for CFD validation [FM, PS, M. Braza]. A key problem in industrial CFD was identified in the sense that grid-converged RANS solutions are still out of reach on cost grounds, for instance in the case of an aircraft with high-lift devices [PS]. Attention was drawn to the high degree of qualitative insight that CFD can already offer on a standard desktop PC [G. Barakos].

4 A Widening Gap between Research and Industry?

As briefly mentioned above, a gap between research and industry is perceived by some to be widening on two key fronts, namely specialist expertise and the availability of advanced methods [WH, MSi]. The existence of such a gap was pointed out to be necessary and indeed an indicator that research is taking place at all. It was however acknowledged that this gap should not grow too large [FM]. A number of counter-examples were given, including the European research programs in which the concept of "industrial observers" has had a positive impact [WH] and direct bilateral partnerships between industry and academia/research [PS, MSt, M. Braza].

The level of specialist expertise required is covered in depth in the next section, however some key contributors to the **"expertise gap"** were identified. Firstly, it was widely acknowledged that the quality of new recruits to industry is diminishing with regard to an understanding of the workings of CFD, writing code and even the equations of fluid mechanics [SF, MSi, PS]. At least a partial cause of this was believed to be the proliferation of commercial software in university teaching, which isolates the student from the inner workings of the solver and encourages the chaining together of such black boxes in a "package approach" [SF, M. Leschziner, S. Rolfo]. A growing tendency away from a necessarily critical interpretation of CFD results was also noted [SF, M. Leschziner]. Secondly, increasing time pressure on industrial engineers enhances the problem of inertia whereby established methods (i.e. RANS) are maintained rather than adopting newer approaches (i.e. hybrid RANS-LES) [MSi, PS].

5 The Level of User Expertise Required for Industrial Application

Although the requirements of a high level of user expertise is an established feature of CFD as a whole ("It's easy to start the engine, but much more difficult to drive." [TG]), the necessary expertise level is seen to be higher still for hybrid RANS-LES methods. This prompted a range of questions by the hosts aiming to draw opinions on the role of best practice guidelines (BPG).

From the point of view of a commercial software vendor, the complexity of CFD and the required expertise are inherent: It is clear that we are far away from a "push-button utopia". Within an industry where the same tools are available to all users of commercial software, this presents an opportunity whereby expertise represents a **competitive advantage**. The observation that certain customers of CFD software consistently produce better results was cited as evidence of this effect [FM].

Current efforts at Boeing to automate the CFD process and reduce the demand placed on users were described, the general approach of which is to incorporate BPG directly in the software. With reference to the comments of the previous paragraph, this **expert system approach** is expected to consolidate and enhance the in-house competitive advantage. Adaptive grids are one key element of this strategy. Although the separation of the user from the inner workings of the solver was widely seen as a problem with commercial software (see previous section), the argument was made that the user should ideally not need to be concerned with such issues. An illustrative analogy with mobile telephones was made: We all successfully make use of them without understanding how they work [PS].

The importance of reliable **adaptive grid strategies** was a widely-held viewpoint [FM, PS, MSt, M. Braza, J. Rokicki], as was a belief that this is a greater challenge for unsteady and LES-based methods than it is for RANS. Adaptation for DES was declared as an ambition, towards which initial steps are underway [PS]. Some success in adaptive grids for RANS was reported, including efforts within the EU ADIGMA project [J. Rokicki] and a 20% improvement in the prediction of drag [M. Braza].

6 Prospects for a "Universal Model"?

A natural outcome of the high level of activity and promise in hybrid method development has been a rapid growth in the number of different formulations proposed, which leads to a bewildering situation for the industrial end-user tasked with selecting an approach for their particular application. Indeed, this mimics the situation in the RANS modelling community of the 1980's and 1990's. Nonetheless, following the initial growth phase (in terms of new proposals) a consolidation towards a small number of industry standards occurred, for example the Spalart-Allmaras and Menter SST models in aerodynamics [SHP]. Indeed, the EU ATAAC project, which started recently, has the stated goal of establishing a small number of recommended RSM and hybrid methods [WH]. This was the motivation for a discussion concerning the prospects for "a universal model".

The concern was voiced that such a "one-model vision" would serve to constrain creativity [P. Doerffer]. Many furthermore expressed the opinion that **a spectrum of approaches** appears to be a necessary feature of hybrid RANS-LES [FM, CM, B. Chaouat]: It is already clear that different families of hybrid RANS-LES methods are forming [FM, CM] each with their respective inherent performance and expense. The most appropriate approach for a particular application depends on the expectations of the engineer and the computational resources he or she is prepared to expend on the problem [B. Chaouat]. Considering a general case of mixed attached and separated turbulent flow, some principle approaches were ranked in order of increasing numerical expense and predictive accuracy as follows: URANS, DES/SAS, WMLES, LES [FM, CM]. Classification of the methods based on their ambitions rather than their formulation was an approach suggested for such analysis. The example of an external aerodynamic flow with large areas of thin, attached boundary layers was cited as a case where DES and SAS are currently the only applicable types of hybrid approach on grounds of computational expense, even though better performance would be expected from WMLES [CM]. As a result, it is clear that the **role of RANS in hybrid RANS-LES** will continue to remain important for many applications, since approaches such as DES and SAS that are "at the mercy" of the underlying RANS model to determine the separation location [SF, CM, M. Leschziner]. The user is hence presented with two near-orthogonal choices, namely the selection of the RANS model and the selection of a hybridisation strategy [PS].

Some discussion regarding the **validity of URANS** was made. URANS was rejected by some on formal grounds (namely, the conceptual difficulty of applying a method calibrated for steady-state flows to the unsteady solutions) [TG, M. Leschziner]. Others however made comments supporting its necessity, which were founded on considerations of computational expense together with the presence of clearly unsteady processes in practical applications (the examples of transonic buffet [PS] and aeroelasticity [M. Morzyński] were cited).

Despite this necessity to maintain a varied toolbox of hybrid methods, the **process of consolidation** within each class of approach was felt to be an important goal. The comments referred to earlier concerning the profound difficulties of assessing the most optimal from a number of different hybrid approaches [I. Mary] represent a fundamental hurdle for such efforts. A further difficulty is seen in that such consolidation runs contrary to the human-nature urge to get one's name on the map [PS].

7 A More Inter-disciplinary Approach?

The suggestion was made that the field of hybrid RANS-LES method development (and indeed turbulence modelling in general) could benefit from the involvement of specialists from other disciplines than numerical fluid mechanics. In particular, the belief that statistical physics could offer an alternative to the eddy viscosity ansatz was expressed. An analogy to the development of feedback control was made, whereby the "desolate situation" ten years ago was portrayed. The involvement of many different disciplines resulted in the establishment of a whole industry of feedback control utilising low-order models. Drawing analogy to the industrial observers in EU projects, the suggestion was made that "outsider observers" could be of benefit [B. Noack]. This sentiment received some support and further examples of successful inter-disciplinary research and cross-fertilisation were given [M. Braza].

A key difficulty was however identified regarding the issue of **communication**. One commentator expressed his difficulty in understanding theoreticians despite his own background in fundamental physics. As such, it was warned that the kind of outsider perspectives proposed can also have the adverse affect of increasing confusion and ultimately hindering progress [TG]. In this spirit, the observation was made that even within the hybrid RANS-LES community we have surprising difficulty understanding each other [CM]. Inherent similarities with the situation in RANS modelling 30 years ago were identified: A difference in perspective exists between engineers/scientists who apply and those who develop models. The latter group can furthermore be subdivided into those pursuing first principles approaches and those employing phenomenological arguments [TG].

An important goal of thematically-specific meetings such as the current symposium was reiterated in terms of bringing us closer together [WH], encouraging the presentation of clear descriptions of the methods as well as their cross-use within the community [PS].

Reflections on RANS Modelling

P. Spalart

Abstract. The article begins with a reminder that the need for RANS modelling in 2009 is not any lower than had been predicted in 1997, when Detached-Eddy Simulation was created against a wave of predictions that Large-Eddy Simulation (LES) would soon make RANS redundant. The CFD community, in large part, has embraced DES and other hybrid RANS-LES approaches for flows other than thin shear layers, while noting their high cost and the uncertainty in arriving at the right numerical resolution and estimating the accuracy of the results. However these approaches leave RANS in charge of the boundary layers. In addition, wall modelling below an LES has the nature of RANS modelling. In short, RANS is in great demand.

Key points are then drawn from a special RANS issue of International Journal of CFD, with the author as guest editor. The articles confirmed the static state of modelling for practical applications since the mid-1990's, and the near-failure of efforts aimed at making RANS modelling more rigorous, using math or DNS data. They also show that in the applied community pure RANS will be tried for many flows, even those where there is little hope for success, just because of computing cost. This is especially the case for a turbulent flow region that is a small part of a larger system; say the antenna on a car. Therefore, even knowing the strong misgivings which apply to Unsteady RANS (URANS) on conceptual grounds and based on poor results with bluff-body flows, there is reason to continue exploring URANS in 2D and 3D, and there would be great value in RANS models which perform better with it. The meaning of Steady RANS solutions in flows which many observers consider "too unsteady" is discussed, using examples such as the circular cylinder and airfoils at very high angles of attack.

Attention then shifts to the treatment of laminar regions in RANS CFD, with emphasis on proper and reproducible use of the models, without hidden influences of inflow values, numerics, or domain sizes. The promise of transition prediction by local equations is examined.

Finally, the issue of acceptance of new RANS models is addressed in terms of how well the scientific publication system and the code industry are prepared to recognize and implant the next good idea in RANS modelling, especially if it comes from a less-established source without a long record and strong resources, and/or if it conflicts with "common wisdom."

P. Spalart
Boeing Commercial Airplanes, USA
philippe.r.spalart@boeing.com

S.-H. Peng et al. (Eds.): Progress in Hybrid RANS-LES Modelling, NNFM 111, pp. 7–24.
springerlink.com

1 RANS still in High Demand

In the 1980's and 1990's there was hope that rapid improvements in computer power and algorithms would allow Large-Eddy Simulation (LES) to displace RANS modelling, which was and is perceived as crude and limited. The move would be away from empiricism, and towards first principles and a sound response to influences such as rotation and curvature, as well as to grid refinement. These are all desirable. A voice of dissent was the paper by Spalart, Jou, Strelets and Allmaras (Spalart et al., 1997 - SJSA) which contained explicit estimates of the number of grid points needed for LES of a complete wing, and the prediction that it would be possible only around 2045, even as a "Grand Challenge". It is too early to prove or disprove this prediction, but it is a fact that in 2009 the wing LES has not been performed. At best, simulations of a narrow slice of wing, at moderate Reynolds numbers, have been possible. Note that the SJSA estimate was generous since it assumed only 20 points in each direction for each cube of size δ in the boundary layer, δ being its thickness. No breakthrough has occurred in LES since 1997 which would make this resolution "overkill," or even truly sufficient. Wall modelling also remains imperfect, and is crucially needed in the 20^3 framework. Other flows of great value than the wing, say over an automobile, may yield a little sooner thanks to lower Reynolds numbers, but pure LES is still many years away from engineering practice. Not only is it too expensive by orders of magnitude, but conclusive grid-refinement studies are essentially impossible for both cost and conceptual reasons, so that establishing numerical reliability is most difficult. We are unable to produce the "error bar," which is disastrous in engineering.

At the same time, the SJSA proposal of combining RANS and LES has been quite successful, and the current wave of methods for separated flows is made of hybrids between the two: DES, LNS, PANS, FSM, SAS, XLES, DDES, OES, ZDES, IDDES, and so on. All these are in principle capable of the mission of what has been called Natural DES, namely, to treat the boundary layer with RANS and the massively-separated region with LES. Another mission is Wall-Modelled LES (WMLES) in which the RANS region is a very thin layer at the wall, and the LES region includes the outer part of the boundary layer. In fact WMLES goes back to the LES pioneers of the 1970's (e.g., Deardorff and Schumann), and subsequent LES conducted without RANS-like wall modelling, or "Quasi-Direct Numerical Simulation," has been criticized (by this author) as a corruption of the LES concept. Note that although a wall model in WMLES has RANS nature, it does not need to contain a complete self-sufficient RANS model, as is needed in Natural DES. Much like a Sub-Grid Scale (SGS) model, a wall model can and should be simple.

Thus the forces which were seen in 1997 as guaranteeing the place of RANS in CFD for decades are still in place. On the other hand, the old complaints about accuracy and lack of intellectual appeal of RANS have not been erased, and if anything they appear worse now in that progress away from empiricism and/or towards precise use of DNS and other new data has been slim. In addition, with CFD increasingly used in engineering practice, it must be said that although

RANS CFD makes grid convergence possible in theory, it is almost never attained in practice, or even at the Drag Prediction Workshops (see Morrison and Helmsch, 2007) with seemingly simple wing-fuselage cases and a large set of very motivated participants. Further, the physical modelling errors are essentially impossible to bound and therefore, RANS CFD is unfortunately *also* unable to produce its own error bar. Most journals ask reviewers a question such as "is the accuracy of experimental or numerical results adequately demonstrated?" and to be rigorous, we would have to stop almost all manuscripts involving turbulent CFD. Regardless of the approach, it is not possible to assert directly that a quantity or another is accurate to within 1%, say, or 5%. Only extensive experience with almost identical flows calculated with almost identical grids, for instance high-speed wings, can bring the confidence in quantitative results to such levels.

2 Special Issue of International Journal of CFD

The author was asked by Prof. Habashi to edit a special issue of his journal devoted to RANS (International Journal of Computational Fluid Dynamics, 2009, v. 23, No. 4). The articles confirmed the absence of definite progress since the 1990's, when transport-equation models with better separation prediction than that from k-ε were introduced. They also confirmed the permanence of RANS for almost all applications.

The reasons for slow progress are many. There is a lack of ideas that are both new and "fortunate." Direct Numerical Simulation (DNS) has provided quality data, but not enough with sufficient Reynolds number and/or pressure gradients. In addition, DNS has provided a "moving target" for refined modelling of the near-wall Reynolds stresses. The reason is that models usually lead to a "law of the wall" for these quantities, giving them a unique behaviour in wall units, but DNS and true turbulence do not. Therefore, a model can achieve excellent agreement on stresses near the wall at one Reynolds number, and not others; in addition, pressure gradients cause even stronger conflicts.

Also at play is the general attitude of modellers that their genuine goal is a single model, as universal as possible within a given class (e.g., one-equation, LEVM, NLEVM, EARSM, and so on). They refuse to specialize their models much at all, and therefore to satisfy one clientele with one version, and another clientele with another version (some will propose two or more models, but in different classes). As an example, they demand that compressibility or curvature corrections be backward-compatible. This makes the intellectual problem of finalizing a "new and improved" version of any model this much harder. Actually, this is excellent discipline; for a while at least, there was a near-industry for some models of "creating a new version per year to deal with a new flow, and travel to a new conference."

Modellers are often asking the community to describe the "principal flaw" in their model: the first one they should address. As mentioned, separation prediction brings less frustration now than with the Baldwin-Lomax or k-ε models, without of course being perfect. The same dynamics of the models controls shock-wave position on high-speed wings. The Special Issue of IJCFD is "rewarding" in that two weaknesses are mentioned which seem to run across models.

The first weakness is too low a level of Reynolds shear stress and mixing after separation; it is seen early in the separated shear layer, and also later in the reattaching layer (Hirsch and Tartinville, 2009; Menter, 2009; Rumsey and Swanson, 2009). The models tend to assimilate the former to a standard mixing layer, and to miss an enhanced level of turbulence which *may* be caused by the neighbouring thick turbulent layer, inherited from the outer region of the boundary layer. The models assimilate the latter flow to a nearly-standard wall layer, and the skin friction is too weak. In both cases, LES brings an improvement, but not perfect agreement with experiment. The experiments are also imperfect; however, the trend is clear.

The second problem area is in corner flows with adverse pressure gradients, typically on airfoils; certainly in wind tunnels, and probably over wing-fuselage junctions. RANS predicts excessive separation, sometimes by very much (Hellsten and Wallin, 2009; Hirsch and Tartinville, 2009; Menter, 2009). Observers tend to blame incorrect stress-tensor anisotropy in linear eddy-viscosity models (LEVM). The reason is that the correct anisotropy creates Secondary Flows of the Second Kind, which propel some fluid towards the corner, for instance in a square duct. Intuitively, this would help reduce separation in corners; however, there is no guarantee whatsoever that the effect is strong enough. It can well be that the corner flow and diffusion from one boundary layer to the other creates the wrong profile for eddy viscosity. Indeed, informal discussions at the Gdansk meeting suggest that non-linear models (NLEVM) do not solve the problem. Stress-transport models will be tried, as will LES. DNS with very good unsteady content in the incoming boundary layers is in order. This does not mean that a perfect corner-flow database would bring about progress in modelling, not knowing how deeply the structure of the models would have to change, if at all. It is possible that a clever sensitisation of simple models to the "corner situation" is all that is needed.

The wing-fuselage corner flow brings up the unfortunate fact that numerical grid convergence is not achieved today, even with grid counts of the order of 10^8 for a half-airplane and structured grids (usually more accurate than unstructured ones). In addition, odd effects are observed, including a fortuitous improvement when using the thin-layer Navier-Stokes approximation (which vaguely supports the conjecture of an unfavourable interference between the two boundary layers) or surprising changes when the correct trailing-edge bluntness is included in the geometry. These strong residual numerical effects still give RANS modellers "a place to hide" from criticism. It is clear that, with grids acceptable today, 3D cases with high-lift devices cannot be grid-converged; an issue which will be clarified with workshops on semi-complex geometries such as the Trapezoidal Wing in the coming years.

3 Steady and Unsteady RANS for Bluff Bodies

Figure 1 illustrates a challenge to CFD which will be long-lived, and highly justified. The reduction of landing-gear noise will be pursued for decades, just as much as the reduction of jet or prop-fan noise, and other airframe noise such as that from flap edges. Unsteady simulations will be instrumental. The photo shows the ultimate goal: a fully-dressed system. The high-frequency but audible noise comes from the small elements and their interactions with other elements. The other picture shows a generic geometry designed by the author as an exercise for unsteady CFD over the next five years or so, with the charter that reasonable agreement between codes would be obtained, although probably not full "grid convergence." Some measures of the flow such as mean and rms wall pressures may agree well, internally and with experiment. Spectra and radiated noise may well lead to (instructive) disagreements. In any case, the geometry is such that all its elements, namely the wheels and the axles and post, can be treated by turbulence-resolving simulation, in simple terms, they can be in the LES region of a hybrid method. This would be within current grid-count limits for LES content, a few times 10^7.

Fig. 1 Landing-gear geometry. Left, ultimate goal; right, level currently perceived as allowing grid convergence

The contrast between the two pictures makes it clear that the smaller elements in the fully-dressed gear will not enjoy LES treatment in the near-to-mid future. Instead, the hydraulic lines and even the braces will be immersed in effectively RANS regions since the grid spacing cannot support LES on their scale, say 1cm and less for the shear layers. The spacing along these slender elements will be much too loose for LES, and the time step too long. Therefore, even as we invest in and praise hybrid methods with a domination by LES after separation, we must understand and do "damage-control" for all simulation modes. As the grid counts gradually evolve from 10^7 towards 10^{12}, each size of component will switch from quasi-steady RANS through Unsteady RANS (URANS) towards LES. These modes are seen in Figure 2.

There is hope for a consensus on the merits of the simulation types for bluff bodies. Steady RANS (SRANS) has bad press, but not always for the right reason. Some observers argue that "this flow is unsteady, and therefore steady CFD cannot be accurate." However all turbulent flows are unsteady and the argument must be that the cylinder flow, as an example, is "more unsteady than others." This does not counter the fact that the long-term average of the cylinder flow is a valid entity, and the perfect RANS model would calculate it perfectly. In addition, we have just seen that industrial CFD does and will attempt such calculations. Thus, there is value is anticipating how good such solutions can be.

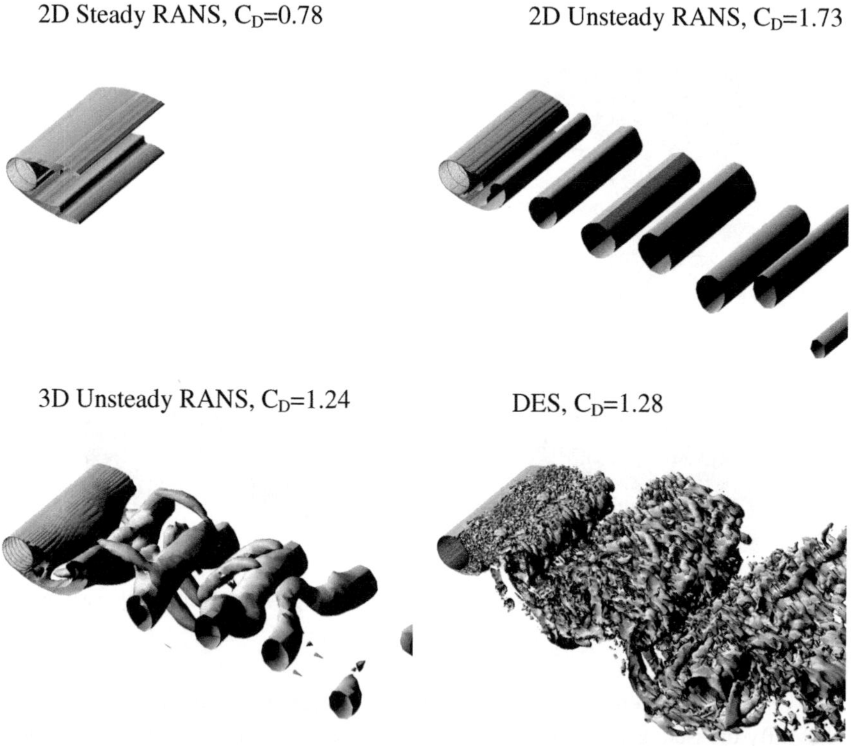

Fig. 2 Four types of bluff-body simulations, with periodic lateral conditions (experimental C_D is in the range 1.15-1.25). Courtesy A. Travin

A less immediate reason against this simplistic argument is that it leads to defending URANS as resolving the (coherent) Karman vortex shedding, and letting the RANS model account for "incoherent" smaller-scale turbulence. This connects with phase-averaging, a practice which is *very* debatable for bluff-body flows, because they are notorious for strong modulations of the shedding. Typically, URANS, at least if 2D, completely misses these modulations. This makes it convenient, but wrong. This may be the reason why 2D URANS is not very accurate; in fact for the cylinder and many other bluff bodies it is less accurate than SRANS (although this depends on the RANS model). It predicts very excessive amplitude of the lift fluctuations. Note that any near-future level of resolution of the fully-dressed landing gear will of course not force strictly 2D URANS on the small tubes, but the lateral grid spacing will exceed the diameter, so that the lateral dependence will be very slow, and thus quasi-2D.

When the lateral grid spacing is well below a diameter, 3D URANS can happen. It tends to be more accurate than 2D URANS, as revealed by the drag coefficients. However, it is a delicate instrument (see Shur et al., 2005). Its results depend on the RANS model; for instance its combination with the SARC model (Spalart-Allmaras with correction for Rotation and Curvature; Spalart and Shur, 1997, Shur et al., 2000) has given odd behaviour. In contrast a reassuring feature of DES and many hybrids is precisely that the dependence on the model is very moderate. The 3D URANS results also strongly depend on the lateral period, much more violently than with DES. Therefore, there is a need to exercise 3D URANS further, in other geometries and other codes, and with independent perspectives.

The figure is favourable to DES, here with the SST model (Menter, 1994) and a fairly fine grid. It is an expensive technique, but one that is relatively free of mysteries, for instance of unexplained differences between codes. The user burden is high, of course, but patient study of flow visualisations such as those in the figure, combined with meaningful grid variations (say by $\sqrt{2}$ in each direction) goes a long way towards ensuring that insufficient resolution in all spatial directions and in time is not destroying any key mechanism.

The next two figures illustrate cross-fertilisation between the bluff-body approaches in Figure 2 by comparison of SRANS, 2D URANS and DES, all based on the SA turbulence model (Spalart and Allmaras, 1994). The DES results are taken as a target, based on rather good quantitative comparison with experiment, qualitative success with the lift modulations, and robustness to grid variations and code differences. SRANS again gives too little drag, consistent with the pressure not being low enough behind the body (see Figure 3). On the other hand, the position of the streamline saddle point is not poor (it is worse with some other models).

The pressure coefficients in Figure 4 reproduce the moderate but definite differences between SRANS with two models (SA and SST) and DES, which is closest to experiment. The large failure of 2D URANS is also clear again, both in terms of drag level and shape of the distribution. A RANS model which would remedy this failure would have value in the field of industrial CFD; however, the direction needed to effect this is far from obvious, as the suppression of the lateral dependence by SRANS and quasi-2D URANS is a very unnatural constraint.

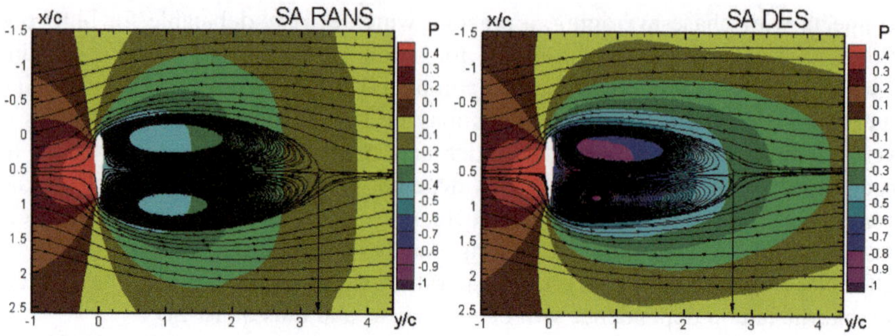

Fig. 3 Streamlines and pressure contours for NACA0012 airfoil at 90° angle of attack

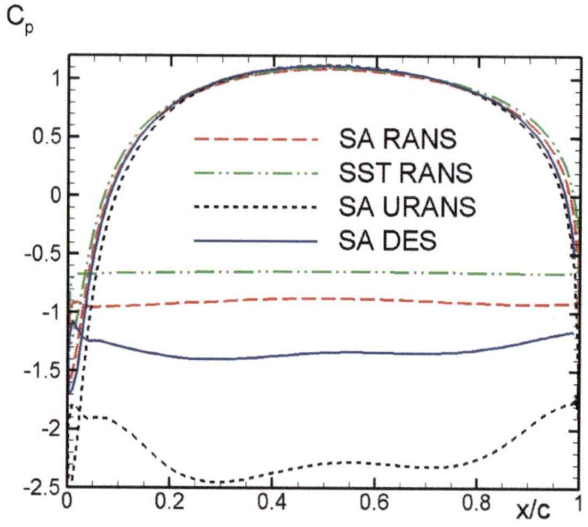

Fig. 4 Pressure coefficient distributions obtained with different approaches. Courtesy A. Travin

4 Treatment of Laminar Regions

In external flows, the turbulent region lives next to a far larger region free of turbulence, and in addition some parts of the boundary layer (BL) may be laminar. Good CFD needs to deal with the first issue without hidden effects and excessive uncertainty. It needs to deal with the second one if the laminar BL is large enough to have much effect on the global flow, and especially if it controls separation.

4.1 Irrotational Region

The interface with the non-turbulent, irrotational region was discussed by Menter (1991), Wilcox (1993), Spalart and Allmaras (1994), and Kok (2000). Even if by design of the model the turbulent region, as it should, dominates the non-turbulent region by diffusion, the values in the latter region retain some influence. Here we summarize work of the author with Rumsey (Spalart and Rumsey, 2007 - SR). A good "default" mode of use of RANS models is the Fully Turbulent (FT) one, in which the model is made to grow turbulence and eddy viscosity in any BL fluid. Such behaviour is shown in Figure 5, where all the BL regions which are thick enough to see visibly contain eddy viscosity. This is really controlled by the Ambient Values of, say, k and ω. These are the values which prevail near the body, and are often much lower than the inflow values; a fact which is not always appreciated. It is common to set very large inflow values and to let deep decay occur; the danger is that this extreme decay is often poorly resolved by the grids. This is revealed in Figure 6, which gives an example of "hidden influence": it is clear that the grid dependence is strong and that with each grid it is possible but not certain that FT behaviour will be obtained. None of the grids get even near the exact solution. Such phenomena are clearly poor CFD practice, and slow both grid convergence and iteration convergence with grid sequencing.

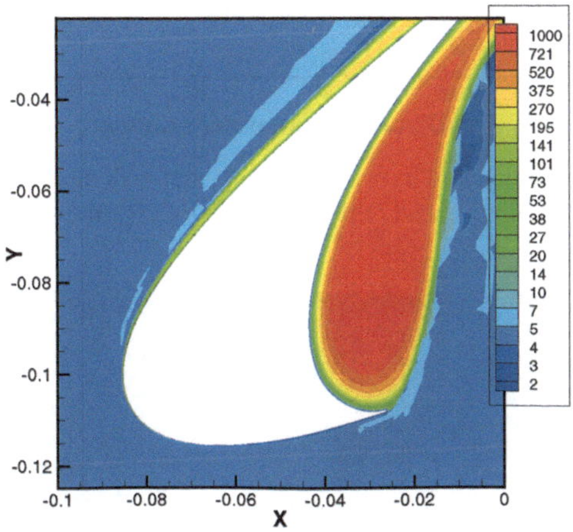

Fig. 5 Fully turbulent behaviour of RANS models. Eddy viscosity normalized with molecular viscosity. Courtesy S. Allmaras

With two-equation models, the decay can continue near the body, to the point that the ambient values near the leading edge and the trailing edge could be noticeably different. This is not the case in Figure 7, with the SA one-equation model. SR's proposal is to obtain the same general behaviour from two-equation models: the inflow and ambient values are the same, and adjusted for reliable FT operation. These models are then as easy to use as one-equation models, or nearly so, because good inflow values still depend on the velocity and body dimensions, whereas the SA model can work simply from the molecular viscosity. This rests on adding Sustaining Terms to the equations, which is easy. Of note is that SR advocated a pragmatic attitude to picking Ambient Values, rather than trying to match wind-tunnel or atmospheric turbulence. Whether Sustaining Terms become widespread or not, the point is that inflow decay is not as innocuous as it appears, and that both model developers and code vendors owe users guidance in this area.

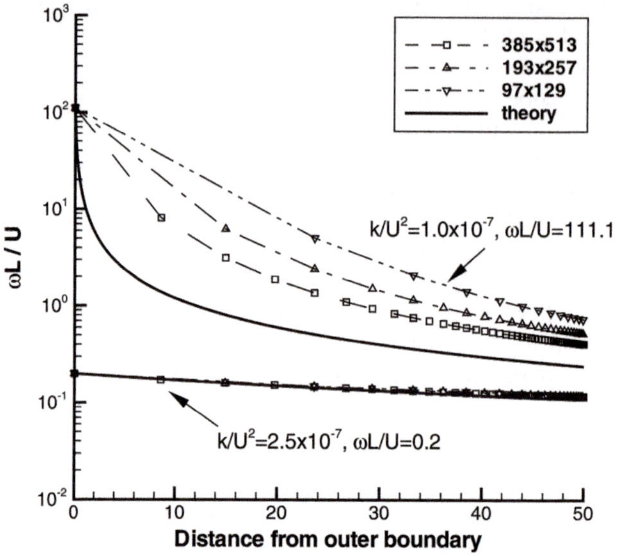

Fig. 6 Decay of ω ahead of airfoil in two-equation model with different grids. Airfoil case with inflow boundary 50 units ahead of body. Courtesy C. Rumsey

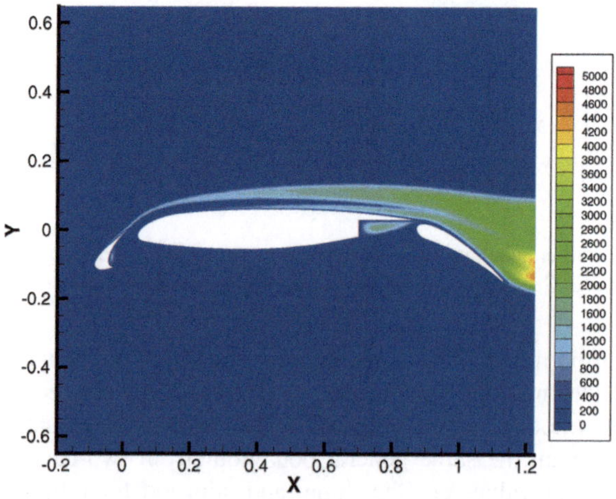

Fig. 7 Turbulence model inside and outside viscous regions. Eddy viscosity normalized with molecular viscosity. Courtesy S. Allmaras

The absence of instabilities or abnormal interactions at the edge of turbulent regions with a particular model does not imply that the situation is ideal. The gradient of the eddy viscosity is discontinuous (except for an extremely small "fillet" due to molecular viscosity), which is a little troublesome numerically, and reduces the formal order of accuracy of the solvers. It has not proven possible to obtain a fillet with a width that scales with the BL thicknesses, as would be preferable.

4.2 Laminar Areas of Boundary Layers

The prototypical flow in which FT operation is not satisfactory is the circular cylinder at sub-critical Reynolds numbers (actually FT operation is far from perfect even at super-critical Reynolds numbers, but that is more subtle (Travin et al., 1999)). It is simple enough to obtain laminar behaviour, by setting low ambient values and in particular an eddy viscosity much smaller than the molecular viscosity. However there is more variety and there are pitfalls in how to initiate the modeled turbulence, in simple terms the transition.

First, transport-equation models reproduce transition by contact with an already turbulent shear layer, at least qualitatively. This is influenced by the details of the model's behaviour at the edge of turbulent regions, which is not finely adjusted, but the phenomenon is basically correct. It is only lost, potentially, when some codes use production blanking to enforce laminar regions.

Second, some models, e.g., the SA model have standard Trip Terms to lift the eddy viscosity away from zero or near-zero. These are meant for use to model actual trips, or transition predicted by other means such as an e^n method. In practice, they have not been widely used. The reasons include the cost of implementation in full 3D, and also some failures in both directions: in some cases transition fails to occur, in others, the transition line creeps forward of the trip line. For both cases, finer grids normally fix the problem, but there is clear need for more robustness.

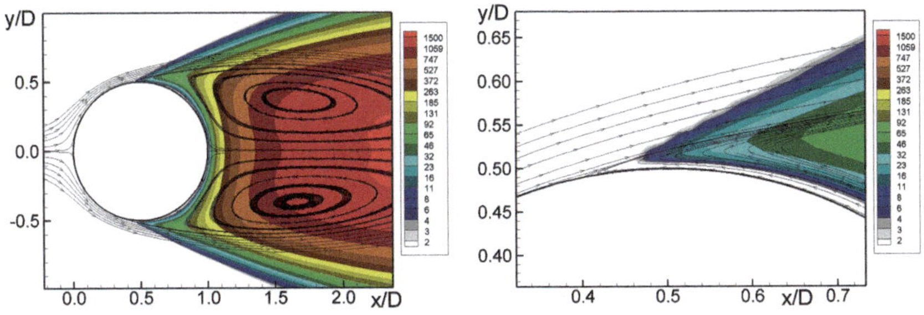

Fig. 8 Streamlines and eddy-viscosity contours for trip-less simulation

Finally, there is the Tripless (TL) mode (Shur et al., 1996), which is illustrated in Figure 8 on a sub-critical cylinder. This is an SRANS, but the treatment of the boundary layer extends to the other techniques of Figure 2. Eddy viscosity is negligible in the attached boundary layer, and is high after separation, as a result of the back-flow carrying it and triggering the separated shear layer which provides vigorous production. Just like for the contamination of one layer by another, this is imperfect, but qualitatively correct. It plays the same role as the mode of "transition caused by separation" which is included in several transition-prediction methods. The key of the TL mode is in the nonzero initial condition for eddy viscosity, combined with zero inflow condition. Alternatively, the simulation can be started with nonzero inflow, leading to an FT state (with late separation), followed by zeroing out the inflow condition and waiting for the propagation of that to the separation line.

A reason to view this TL mode as imperfect is that the cylinder has a wide range of Reynolds numbers for which separation is laminar, but transition appears to mingle with it, leading to the drag crisis. This is too subtle for RANS modelling, as of now.

4.3 Transition Prediction

The classical e^n methods are well-adapted to boundary-layer calculations, but quite difficult to couple with Navier-Stokes codes, and this has been done only for simple topologies such as a high-aspect-ratio wing, with strips. The CFD user's "dream" is a method that requires as little input as the RANS model, and works seamlessly in the entire domain. RANS models have made some progress since the 1980's towards this goal, but only for by-pass transition, which is a more subtle form than the transition by contact mentioned above It is still far less subtle than boundary-layer specific transition through extended instability growth, with all its sensitivity to pressure gradients, whether adverse and inducing TS waves, or favourable and inducing cross-flow vortices. These effects are strongly reflected in the shape factors and Reynolds numbers of the boundary layer; these are global measures, but the code (especially unstructured) writers and users demand a local measure. In fact the shape factors and thicknesses are proving difficult to extract from typical Navier-Stokes solutions with much robustness, even for casual plotting.

It appears the task of creating a transport-equation transition-prediction method is not as hopeless as it seemed, as Menter et al., 2006 and Langtry et al., 2006 have produced a surprisingly powerful method of that type. It is based on adding two extra transport-equations to the SST model and is local. The derivation of the method is quite mysterious to the author, who had personally made a similar attempt, also starting from a local Reynolds number based on vorticity and squared distance to the wall. The method has capabilities in the realm of both contamination and separation-induced transition, which are not unique, and also in the realm of pressure gradients and TS waves, which are unique. The latter is illustrated in Figure 9. The agreement with experiment and with the well-established (and boundary-layer based) Xfoil method is astounding; the "laminar bucket" is almost perfect. In each method, one parameter with the nature of a "turbulence level" was adjusted.

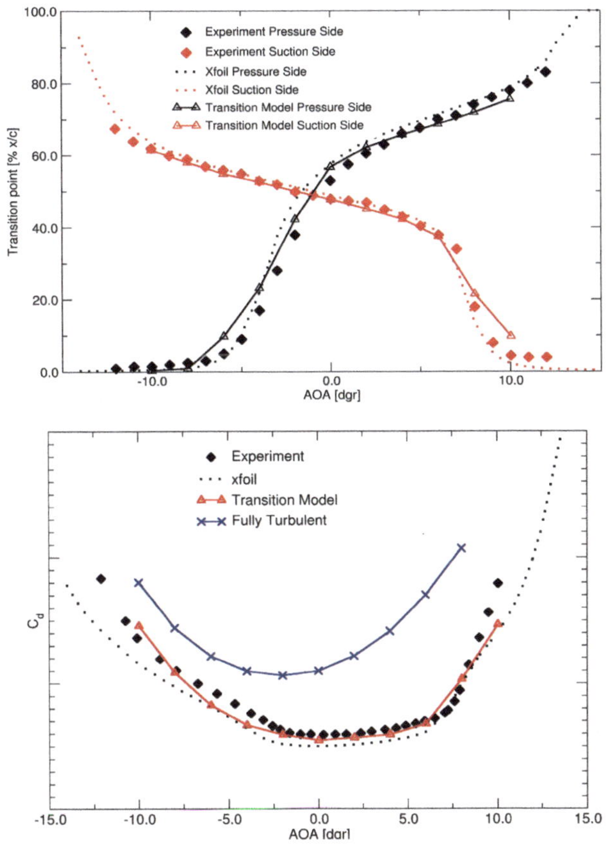

Fig. 9 Langtry-Menter transition model applied to airfoil. Courtesy F. Menter

This method was published incompletely, two functions being treated as proprietary to the CFD company. This initiated a stream of papers attempting to reproduce them, not with full success. However, Langtry and Menter have now submitted the complete formulation for publication (personal communication, 2009).

The method is not yet fully general; to begin with, it does not address the cross-flow instability. Still, it is likely to be the forerunner of the methods which will dominate for a long time. The user will have to provide more information than for the FT mode; some material on how turbulent and noisy the freestream is, and how rough the surface. This is similar to providing a transition n factor.

To summarize this section on laminar regions, our principal purpose was to raise awareness of both some CFD Quality Control issues and the physical possibilities for more refined modelling of flows. While the trend towards an ultimate "black-box" tool is here, this must be balanced with user competence and attitude, in order for high-quality predictions to come out.

5 Acceptance of New Models

The scientist's life includes watching the conferences and journals for progress - and reviewing numerous manuscripts every year. Both are duties, and can bring intellectual pleasure. During this decade, there has been some creativity, but no new model has met with true success, if measured by wide implantation into codes and use. The barriers to entry in the field must be remembered. New models are published with a small number of cases and within the page limit or guideline of the journal. There is no consensus set of cases which a model needs to be good in before it "is taken seriously." Reviewers are not likely to bar a new model, unless the set of available results is meaningless (for instance, only homogeneous turbulence) or the model violates basic requirements such as dimensional analysis or invariance. A less clear-cut but essential property is good response to ambient values, and ascertaining this requires some mathematics, and some numerical testing. The difficult next step after this initial publication is an independent implementation in a commercial or government CFD system. These institutions are understandably intent on spending their resources wisely. This independent use is crucial in order to prove the completeness of the formulation, indicate robustness, and provide an impartial opinion.

An AIAA Discussion Group led by Dr. B. Smith of Lockheed-Martin is bringing some order to the process of model assessment and diffusion with a rating scale for maturity, and suggested test cases. As part of this, Dr. C. Rumsey of NASA is recording the formulation of models at http://turbmodels.larc.nasa.gov/index.html. There is also a survey of CFD users and experts, the results of which are not yet available. In general, the community of model developers is not very secretive, and in the long run there is a good level of effective collaboration and cross-fertilization.

When trying to improve the prospects for new ideas, we need to remember also that in the past models have been successful that "should not have worked." For instance, 20 years ago it was "known" that one-equation models would never be useful, and 4 years ago that transition prediction from local quantities would never respond to pressure gradients. We need to be ready, and hope, for deep surprises.

6 Outlook

Although it is more useful than ever, RANS modelling inspires impatience in most of the people involved in it. We must hope that far better turbulence models are waiting to be born, and imagine what new approach could enable this. In spite of some predictions, often coming from Physics, it is not clear that these models will result from a more scientific thought process than the very pragmatic one that has paid off so far. Another possibility is a pragmatic process, but one based on automatic optimization making use of a very wide experimental and DNS database. The structure itself of the model would be left to optimize, instead of only the constants, in particular the exact choice of the length-scale variable, and also maybe the more obscure structure of non-linear constitutive relations, or sophisticated pressure-strain terms. Optimization would surpass intuition. This is not to suggest that a truly general and non-arbitrary structure for a RANS model can be defined, but subsets of this problem may be able to, and the prospect is fascinating (although the selection of the set of test cases remains arbitrary). Models optimized only to play a role in Natural DES may also be created.

The Reynolds-number-related "Moving Target" issue mentioned above may be pondered, so as to avoid spending too much effort obtaining perfection on a set of cases "of measure zero." Similarly, the author's opinion on the relevance of homogeneous turbulence to model calibration is no secret: is it of much value? It is of value to knowing the inertial range, but the inertial range never covers all the energy of the turbulence. RANS models are needed to provide the Reynolds stresses, which necessarily involve non-universal eddies. In homogeneous turbulence, the driving (and often driven by some plausible device) non-universal eddies are artificial.

The fundamental contradiction of RANS modelling is the demand for local Partial Differential Equations to render turbulence, which is not local in nature. Local equations are merely convenient for CFD, and in this strategy, non-local effects are mimicked by transport and diffusion terms. There is no physical reason to expect that any elegant solution to this demand of CFD exists. The perennial debate over whether it is "proper" to use wall distance or wall-normal direction is a symptom of this contradiction.

A similar topic of dissent between the author and much of the community concerns "unsteady flows." Many argue that modelling steady flows is somehow easier, so that models would need extra credentials before they perform well in unsteady flows. However, the fact that the Eulerian derivative $\partial/\partial t$ is zero, when the material derivative $(\partial/\partial t + U_i \partial/\partial x_i)$ is not zero, means very little physically to the turbulence. Imagine turbulence going through a "steady" shock wave. The steady or unsteady character of a flow depends on the reference frame, and therefore is not a deep feature of the flow; in fact, transport-equation models do not and should not recognize it. Unsteady boundary conditions with frequencies higher than the frequency scale of the turbulence do create new phenomena, but it's not unusual for them to be quasi-laminar, and therefore relatively easy to calculate given enough resolution. An example is a periodic Stokes layer.

A puzzle remains why elaborate models up to Reynolds-Stress Transport have not found a widely acknowledged niche: a class of flows in which linear eddy-viscosity models are much too crude, yet flows which are not "almost obviously" unsuited for RANS, such as the cylinder or stalled airfoils. Moderate rotation and curvature effects quickly come to mind, which are probably more present in internal flows than in aerodynamic flows. However in the Special Issue, outside of the article devoted to them (Hellsten and Wallin, 2009), Algebraic Reynolds-Stress models were tested only twice, by independent teams, in semi-complex flows (and Reynolds-Stress Transport models not at all). The first time (Rumsey and Swanson, 2009) they had the worst agreement with experiment, and the second time (Zingg and Godin, 2009) they returned an essentially unphysical behaviour of the shear stress. Several of the authors in the issue know internal flows far better than this author, so that a bias towards aerodynamics is not apparent. Still, this paragraph is not construed as a review of the recent literature. It is possible the elaborate models have lacked scientific stability (not just numerical stability); frequent publications and "tweaks" by rival teams would have discouraged the code developers. This gives food for thought over a possible similar evolution for hybrid methods, marked by unproductive rivalries.

RANS providers have complained of poor funding and lack of interest by government institutions. There is no doubt that the humanpower devoted to modeling has long been dwarfed by what is applied to other aspects of CFD: solvers, grid generators, visualization, GUIs, user support. It just seems to naturally be a small-team endeavor, and to support only a few teams or even individuals world-wide. The quality of collaboration is very modest. In the last few years, there has in addition been a "Brain Drain" towards hybrid methods. On the other hand, programs such as Flomania and DESider (Haase et al., 2006; Haase et al., 2009) are indeed turbulence-modelling research, well-funded and well-run; a generous act by the EU. They have served the needs of some of the creative minds.

A durable concern as CFD extends its reach into real-life flows will be grid resolution. Improved CFD packages make it possible to obtain "solutions" for complex geometries. Let us take a Formula-1 race car as an example. A CAD file is certainly available with extreme detail. A "very large" grid is generated, say with 10^8 points for half the car. Does this make the output a solution for that flow? We cannot tell. Reports that only steady RANS is used suggest poor accuracy, considering the physics of open wheels. It seems certain that a real step in grid refinement, say by a factor of 2 in each direction, would strongly alter some parts of the flow; therefore, this is not a true numerical solution. F-1 aerodynamics are dominated by vortices which create interactions between distant elements such as wing fences and "barge boards", and the current non-adaptive grids do not track the vortices and keep them tight. The diffuser flow is extremely sensitive and fed by turbulent flow between the long bottom plane of the car and a rough road, with massive separation around it. Considering the open-wheel flows, this is a perfect candidate for hybrid treatments, but these do not make grid design and Quality Assurance any easier. Now this is a field with numerous repeats of almost identical runs, which as we have seen normally builds confidence, and the collaboration with wind-tunnel testing is obviously intimate. However, it appears possible that the relatively minor rule changes for 2009 (higher, narrower rear wing, wider front wing, different diffuser) were sufficient to throw some of the teams into a slump which their CFD was unable to predict or prevent.

The situation is similar with the complex 3D flow fields around aircraft with high-lift devices and landing gear deployed. We have CFD output, but do we have true solutions? Distant interactions are also present, for instance as the vortex from the nacelle chine controls the flap region after climbing over the wing. An additional obstacle is that, as we have seen, SRANS solutions are often unstable around bluff bodies, so that the true "RANS answer" is URANS with vortex shedding. This often leads the codes to a (locally) poorly converged SRANS solution (poorly converged in iterations rather than grid spacing, here), which is not a sign of good CFD ethics. It also probably leads users to choose more dissipative algorithms, which in turn degrades the resolution of vortices and other fine flow features. If the computing power suddenly increased by a factor of 100, it would be used by 3D turbulent CFD tomorrow, and with good reason. This would begin to shine light on the deficiencies of the models with more rigour than we can today, assuming of course a concurrent improvement in the reference database.

The future of RANS is very secure, but orderly and "on-demand" progress is not likely.

Acknowledgements. Dr. M. Shur provided extensive help in writing this article, and Dr. M. Strelets made insightful comments.

References

Haase, W., Aupoix, B., Bunge, U., Schwamborn, D. (eds.): FLOMANIA - A European Initiative on Flow Physics Modelling. Notes on Numerical Fluid Mechanics and Multidisciplinary Design, vol. 94. Springer, Heidelberg (2006)

Haase, W., Braza, M., Revell, A. (eds.): DESider - A European Effort on Hybrid RANS-LES Modelling. Notes on Numerical Fluid Mechanics and Multidisciplinary Design, vol. 103. Springer, Heidelberg (2009)

Hellsten, A., Wallin, S.: Explicit Algebraic Reynolds Stress and Non-Linear Eeddy-Viscosity Models. Int. J. of Computational Fluid Dynamics 23, 349–361 (2009)

Hirsch, C., Tartinville, B.: Reynolds-Averaged Navier-Stokes Modelling for Industrial Applications and Some Challenging Issues. Int. J. of Computational Fluid Dynamics 23, 295–303 (2009)

Kok, J.C.: Resolving the Dependence on Freestream Values for the k–ωTurbulence Model. AIAA J. 38, 1292–1295 (2000)

Langtry, R.B., Menter, F.R., Likki, S.R., Suzen, Y.B., Huang, P.G., Volker, S.: A Correlation-Based Transition Model Using Local Variables Part II - Test Cases and I Industrial Applications. J. of Turbomachinery 128, 423–434 (2006)

Menter, F.R.: Influence of Freestream Values on k–ωTurbulenceModel Predictions. AIAA J. 30, 1657–1659 (1991)

Menter, F.R.: Two-Equation Eddy-Viscosity Turbulence Models for Engineering Applications. AIAA J. 32, 1598–1605 (1994); also AIAA Paper, AIAA-93-2906

Menter, F.R., Langtry, R.B., Likki, S.R., Suzen, Y.B., Huang, P.G., Volker, S.: A Correlation-Based Transition Model Using Local Variables Part I – Model Formulation. J. of Turbomachinery 128, 413–422 (2006)

Menter, F.R.: Review of the Shear-Stress Transport Turbulence Model Experience from an Industrial Perspective. Int. J. of Computational Fluid Dynamics 23, 305–316 (2009)

Morrison, J.H., Helmsch, M.J.: Statistical Analysis of CFD Solutions from the Third AIAA Drag Prediction Workshop. AIAA Paper, AIAA-2007-0254 (2007)

Rumsey, C.L., Swanson, R.C.: Turbulence Modelling for Active Flow Control Applications. Int. J. of Computational Fluid Dynamics 23, 317–326 (2009)

Shur, M., Spalart, P.R., Strelets, M., Travin, A.: Navier-Stokes Simulation of Shedding Turbulent Flow Past a Circular Cylinder and a Cylinder with Backward Splitter Plate. In: Proceedings of the Third ECCOMAS Computational Fluid Dynamics Conference on Computational Fluid Dynamics 1996, pp. 676–682 (1996)

Shur, M.L., Strelets, M.K., Travin, A.K., Spalart, P.R.: Turbulence Modeling in Rotating and Curved Channels: Assessing the Spalart-Shur Correction. AIAA J. 38, 784–792 (2000)

Shur, M.L., Spalart, P.R., Squiers, K.D., Strelets, M.K., Travin, A.K.: Three Dimensionality in Reynolds-Averaged Navier–Stokes Solutions Around Two-Dimensional Geometries. AIAA J. 43, 1230–1242 (2005)

Spalart, P.R., Allmaras, S.R.: A One-Equation Turbulence Model for Aerodynamic Flows. La Recherche Aerospatiale 1, 5–21 (1994); also AIAA Paper, AIAA-92-0439

Spalart, P.R., Jou, W.-H., Strelets, M., Allmaras, S.R.: Comments on the Feasibility of LES for Wings, and on a Hybrid RANS/LES Approach. In: Liu, C., Liu, Z. (eds.) Advances in LES/DNS, First AFOSR International Conference on DNS/LES. Greyden Press (1997)

Spalart, P.R., Shur, M.L.: On the Sensitization of Turbulence Models to Rotation and Curvature. Aerospace Science and Technology 1, 297–302 (1997)

Spalart, P.R., Rumsey, C.L.: Effective Inflow Conditions for Turbulence Models in Aerodynamic Calculations. AIAA J. 45, 2544–2553 (2007)

Travin, A.K., Shur, M.L., Strelets, M.K., Spalart, P.R.: Detached-eddy simulation past a circular cylinder. Int J. Flow, Turbulence and Combustion 63, 293–213 (1999)

Wilcox, D.C.: A Two-Equation Turbulence Model for Wall-Bounded and Free-Shear Flows. AIAA Paper, AIAA-93-2905 (1993)

Zingg, D.W., Godin, P.: A Perspective on Turbulence Models for Aerodynamic Flows. Int. J. of Computational Fluid Dynamics 23, 327–335 (2009)

Flow Modelling in Aircraft Design – An Engineering Perspective

M. Sillén, E. Amundsson, S. Arvidson, I. Persson, and M. Säterskog

Abstract. Computational fluid dynamics is commonly used in the aerodynamic design process at Saab. The improvements in flow simulation capability and not least computer capacity over the last years have enabled aerodynamicists and propulsion engineers with an efficient design tool. In the development of the Gripen fighter aircraft numerical simulations are performed to guide the design and to make trade off studies ahead of wind tunnel tests. In this paper several examples are presented covering overall design, inlet design and weapon integration where numerical simulations have played an essential role in the development.

1 Introduction

Computational Fluid Dynamics (CFD) plays an important role as a tool for design and analysis in the aerospace industry today. Typical characteristics of the industrial CFD environment are complex geometries, the frequency of large computationally heavy problems, the requirement of high fidelity results and the short time schedule for producing results. A variety of methods are used from preliminary design studies, where short problem turn-around time is crucial, to detailed component analysis where accurate results are required. In order to further improve the confidence in aerodynamic simulations, advanced CFD methods, i.e. Reynolds Average Navier-Stokes (RANS) based 3D methods, must be deployed at an even earlier stage than today, c.f. Sillén (2006). It also requires the introduction of improved methods for unsteady flow e.g. separated flow regions. As a result, focus now lies on CFD development towards efficient numerical algorithms and accurate modelling of flow physics. Even though dramatic improvements in simulation capability have been made over the past ten years we are not able to exclude wind tunnel tests from the design process. Today, simulations serve as guidelines in the design loop and complement experimental results when generating aerodynamic data bases for stability and control, performance, aerodynamic loads and external store integration.

Aircraft design is truly a multidisciplinary challenge. A significant reduction in turn around time of a design can be accomplished by carrying out the design

M. Sillén · E. Amundsson · S. Arvidson · I. Persson · M. Säterskog
Saab Aerosystems, Sweden
mattias.sillen@saabgroup.com

S.-H. Peng et al. (Eds.): Progress in Hybrid RANS-LES Modelling, NNFM 111, pp. 25–39.
springerlink.com

activities simultaneously in each discipline using simulation and optimization tools. In order to accomplish this, numerical simulations are essential to establish the initial mathematical model of the product that all disciplines can continue working on. The concept of using a high fidelity mathematical/numerical representation of the physical properties and the functions of the product is often called a *virtual product* (VP). Critical to the success of the VP is the capability for rapid generation of high fidelity information from all disciplines, which recent advances in information technologies and high performance computer systems now have enabled. In this way the development can proceed in parallel to reduce the development risk and time as the different disciplines are matured iteratively. This method of building a numerical model of the product and using it for simulations and as information carrier throughout the development process is often called Model Based Design and is a key component in the VP vision. Using models and simulations, product requirements can be verified at an early stage and reduce the number of design cycles, the development risks, the number of flight tests, the cost and shorten the time to market.

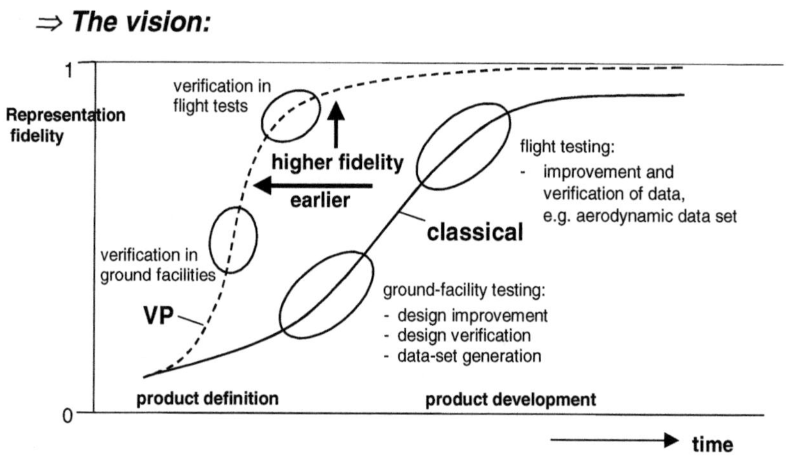

Fig. 1 Illustration of the benefit of the virtual product in aircraft design, namely higher fidelity early on in the design cycle, from Hirschel (2000)

Aerodynamics is one of the disciplines in the development chain that provide information to a number of disciplines. The output from aerodynamic activities is e.g.

- Support in design of external shape to the design engineers
- Aerodynamic data regarding stability and control for handling quality judgment and flight control development
- Aerodynamic data bases for performance evaluations
- Duct design and air flow quality analysis for engine integration and engine performance
- Aerodynamic loads data bases for structural design

A schematic view of this iterative process is seen in Fig. 2 where the different enablers of information (tools and tests) are given at the left side feeding information to the models (aerodynamic data bases) describing the product. The models are used in the analyses and to support other disciplines with information. The findings from the analyses and feedback from other disciplines initiate design changes and the need for updated or refined information. As the development process progress the design matures and more elaborate methods are used to feed more comprehensive information into the databases. At the end of the development, flight tests are used for verification of the models used in the design as well as in simulators and to close product requirements.

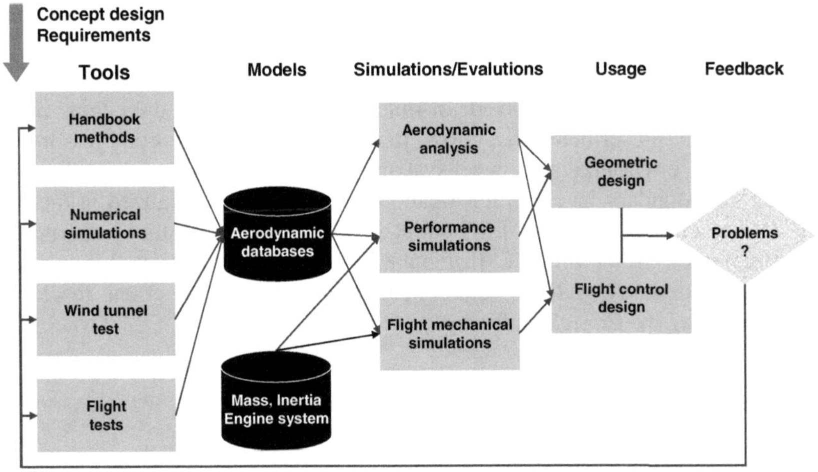

Fig. 2 Schematic processes of aerodynamic design and modelling

2 Design Environment

The design environment at Saab is based on Catia. For flow simulations the Catia models are imported into ICEM CFD package by using the direct CAD interface add-on module. Minor simplifications or modifications of the CAD data are often necessary to obtain good quality of the final computational grid. The usual type of grid is unstructured with prismatic elements in the viscous layer and tetrahedral elements elsewhere. In case of in-viscid flow only tetrahedral or in same cases hexahedral elements are used. Hybrid grids with mixed tetrahedral and prismatic elements are generated using the CFD Tetra / Prism grid generator.

Aerodynamic simulations are performed with the Edge flow solver developed by the Swedish Defence Research Agency (FOI), in detail described in Eliasson (2002). The Edge code is an edge- and node based Navier-Stokes flow solver applicable for both structured and unstructured grids. Edge is based on a finite volume formulation where a median dual grid forms the control volumes with the unknowns in the centres. The governing equations are integrated in time by a multistage Runge-Kutta scheme to steady state and with acceleration by FAS agglomeration multigrid. A pre-processor creates the dual grid and the edge based data structure, which is also employed to agglomerate coarser control volumes for the multigrid. For load balancing on distributed computer architectures the Metis graph partitioner is used, cf. Karypis (1998). The Edge solver is able to provide different options such as different spatial discretisation techniques, turbulence models, low speed preconditioning and high temperature extensions. In the applications presented here, the code has been used in both in-viscid as well as in viscous modes. In the in-viscid mode, it solves the Euler equations and in the viscous mode it has been used to solve the fully turbulent thin shear layer version of the Navier-Stokes equations. To model turbulence, different models based on k-ω equations are used. Boundary conditions are imposed on the far-field using Riemann invariants. On solid surfaces, slip or no-slip conditions are enforced depending on the physics involved. In full aircraft simulations the engine inlet is modeled by a prescribed mass flow just upstream of the compressor disc. Likewise, the environment climate system (ECS) intakes located in the space between the boundary layer splitter plate and the fuselage is realized in the same way. On the engine outlet and ECS outlet surfaces, conditions that preserve their respective mass flows are set. Computations are performed on Linux-cluster whereas the detailed analysis and visualization of the numerical results are performed using the EnSight software.

3 Applications

The ongoing Gripen development comprise of two parallel tracks. The first is the successive upgrades of the Gripen C/D version delivered to the air force every third year adding new functions and external stores. The second is a more long term development where the basic aircraft is redesigned with the objective of increasing the overall performance of the aircraft e.g. increasing the payload capacity and the range of the aircraft. A demonstrator aircraft is also developed to study several of the modifications and reduce the development risk for the future Gripen version. In the redesign several major changes are introduced affecting the external shape and the aerodynamics. New fuel tanks are introduced to increase the range of the aircraft by relocating the main landing gear to underneath the wings instead of installed within the fuselage. New pylons are introduced to carry additional external stores. All these changes increase the weight of the aircraft and a new engine with higher thrust is required to maintain and increase the performance of the aircraft. In this paper we study a few examples of aerodynamic simulations and analyses performed within the development process.

The aerodynamic design and analysis are generally performed according to the scheme in Figure 2. Regardless if the task is to develop a new configuration or make changes to current design the work starts with a design idea, a concept. This is numerically analysed and geometrically refined until the confidence in the design has matured. Along the way the aerodynamic databases is updated and preliminary data on aerodynamics, flight dynamics and loads as well as performance characteristics are judged against the requirements. These numerical studies serve as guidelines for the shaping ahead of the wind tunnel tests. The numerically obtained data is however not yet of sufficient quality to make detailed judgements of handling qualities or performance evaluations. For this wind tunnel data is still needed.

The design and analysis activities are often planned from a given point in time when a final geometry definition is needed. The geometry is often frozen in steps where the overall external shape is at first. After that the air inlet and duct as well as external sensors and pylon configurations are defined. Fig 3 gives a schematic master time schedule for the redesign project. The first 10 months of design and analysis activities focus on the overall configuration. In total, 10 configurations were evaluated with modifications in wing span, wing and fuselage changes due to increased internal fuel and changed position of main landing gear. Before the overall external shape is fixed a wind tunnel test is performed. This allows for last minute changes before the structural development goes into detailed design and changes become costly. The second phase of aerodynamic design and analysis focuses on internal flow and details e.g. sensor integration, pylon configurations and air duct design. A verifying wind tunnel test is performed after the complete configuration is fixed to generate aerodynamic data bases for stability and control as well as performance and aerodynamic loads. These data bases are used further down the development chain in structural design, in flight control design and for verification of performance.

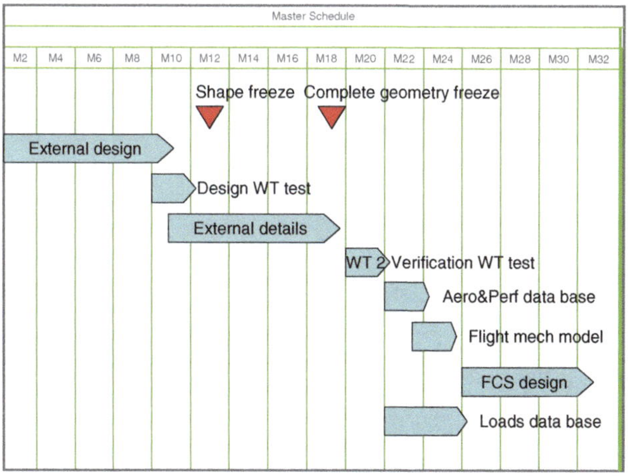

Fig. 3 Master schedules for aerodynamic design in Gripen development project

3.1 Development of the External Shape

The aerodynamic analyses performed on complete aircraft level are dominating early on in the design process as the general layout is defined. Other subsystems are however not far behind in the design process studying e.g. effects on external installations as sensors or specific requirements on local flow parameters. These studies are initially performed using in-viscid flow modelling when comparing different geometry alternatives. As the geometry matures viscous modelling is introduced to further improve the results.

In the development of the future Gripen different wing span alternatives are studied. A larger wing span is due to several reasons considered. For example, a larger wing area enables a higher take-off weight, greater separation of wing pylons and allows for an introduction of a wider fuselage to ease the maintenance of the engine. In combination with the introduction of a 'backpack' to meet the requirements on increased internal fuel volumes as well as a modified dorsal fin this is the largest modification made of the external shape design when comparing a Gripen aircraft to its future design concept. The CFD studies made within this project has indeed served as a favourable tool when generating aerodynamic data sets for flight performance studies and deciding on which configurations to take into the wind tunnel. For each configuration both clean aircraft as well as different external store combinations are studied. Initial studies are performed with in-viscid flow modelling whereas viscous simulations are applied in the second iteration when more detailed information is required and higher accuracy is needed in e.g. aerodynamic load predictions.

In Figure 4 the surface grid used for simulations on a configuration with two external fuel tanks as well as pressure contours and the shock surface are presented. The grid to the left, for in-viscid simulations, contains 3.3 Mnodes whereas the grid for viscous simulations (right) has the same surface grid but consists of 19.6 Mnodes with a boundary layer resolution of $y^{+}=1$ and about 40 prism layers. Typical turn around time for an in-viscid analysis is one week including geometry preparation, grid generation and 20-40 flow cases. A viscous analysis has a typical turn around time of 2-4 weeks but contains less flow cases.

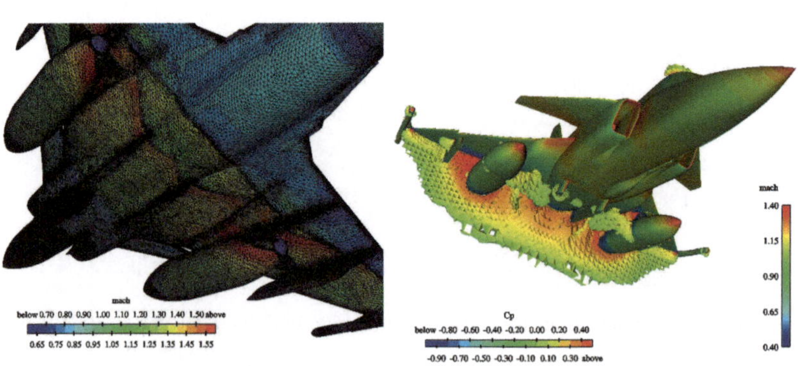

Fig. 4 Surface grid and pressure distribution in transonic flow, M=0.9, $\alpha=0.54^{\circ}$

Another example where detailed flow information is needed early on in the design process is when considering the design of the fuel system and in particular the placement of the ventilation outlet. The Gripen fuel system consists of several tanks distributed across the fuselage and wings. The system is pressurized to be operational throughout the complete flight envelop and fuel is transferred between the different tanks in order to maintain the centre of gravity. One of the important design aspects of the fuel system is to place the ventilation outlet in an area where the surface pressure is relatively stable and close to the free stream pressure - throughout the flight envelop and for all external store configurations.

During the development of the Gripen demonstrator aircraft the ventilation outlet placement below the right main wing was found to be less favourable. This as shock wave patterns on the aircraft in flight could cause such big relative pressure differences that the proper function of the fuel system at times could be limited. In order to avoid a malfunctioning system a restriction of local pressure variation $p- p_{\infty}$ of +/- 5kPa in level flight is required. The design challenge for the future Gripen aircraft is thus to find a proper location for the outlet, and due to fuel system installation reasons preferably in the rear fin area. Of certain interest for this analysis are transonic flight cases where shock waves are present. Further, the proximity of airbrakes and rudder also has to be taken into account as the extended airbrake and deflected steering surfaces when manoeuvring the aircraft alters the relative surface pressure distribution on the fin. As the outlet is preferably placed on one side of the fin, it is also important to determine the effect of angle of sideslip cases, as this influence is much greater here than if the ventilation outlet were to be located below the wing (where the effect of angle of attack is dominating).

The analysis was performed using in-viscid flow modelling on three different aircraft geometries and about 40 flight cases in the transonic and supersonic flight regime with various angle of attack and sideslip. Typically, the grid models consist of about 6 Mnodes. The computed relative surface pressure distribution on the aircraft during a transonic flight case subjected to an angle of sideslip is shown in Figure 5 below. In the illustration, both the effect of a shock wave and that of an angle of attack on relative surface pressure, dfsp [kPa], are clearly indicated.

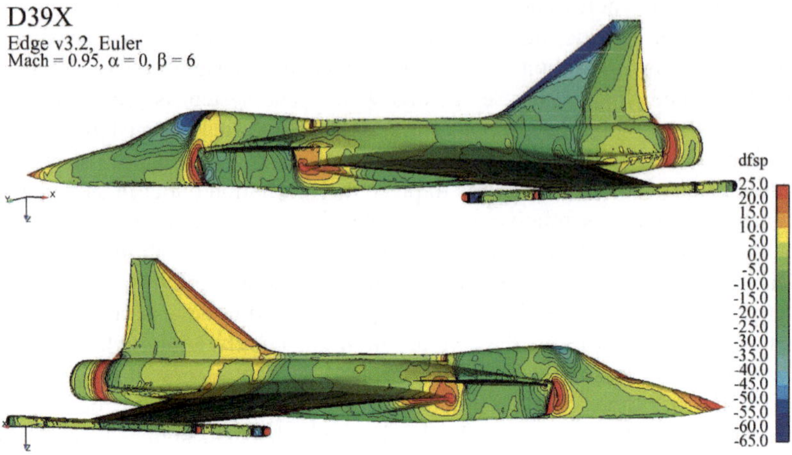

Fig. 5 Surface pressure distribution, M=0.95, α=0°, β=6°

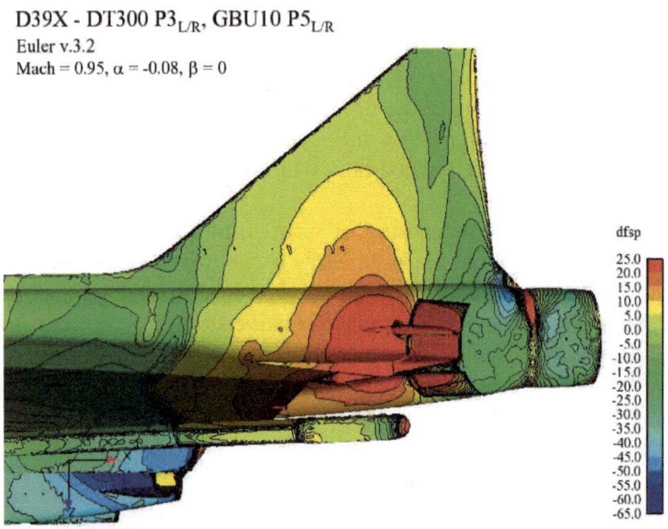

Fig. 6 Surface pressure distribution, M=0.95, α=0°, β=6°

As stated, the area of interest is in the vicinity of the airbrakes and flow cases with both retracted and extended airbrakes where thus considered, see example in Figure 6. Several fin placed positions were evaluated with respect to the limitations stated by the fuel system. As expected, the positions at the fin root are exposed to large pressure variations with airbrake extended and the influence becomes weaker towards the top of the fin. This analysis has shown that, from an aerodynamically point of view, as well as when considering leaking fuel and fire hazardous risks, the preferred location of the ventilation outlet is at the top of the fin, just above the rudder. This area has a relatively stable pressure environment and only weak influence of airbrake deflection. The shock wave pattern will at one point or another pass above this position too, but the indicated drop in relative surface pressure is within the limitations of what the fuel system is able to handle during a short period of time.

3.2 Inlet Duct Analysis

One of the key issues in the future Gripen is to replace the GE F404 engine with a GE F414 engine that delivers 20 % more thrust to compensate for e.g. higher take off weight. Compared to the current Gripen aircraft the inlet duct must be redesigned to account for a higher mass flow rate to the engine.

Whilst designing the air inlet and the duct there are a number of aspects to take into account. The inlet and duct must be able to operate throughout the flight envelop and provide the engine with air flow of sufficiently high quality in terms of total pressure losses and distortion. The size of the inlet is set by the maximum mass flow required by the engine. Nevertheless, when designing an inlet a number of important design parameters must be considered, e.g., throat location, inlet duct area distribution, inlet capture ratio and lip profile. Since the flow around the inlet opening is strongly three dimensional numerical methods has to be applied in the analysis.

In this study the effect of inlet duct area distribution and lip profile design is analyzed regarding inlet performance and inlet spillage drag. Since the flow inside and around the aircraft inlet is strongly viscous, the computational meshes have to be very well resolved, i.e. $y^{+}=1$. The standard meshes used for inlet analysis comprise of the aircraft fore body, inlet lips and inlet air duct with a typical size of 10 to 15 Mnodes. The grids are unstructured and generated with ICEM Tetra and Prism, see Figure 7.

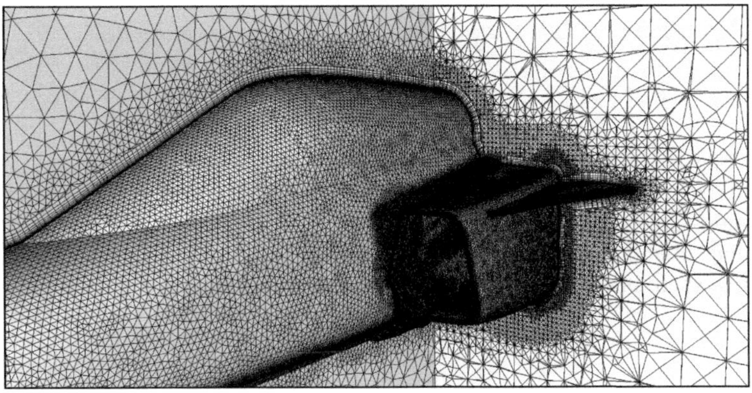

Fig. 7 Mesh used in the CFD calculations

Most of the CFD analyses on inlets are performed using RANS-methods due to the short lead times required in the industry. This restricts the analysis and dynamic distortion or time dependent flow phenomena inside and around the inlet cannot be analyzed with this methodology. However, the RANS approach is very useful and results agree well to wind tunnel data for flow cases where none or only small separations occur. A comparison between duct wall pressures measured in wind tunnel and wall pressures calculated using CFD is presented in Figure 8. It can be seen that the agreement between the data is good which shows that the flow field inside the inlet is well predicted as well as the losses achieved in front of the inlet opening.

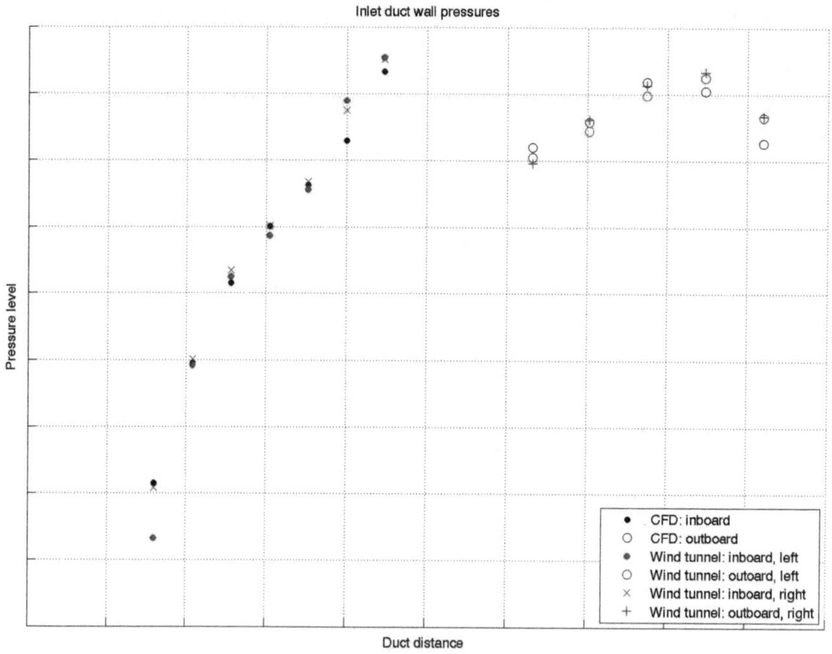

Fig. 8 Wall surface pressure

Inlet spillage drag can be well predicted with RANS methods while inlet mass flow ratios are sufficiently large. Therefore, CFD is a very important tool when designing inlet lip profiles and the inlet opening. An example can be seen in Figure 9 where inlet spillage drag has been calculated for different inlet mass flow ratios.

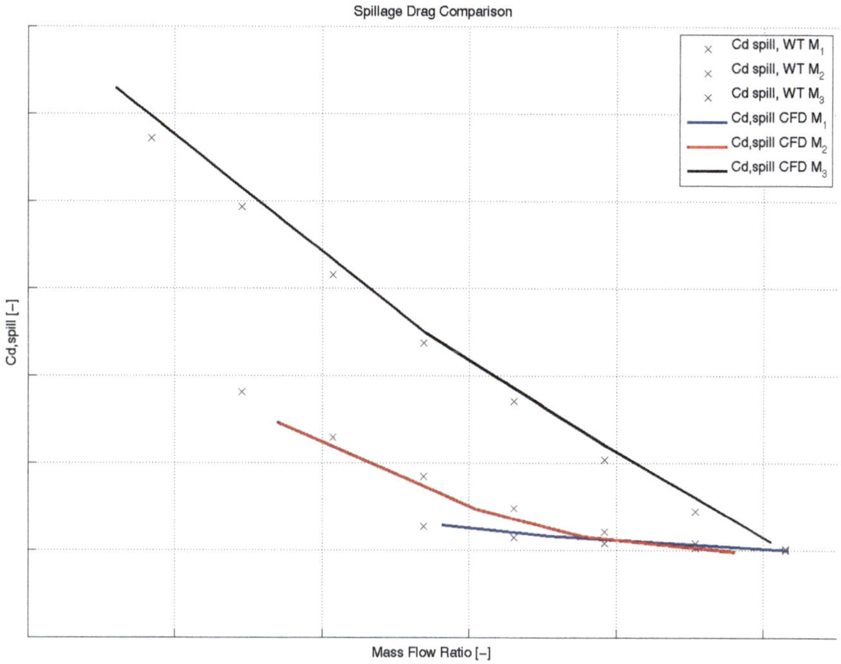

Fig. 9 Inlet spillage drag

3.3 Store Separation Analysis

Integration of new external stores/weapons is a recurring activity as new customers' acquire the aircraft or new weapons become available. On a military aircraft, each external store must be able to be released in a safe and well predicted manner. To predict the store trajectories, a 6-DOF simulation model is used that describes the movement of the aircraft and the store relative to each other as well as an inertial system. The simulation model includes free flying store aerodynamics as well as interference aerodynamics due to the disturbance from the aircraft when the store is within close proximity of the aircraft. The interference aerodynamics is dependent on several parameters such as the flight condition at the time of the store release, the store position and the overall configuration layout, e.g. what adjacent stores are present. This amount of parameters in the aerodynamic data base often requires complicated and extensive wind tunnel tests using e.g. a two-sting-rig. Time and cost aspects have led to an increased usage of numerical simulations in the store certification process at Saab during the last 5 years. Much of its success has been dependent on the advances made within automated grid generation, making turn-around times for the CFD process a very interesting complement to wind tunnel tests. As store separation analysis to a large extent still are based on older tests of geometrically similar stores, CFD has also provided key parameters to existing aerodynamic data when traditional scaling can not be applied. Such an example is demonstrated within this study, further described in Persson (2008), i.e. predicting the correct captive store loads gives the store separation model the initial condition it needs to deliver a trustworthy separation trajectory.

Transonic store separation of an Mk82LD bomb from the right side inner wing pylon of the Gripen aircraft has been studied by means of computational aerodynamics. The studied cases consist of two configurations. One configuration has a 300 gallon drop tank (DT300) attached to the central pylon below the fuselage. This configuration has been wind tunnel tested. The other configuration has a camera pod (FUNK) attached to the central fuselage pylon. This is a flight tested configuration where photometric data from the separation phase has been collected. When store separation simulations were carried out on the FUNK configuration using aero data from the DT300 wind tunnel tests, the results were not good. Obviously, the flow fields below the aircraft are very different in the two cases thus generating completely different captive loads on the Mk82LD bombs. Therefore, computational fluid dynamics was used to compute both cases, with the bombs in their captive position, in order to provide aerodynamic load differences that could be added to the existing wind tunnel data. The computational models are generated with ICEM CFD and typical grid sizes for the tetrahedral grids were approximately 4 Mnodes and for the mixed viscous grids around 22 Mnodes, see Figure 10 for grid details of the Mk82 bomb in captive position.

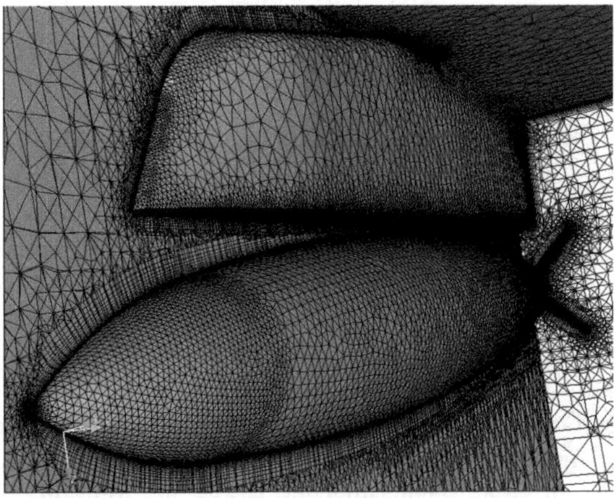

Fig. 10 Grid detail of a pylon with Mk82LD in captive position

For a separation study based on CFD, an initial carriage solution has to be established as a first step. This gives the initial conditions needed for the store separation simulation. The studied case has a free stream Mach number of 0.9 and a 1.9 degrees angle of attack. In Figure 11, the computed carriage forces and moments are compared for solutions based on both the Euler as well as the Navier-Stokes equations. Note the significant difference in side force coefficient and yawing moment coefficient of the Mk82 store depending on what store is attached to the central fuselage pylon. It is the basic explanation as to why the store separation simulation is unable to simulate the flight tested configuration using uncorrected wind tunnel data.

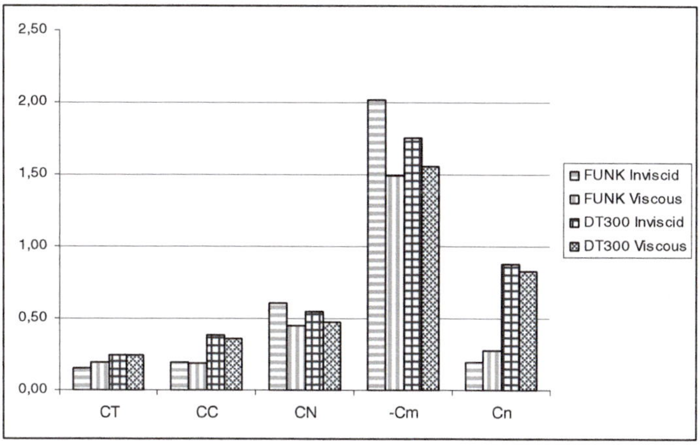

Fig. 11 Computed carriage loads when the store is attached to pylon in position 3R with different stores attached to pylon 5

The store release simulation using wind tunnel data for a configuration with a DT300 drop tank attached to the fuselage centerline pylon (P5) is presented in Figure 12 left. This clearly gives inadequate results when compared to flight test data with a FUNK camera pod attached to the P5. The drop tank exerts a much stronger influence on the Mk82 bomb than a camera pod, making the store simulation fail its attempt to capture the actual drop test. Computing both cases with CFD and adding the difference to the existing wind tunnel data gave a significant improvement as depicted in Figure 12 right.

Fig. 12 Store separation (simulation red/green, flight test grey), Simulation time is 0.45 s. Left figure compare flight test with simulations based on wind tunnel results from DT300 configura-tion. Right figure compare flight test with simulations based on correlated data for FUNK configuration

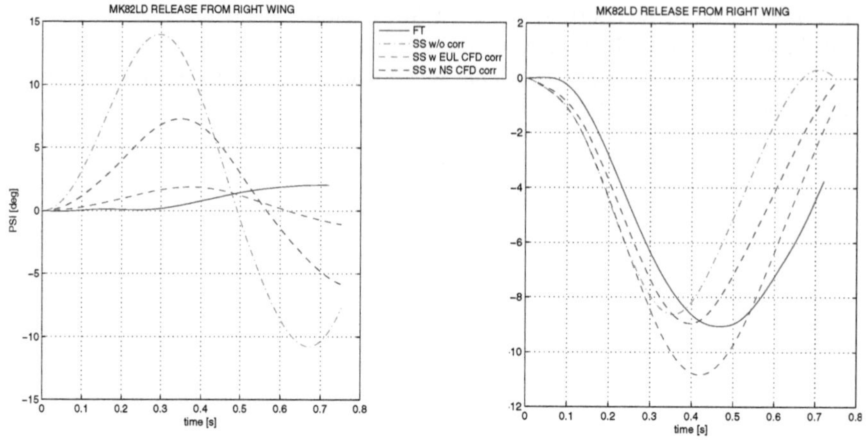

Fig. 13 Relative Y displacement (left) and relative pitch angle (right)

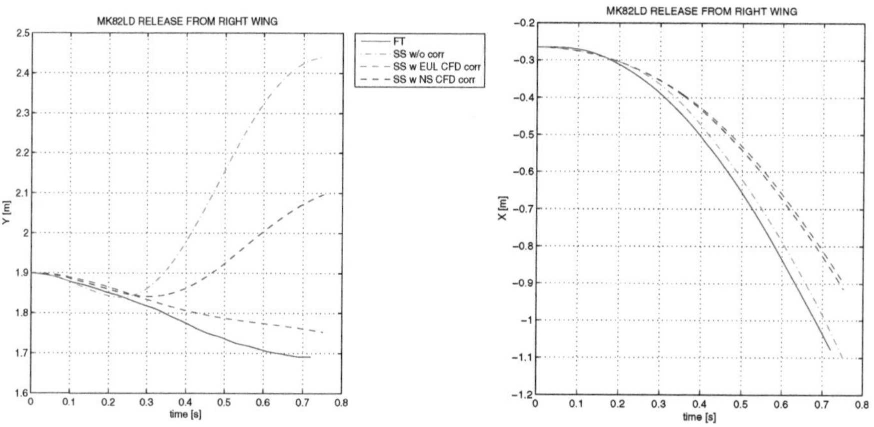

Fig. 14 Relative yaw angle (left) and relative X displacement (right)

Comparing the translation and orientation at greater detail from Figures 13 and 14, it can easily be seen how well the Euler corrected simulation corresponds to flight test data. Simulation of the Y translation is now following the flight test instead of diverging at 0.25 s as before. The yaw orientation is of correct magnitude, although the yaw starts earlier than what flight test shows. The pitch orientation is better in phase but the corrected simulation still predicts a premature nose down movement of the bomb. This causes the pitch recovery to be predicted 0.05 s too early. The only data that becomes impaired by the correction is the X displacement. Investigating the corresponding viscous corrections, it is noted that these lead to slightly less accurate results. In particular, the store separation simulation indicates that the Y displacement is diverging after 0.31 s and that the yaw orientation is clearly from the start already over-predicted. The pitch orientation is somewhat less accurate in phase, predicting pitch recovery 0.07 s too early, but the magnitude of the pitch angle is very well predicted. For both Euler and Navier-Stokes corrected simulations, they follow flight test data well when considering roll orientation and Z displacement.

The Euler grids were decomposed into 16 domains whereas the Navier-Stokes grids were partitioned into 32 domains. An in-viscid solution was established in 4 hours and the viscous computation took approximately 70 hours However, the results in this study do not motivate the use of this model complexity. Our results indicate that the Euler equations are fully sufficient. From this test case, it has been concluded that computational aerodynamics indeed is a useful tool when using it as a correction method.

4 Conclusions

CFD is used by engineers throughout the aerodynamic design process to evaluate and generate data bases. Today, the simulations are complementary to wind tunnel tests and increase in significance. The examples presented in this paper demonstrate challenges both in terms of project planning and physical modelling that still remain to be solved. The combination of well resolved models with the appropriate physical modelling is often too costly to be applied in a tight project schedule. At the same time, by neglecting the correct physics in the simulations introduce technical risks and a continued dependence on experimental testing.

Only by enhancing the physical modelling the numerical simulations can start replacing wind tunnel tests. To accomplish this further development must be paid to both physical treatment to model the correct physics and to numerical treatment of the equations to reduce simulation cost and complexity. Not the least – both engineers and project management must build confidence in the enhanced simulations.

References

Eliasson, P.: EDGE, a Navier-Stokes solver for unstructured grids. In: Proc. To Finite Volumes for Complex Applications III, pp. 527–534 (2002)

Hirschel, E.H.: Towards the virtual product in aircraft design? In: Proceedings of NDF 2000–Towards a New Fluid Dynamics with its Challenges in Aerospace Engineering, CNRS, Paris (November 2000); Champion, M., Periaux, J., Pironneau, O., Thomas, P. (eds.) NFD 2000: Towards a new fluid dynamics. CIMNE Handbooks on Theory and Engineering Applications of Computational Methods, Barcelona (2000)

Karypis, G., Kumar, V.: Multilevel k-way Partitioning Scheme for Irregular Graphs. J. Parallel Distributed Computing 48(1), 96–129 (1998)

Persson, I., Lindberg, A.: Transonic store separation studies on the Saab Gripen aircraft using computational aerodynamics. In: Proceedings of the 26th ICAS Conference, Paper 40 (2008)

Sillén. M.: Application of Parallel Computers to Enhance the Modelling Capability in Aircraft Design, Linköping Studies in Science and Technology, Thesis 1246 (2006) ISBN 91-85523-82-8

Investigation of Practical Flow Control Methodologies with RANS/LES Hybrid Methods

S. Fu, Z.X. Xiao, J.B. Huang, and Y.F. Zhang

Abstract. Passive flow control methods, such as vortex generators (VG) on supercritical wing and before a cavity, spoiler in a large expanded diffuser, boat-tailed or multi-step afterbodies are numerically investigated with RANS or RANS/LES hybrid methods. The basic turbulence models are two-equation k-ω models, such as shear stress transport (SST) and with Durbin's three-dimensional correction (WD+) models. Before applying the flow control methods, the fundamental flows are explored and validated firstly. The flow-fields with flow control methodologies are compared with those of fundamental flows. The results show that the VGs can effectively restrain the separations induced by shock wave and boundary layer interaction on the supercritical wing. VG or VG pair can affect the shear layer instabilities after the leading edge of the cavity. The spoiler appropriately located can eliminate the separations in a diffuser. The boat-tailed and multi-step afterbodies can effectively increase the pressure at the base showing a significant reduction in the base drag.

1 Introduction

The flow control methodologies are widely used in modern vehicles. They can provide an expanding degree of freedom in the design and optimization process. The performances of economics and environment protection can be greatly improved due to the flow control methodologies, which can suppress the separation, reduce the drag, control the noise, and so on.

The flow control methodologies include active and passive methods. Active methods can effectively control the flow for the design purpose. However additional devices and control system, which may lead to increase the weight, cost and complexity, are often needed. Passive flow control methodologies, such as vortex generator (VG), spoiler and multi-step, basically keep the baseline configuration, but often offer good remedy on overall performances. The passive approaches will be investigated in this work.

Although there are many unanswered questions including the turbulence modelling uncertainty, solving Reynolds-averaged Navier-Stokes (RANS) equations with turbulence models is widely employed as a aero-vehicles design option. The RANS methods had been developed to predict many of the important

S. Fu · Z.X. Xiao · J.B. Huang · Y.F. Zhang
School of Aerospace, Tsinghua University, Beijing, China, 100084
fs-dem@tsinghua.edu.cn, xiaotigerzhx@mail.tsinghua.edu.cn

S.-H. Peng et al. (Eds.): Progress in Hybrid RANS-LES Modelling, NNFM 111, pp. 41–58.
springerlink.com

mean flow characteristics, such as the force, moment and velocity, etc., but it was not intended to simulate the complex unsteady flows, pressure fluctuations or turbulent stresses.

Large eddy simulation (LES) is a powerful tool for resolving the large, energy-containing scales of unsteady motion that are typically time and geometry-dependent. However, the sub-grid scale models for the boundary layer and compressible flows are not sufficiently well-developed to accurately predict the high Reynolds number and compressible flows. At the same time, LES require almost the same grid cells as direct numerical simulation (DNS) for high Reynolds number flows near the wall.

Limited by computational resource, the combination of LES with RANS can achieve both reasonably well with high efficiency and numerical accuracy when predicting the flow with massive separation. An alternative modelling strategy of turbulent flows, often called as RANS/LES hybrid methods (proposed by Spalart, 1997, 2006; by Strelets, 2001; by Menter, 2003), has recently been proposed to predict the unsteady and geometry-dependent separating flows. Such hybrid methods combine a high-efficiency turbulence model near the wall, where the flow is dominated by small scale motion with an LES-type treatment for the large scale motion in the flow region far away from the wall. The fundamental turbulence models still significantly affect the predicted flow characteristics for many applications. To construct the hybrid RANS/LES methods rationally, one generally hopes that the turbulence model has good numerical properties and has a *low-Reynolds-number* capability to resolve the near-wall turbulence characteristics.

In this article, two k-ω (Wilcox, 1988) models, shear stress transport (SST, Menter, 1994) and weakly nonlinear correction (WD+, Durbin, 1996), are chosen as the fundamental turbulence models. URANS, DES (Strelets, 2001) and DDES (Menter, 2003) based on the SST or WD+ model are applied to predict the unsteady turbulent flows. Roe scheme (Roe, 1981), which is one of flux-difference splitting (FDS) methods with a 3rd-order MUSCL or a 5th-order WENO (Jiang, 1996) interpolation, is applied to disperse the convection of Navier-Stokes equations. The diffusion terms are dispersed using central differencing. The RANS and turbulence model equations are solved by LU-SGS implicit methods in time-stepping (Yoon, 1987). To achieve higher temporal accuracy, several sub-iterations are applied in pseudo-time.

Three kinds of flow control devices are here investigated: VG or VG pair or VG-in-line (on the plane, before a cavity, and on the supercritical wing), ring rod (in a diffuse), boattailed (Hermin, 1994a) and multi-steps (Viswanath, 2001) afterbodies. In these cases, they consist of very complex flow physics, such as boundary-layer separations, shock wave/boundary layer interaction, vortex shedding, recirculations, etc. In our studies, RANS methods are used to acquire the overall effect of flow control devices and DES or DDES is then implemented for detailed investigation of flow structures and unsteady features.

2 Passive Flow Control

2.1 VGs

VGs are flow control devices which can generate the longitudinal vortices giving rise to local mixing of mean flow, energizing the boundary layer and delaying or preventing the turbulent boundary layer separation. It can be used to improve the performance or to optimize the design when the designer expect to reduce the weight, drag and cost. Therefore, they are very efficient and widely used in both external and internal aerodynamics. Experimental studies as well as computations have shown the ability of controlling separation with VGs.

VG on a flat plane: validation of RANS based on SST model. Yao, Lin and Allan (2002) presented the detailed flow-field measurements downstream of a vane-type low-profile VG in the Langley Shear Flow Tunnel. Here, the flow fields around the same VG are predicted to validate our code.

The deflection angle from the main flow of the VG is 23° and the streamwise velocity is 34m/s. The VG's height (H) is 7 mm and the Reynolds number based on it is about 20,000. The baseline boundary layer flow is attached without the VG. The boundary layer thickness (δ) at the VG location is 35mm, the VG's height is about 20% of δ. It means that the VG is submerged in the boundary layer. It indicates that RANS may be the suitable turbulence modelling method. Of course, many cells are needed with LES because boundary layer should be well resolved at such high Reynolds number (2.843×10^6/m).

About 16 million cells, shown in figure 1a, are generated around the VG with point-to-point (P-to-P) structured grids. From this figure, we can see that the grids in the wake are clustered to capture the flow details.

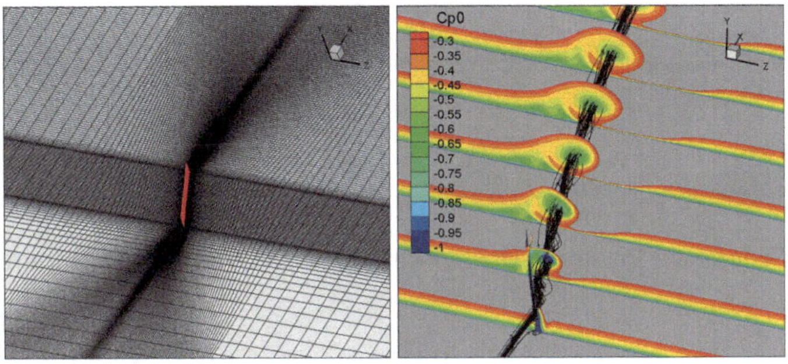

Fig. 1a Grids and flow near the single VG

The vortices past the VG are given in the same figure of 1a. Due to the large deflection angle, the flow can not keep attached on the leeward of the VG. Strong vortices are detached from the leading edge of the VG and they will introduce the high energy flow at the outer boundary layer into the inner low momentum boundary layer.

Comparisons on the streamwise velocity U and vortices at three streamwise sections, where x are 15.2, 27.9, 53.3 and 104 cm after the trailing edge of VG, between the measurements and computations are plotted in figure 1b. From this figure, the vortices generated by the VG can give a continuous influence on the main flow to a very long streamwise distance, such as at section x =104 cm. The streamwise-vorticity (Ω_x) is very high and the position of the maximum is corresponding to the vortex core.

These comparisons show that the numerical methods (such as the spatial discrete schemes, turbulence modelling methods and implicit time-marching methods) are validated through predicting the flows around a low-profile single VG on the flat plane.

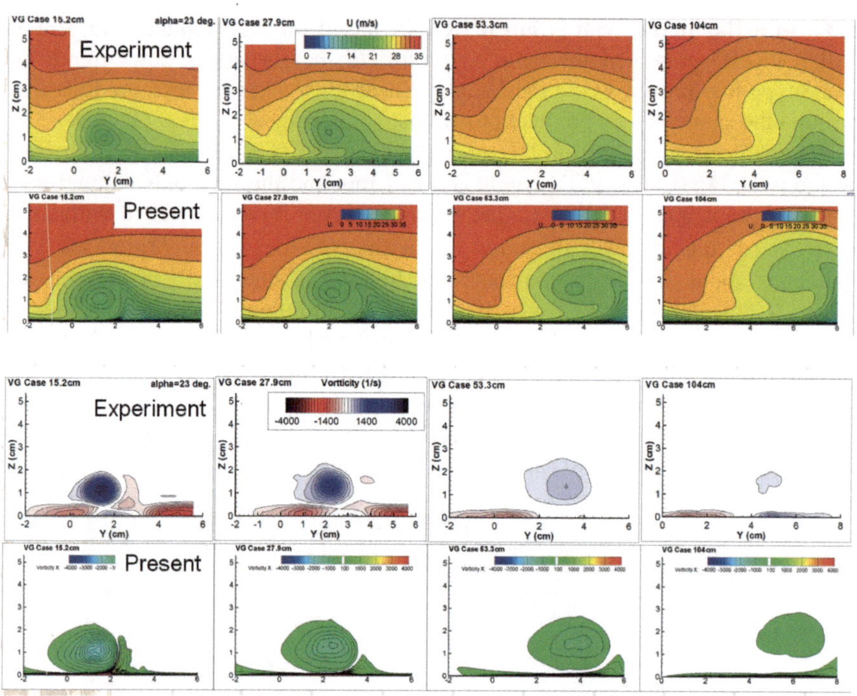

Fig. 1b Comparison for streamwise velocity (m/s) and vorticity components((1/s) using RANS

VGs on a supercritical wing by RANS based on WD+ model. On the upper wing surface of the Boeing-737 or 757 or 777 new generation airplanes, a row of VGs are installed near the leading edge of the wing. Here, the transonic flow around a supercritical wing with VGs mounted on the upper wing surface is investigated to study of the effect of these VGs.

The baseline configuration and the grids are presented in figure 2a. In the case of baseline wing, two-block P-to-P grids ($151\times101\times81+66\times11\times26 \cong 1.25$ million cells) are applied to depict the original configuration. In the case of wing with 3 and 7 VGs, about 3.22 million quasi-P-to-P grids are generated to depict the small size of VG and to save computational resource, which means that the grids around the VGs are much denser in both streamwise and spanwise directions than the region without VGs. The VGs are located at 2z/B=48.4%, 57.3% and 66.2% for 3-VGs case, and from 43.9% to 70.6% with equal interval for 7-VGs case, respectively. The size of each VG is equal and the VG's length is about 1.5% mean aerodynamic chord (MAC). The ratio of height over length is 1:4 and the thickness is taken as 0. The deflection angle is taken as 25°.

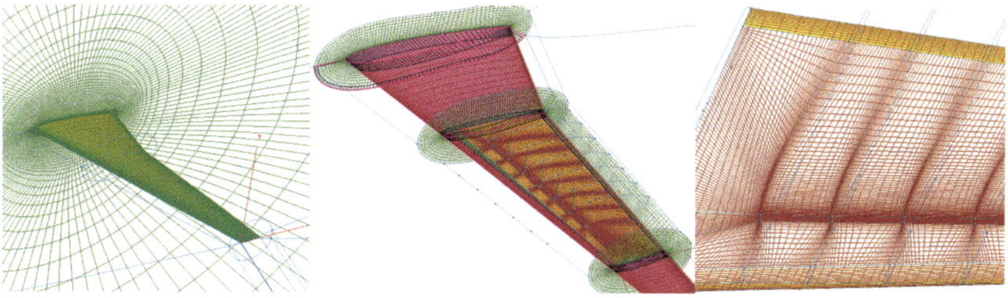

Fig. 2a Grid around the supercritical wing and VGs

The flow is transonic flow with a cruise Mach number M_{crs} at about 0.8 and a Reynolds number 5.94×10^6/m. To explore the VG effect on the shock wave/ boundary layer interaction, the angle of attack is taken as 4°.

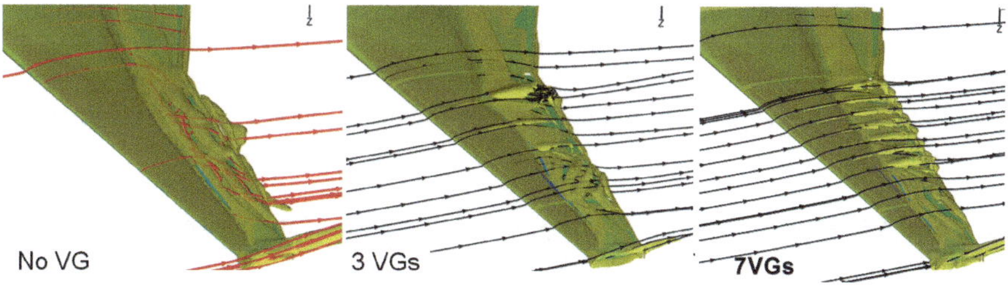

Fig. 2b Vortices and streamlines over the wing

The spatial vortices and streamlines are presented in figure 2b. The spanwise flow is very intense and the flow is almost separated in the wing middle-span region. The spanwise vortices accumulating to the wing-tip region lead to very strong vortex detached from the wing tip. It indicates that the induced drag is remarkable.

The flow around the wing with three VGs is very different from that without VG. Firstly, the inner VG seems to hold up the spanwise vortex development and an obvious vortex bump exists after the separation. However, flows past the second VG seem to show no separation, even with no spanwise flows. Secondly, between the outer VG and wing-tip, the spanwise flow is relatively weakened. Lastly, the wing-tip vortex becomes very small due to the three-VGs.

The flow around the wing with 7-VGs is also different from that without VG and with 3-VGs. The spanwise flow is effectively controlled as viewed from the streamlines and spanwise vortices. The large and strong spanwise vortices are significantly weakened and no saparation are present in the VGs region. Again, the wing-tip vortex is much weaker than that without VG.

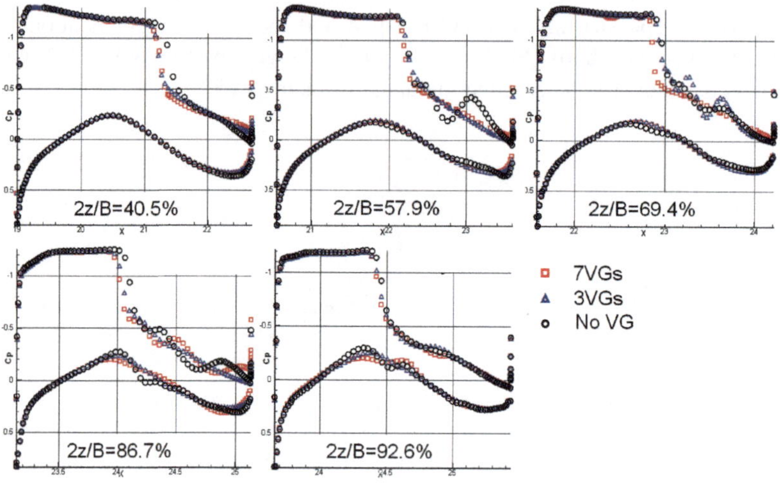

Fig. 2c Pressure coefficients on the wing surface

Comparisons on pressure coefficients at five spanwise sections for the three cases are presented in figure 2c. Sections of 2z/B = 57.9% and 69.4% are located in the VGs region. Although the shock wave position is somewhat more upwind in the case with VG, the separation induced by SW/BL interaction is almost removed by the 7-VGs.

VG and VG pair before a 5:1 cavity. Cavities, such as landing gear bay of a civil airplane, bogie of the high-speed train or car with open window, can generate unsteady vortex shedding, interactions and strong self-sustaining oscillation which in turn can generate acoustic resonance phenomena. Resonance in this kind of open cavities arises from the feedback mechanism between the growth of shear layer instabilities (Kelvin-Helmholtz instability) after the cavity edge and the periodic impingement of high speed flow on the rear wall. Pressure waves emanating from the rear wall trip the former leading to undulations in the shear layer with further high speed impingement on the rear wall. Resonance always occurs when the frequency and the phase of the acoustic energy coincide with the instabilities of the shear layer.

In the case of civil airplane, very strong noise generated at the phase of taking-off and landing badly pollutes the environment of airport and flight path. In the case of a car or a high-speed train, the window or the bogie produces very loud and uncomfortable noise when they operate at specific speeds.

Through analyse of the noise generation mechanism, the following counter-measures are considered effective for cavity noise reduction: 1) Weakening vortex shedding from the edge; 2) Deflecting shear layer between the outside and inside of a cavity for vortices not to collide with the downstream wall; 3) Weakening pressure pulse generation even if vortices collide with the downstream edge. Some devices with low-profile, such as ramps, spoilers, rods in cross flow, or VGs can ameliorate these resonant phenomena. This article tries to explore the VG or VG-pair effects on the vortex shedding and shear layer instabilities to realize the first or second countermeasures.

The cavity geometric configuration is represented in figure 3a, where the length over depth ratio is L/D=5. The cavity height D was used as the reference distance for nondimensionalization. The spanwise width is also D. Thus, the ratio of length/width/height is 5:1:1.

The freestream Mach number is taken as 0.2.The Reynolds number based on D is 1.36×10^{6}. The turbulence modeling methods are URANS and DES based on k-ω-WD+ model.

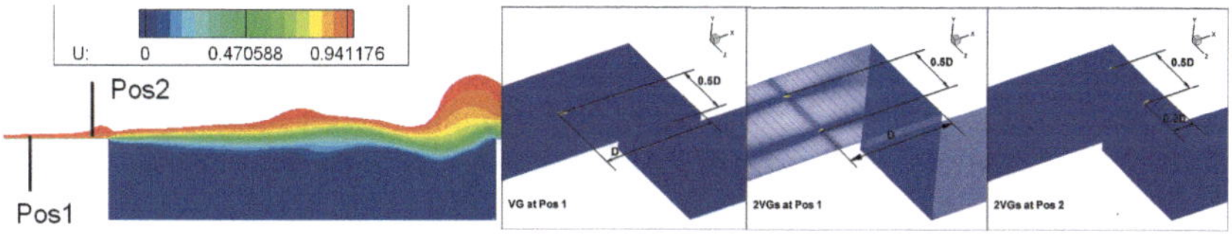

Fig. 3a Positions of VG pair

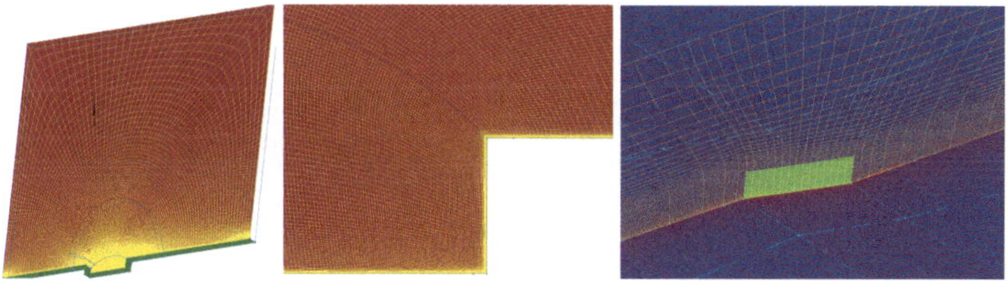

Fig. 3b VG and VG pair position, grids (overall, near rear wall and VG)

The VG and VG pair are installed before the cavity leading edge. The size and position are chosen through the clear cavity flowfields from figure 3a and CASE 2.1.1. The parameters of low-profile VG or VG pair are given as following. The deflection angle is 16°. The height (*H*) of VG is 0.015D, the length (*L*) is 0.06D and the thickness is taken as 0. Two installation positions of VG or VG pair are chosen. One is 1 D (named as Pos 1) and another is 0.2 D (named as Pos 2) before the cavity leading edge. In the case of a single VG, the VG is located at Pos 1, while in the case of VG pair, the VG pair is located at both Pos 1 and Pos 2 to explore the effect on the streamwise development of the VG vortices.

The computational grids are presented in figure 3b. The grids consist of 301×201×201(totally about 12 million) cells in streamwise, normal and spanwise directions, respectively. The grids of this topology can save the computational resource and keep the grid quality. The grids near the VG pair are clustered in both normal and spanwise directions.

In the case of a single VG, URANS and DES are applied to investigate the characteristics of the cavity flow. From figure 3c, DES can predict relatively finer scale structures than that of URANS. The vortices detached from the VG can affect the main flow very long in the streamwise direction. This vortex is very concentrated over the flat-wall before the cavity and the spanwise structures after the leading edge are split into two segments. Moreover, the concentrated vortex can continuously perforate the spanwise quasi-three-dimensional structures to about 1.5D (i.e. about 166*H*) distance after the leading edge of the cavity, labeled as red dashed line.

Of course, the difference between without and with VG or VG pair is our research interest. The flow without and with VG pair at Pos 1 are investigated using DES. In fact, the iso-surface of Q (= S^2-Ω^2) in figure 3d is different from those in figure 3c, because no coherent structures without VG pair are shown if the same iso-Q is plotted. From figure 3d, quasi-two-dimensional structures without VG pair are very obvious and the spanwise three-dimensional structures are well predicted with VG pair. Morever, the vortices detached from the two VG look merged after the cavity leading edge. From comparisons on the frequency VS pressure level, the pressure level without VG is almost larger than that with single VG or VG pair, although the primary frequency is a little higher that that with VG.

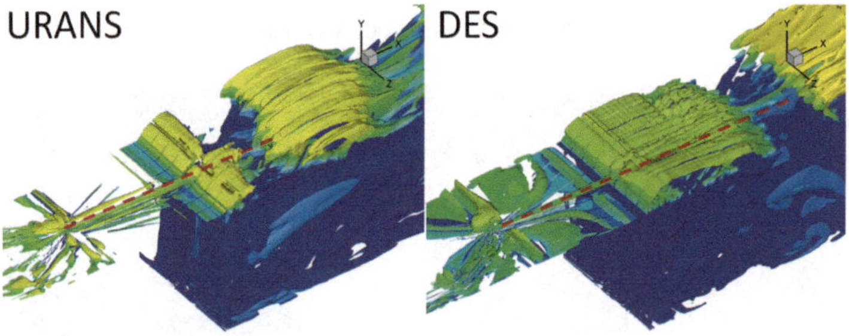

Fig. 3c Flow with a single VG by URANS and DES near the cavity leading edge

The effect on the different locations of VG pair is presented in figure 3e. From this figure, the VG pair effect on the shear layer is very different at the two locations. The vortex generated from the VG pair at POS 1 is almost fully developed; however, the vortex from VG pair at Pos 2 is underdeveloped. After analyzing the flow, the main reason seems that VG pair of the same size is relatively much smaller than that at POS 1. The vortex generated from the VG pair can not be well developed in the thicker boundary layer thickness at Pos 2. From the figure of frequency VS pressure level, the pressure levels about the VG pair at Pos2 are higher than those of VG pair at POS 1, where four pressure sample points on the bottom wall of the cavity are located at (0.15, 0, 0.061), (0.45, 0, 0.061), (0.75, 0, 0.061) and (0.95, 0, 0.061). It should be noted that upper-left point on cavity bottom is taken as the origin point O(0,0,0).

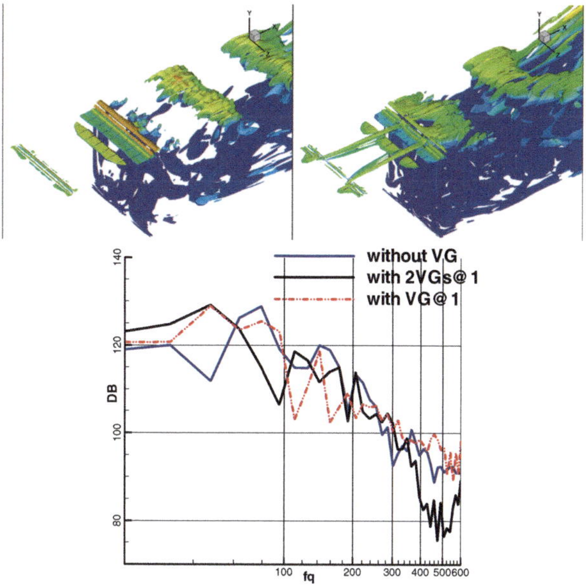

Fig. 3d Cavity without or with VG pair

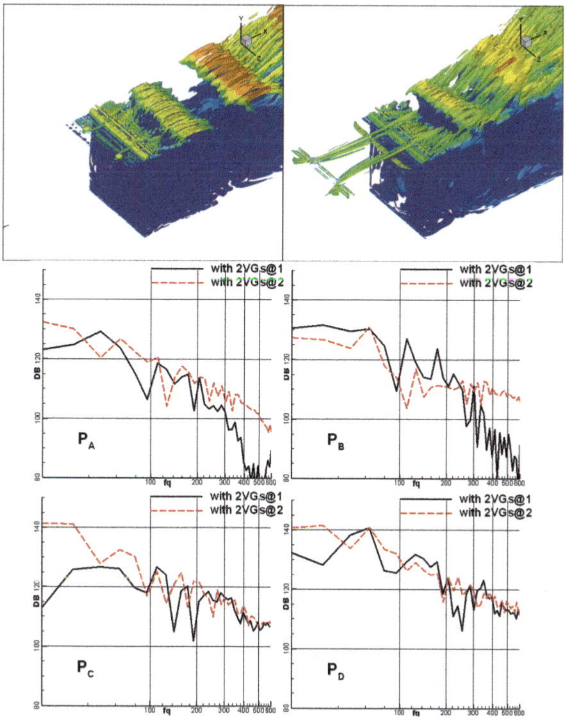

Fig. 3e Effect on positions of VG pair

2.2 Ring Rod in a Diffuser

Diffuser is widely used in engine system. The performance of engine strongly depends on the flow quality of the diffuser from the inlet to the engine. To obtain higher efficiency of the engine, the flow separation is undesirable and causes the total pressure distortion. The sub-merged inlet of the modern fighter or high performance bomber in the fuselage is designed to minimize the radar cross section (RCS). At the same time, small inlet is hoped to save space for more fuel and longer range. Thus the divergence angle and expansion ratio become large. It indicates that massive flow separation appears in the duct, and strong velocity and pressure fluctuations exist in the flow field. Kaltenbach et al. (1999) investigated the flow in a planar diffuser with a 10° divergence angle using LES, and strong fluctuations were presented.

In present, the flow past a diffuser with a 14.57° of divergence angle and 3.5 times of expansion ratio is numerically studied. The geometry is presented in figure 4a. The radius of inlet is 0.133m, the radius of the outlet is 0.25m and the length of the diffuser is 0.45m. The total temperature is 930K and velocity is 221m/s, the inlet Mach number is about 0.45.

The flow keeps subsonic in the diffuser due to geometric expansion. A large separation bubble locates near the wall around the expansion region.

A simple flow control device, a ring rod, is installed at the beginning of the expansion to suppress the flow separation. We hope that the unsteady vortex shedding triggered from the ring can introduce high energy flow into the low momentum recirculation region and to sppress the separation.

The ring's radius is 0.01m. The position of the ring is also shown in figure 4a. The grids around the ring are presented in the same figure and the total cells in the diffuser are 10.5 million. In this section, SST model is taken as the fundamental turbulence model.

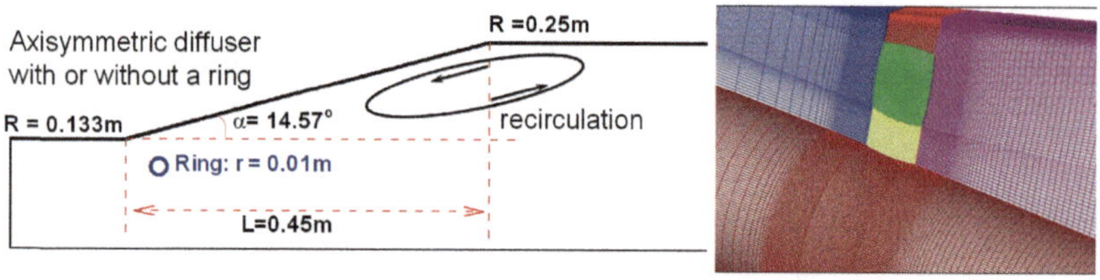

Fig. 4a Geometry of the diffuser and the grids near the ring

Firstly, the baseline flow without the ring is investigated using URANS. The flow on the symmetric plane is shown in figure 4b (Left). Large scope of recirculation approaching to the expanded wall is very distinct and high energy flow is clustered to the central region. The total pressure distortion at the outlet is very strong. The time-averaged flow field with the ring by URANS is also presented in this figure (Right). Although low Mach number region exists, separation disappears. The total pressure recovery of the controlled diffuser is better than the uncontrolled one. It indicates that the purpose is achieved through the ring.

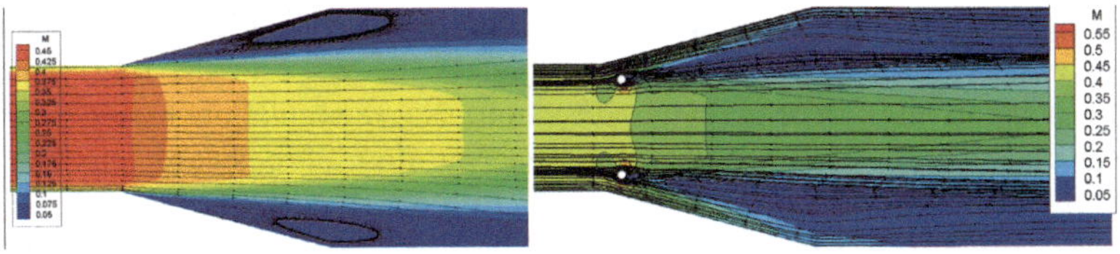

Fig. 4b The baseline flow and the time-averaged flow with ring (by URANS)

As we known, URANS has very large eddy viscosity. Small turbulent structures are easily removed when the unsteady turbulent flows are simulated with URANS and only large scale structures are kept. From the left figure of 4c-1, circumferential flow structures are presented in series and no small structures are shown by URANS. DES or DDES, which is one kind of RANS/LES hybrid methods, was developed to predict the massive separations and obtain more detailed flow features through changing the length scale of the turbulence structures. It leads to lower turbulent dissipation in the flow fields. Therefore, little turbulence structures are easily predicted using this kind of hybrid turbulence modelling methods. From the middle and right figure of 4c-1, plenty of small scale structures after the ring are presented through the iso-surface of vorticity. The circumferential continuous structures near the ring keep a very little distance and they split to form several small scale three dimensional structures.

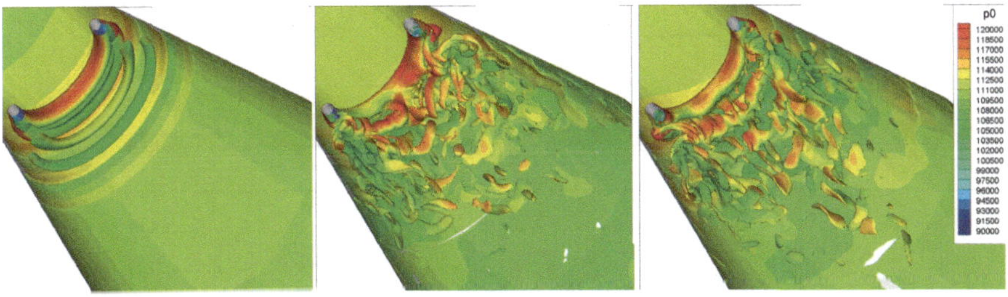

Fig. 4c-1 Comparisons on isosurface of instantaneous vorticity (coloured by total pressure)

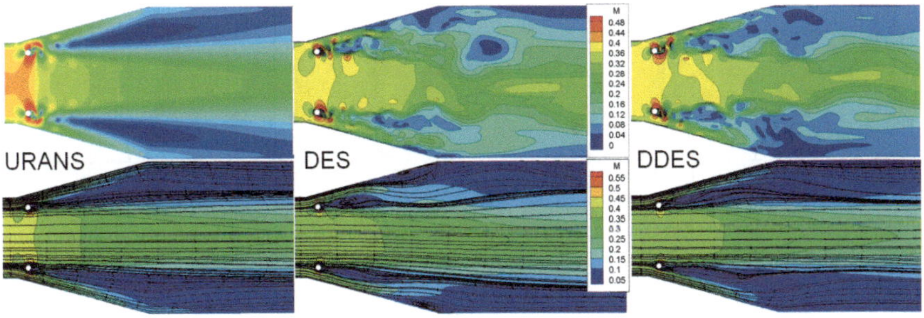

Fig. 4c-2 Comparisons on time-averaged and instantaneous Mach contour

From figure 4c-2, the flow features on the symmetric plane are presented. URANS shows clearly Karman vortex street in the wake of the ring, which is synchronous in the circumferential direction. No recirculation exists from the time-averaged flow by URANS. High energy flows are swept into the low momentum boundary layer. DES can predict the small scale turbulent structures. However, a relatively small scale recirculation exists near the wall from the time-averaged results. DDES, from our experience, can give a more reasonable time-averaged flow fields (i.e. no recirculation) and instantaneous three-dimensional turbulence structures.

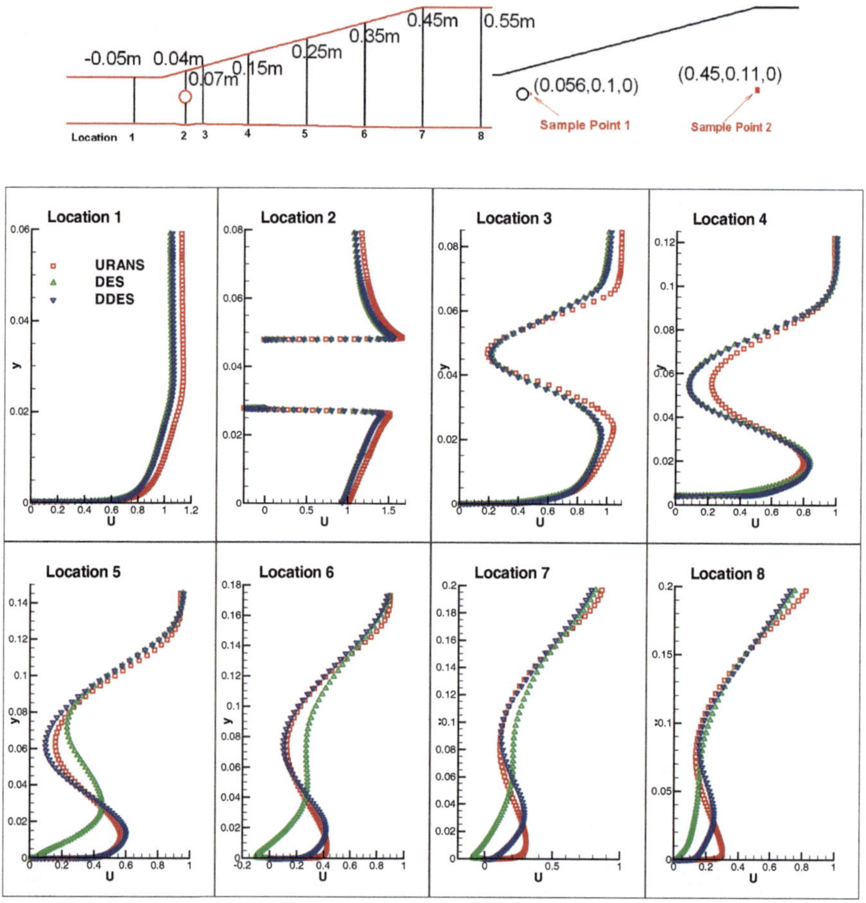

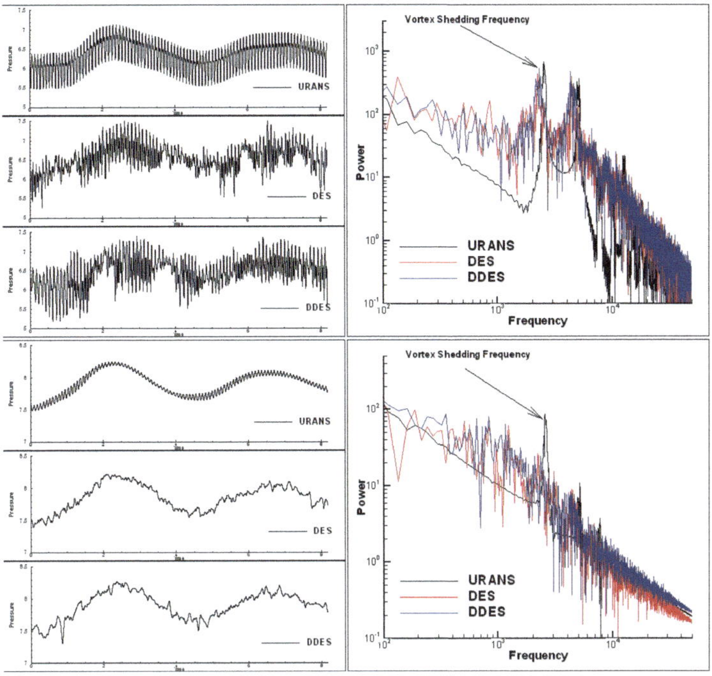

Fig. 4d Averaged u-velocity profiles, pressure fluctuation and frequency

Time-averaged streamwise velocity profiles at eight positions, where the beginning of the expansion is taken as the original point, are presented by URANS, DES and DDES. The details of positions are listed in figure 4d. When x is less than 0.15m, velocity profiles by URANS, DES and DDES show little difference, although the velocity by URANS is a little larger than DES and DDES. At the rear of the expansion, the results by URANS looks like those of DDES. DES predicts separation near the wall because of the u-velocity is negative. The low velocity region predicted by DES looks much larger than that of both URANS and DDES.

The pressure fluctuations at two samples points are investigated. Point A (0.056, 0.1, 0) is located in the wake of the ring and the Point B (0.45, 0.11, 0) is located at the end of the expansion. Because Point A is approaching to the ring, the pressure coefficients look like high-frequency sinusoidal fluctuation plus low-frequency wave. The dimensionless time 1 means 1/221 second in physical time. The main frequency is about 2400Hz, which is corresponding to Strouhal number 0.2 of the Karman vortex shedding. Three turbulence modelling methods can predict the main frequency, although the frequencies by DES and DDES are about 2200Hz. The second frequency about 5000Hz is also predicted by three methods, which is the effect on the vortex pair shedding from the upper and lower surface, respectively. At point B, the main frequency is kept only by URANS. However, both DES and DDES can not demonstrate an obvious primary frequency.

2.3 Boattailed and Multi-steps Afterbodies

Base flow (Herrin and Dutton, 1994b) is one of the fundamental fluid mechanics problems occurring in flows around bluff bodies, such as projectiles, missiles, rockets, spaceships, or ground vehicles like cars, trucks, trains, *etc*. The aerodynamic characteristic at the base strongly influences the drag and stabilities of these moving objects. The low pressure behind the base due to the recirculation commonly existing there can cause significant amounts of total drag (40% or more), especially for a missile in unpowered flight situation, where the high pressure and high momentum jet is unavailable. If the base drag is effectively reduced, the flight range or the payloads can be significantly extended.

Among the passive methods, the B.T. or conical concept is very popular and proved very effective in reducing base drag. However, to achieve more drag reduction, the length of conical afterbody has to be sufficiently long because a short B.T. afterbody does not offer necessary pressure recovery on the base surface. Application of shorter, lighter and lower cost afterbody with possibly less base drag is the perfect objective for designers. M.S. afterbody is another concept which can effectively reduce the base drag through controlling the separation before the base. It can offer less base drag, shorter afterbody length and lighter structure with careful optimization of the afterbody geometry. They can also achieve friction reduction due to the recirculation behind the sub-steps. Both B.T. and M.S. afterbodies are easy to implement in practice and to achieve the drag reduction effectively.

Two-block grids ($111\times97\times161+25\times25\times161\cong1.83$ million cells for Cy. and B.T. cases and $161\times97\times161+25\times25\times161\cong2.61$ million cells for M.S. case) are applied to avoid the singularity of the base centre, as shown in figure 5a. The base area of M.S. is the same as the B.T. one and the ratio of length over height is the same as each other. The ratio of length over height of each sub-step is arctan (5°) (i.e. $\eta = l / h = \arctan(5°) \cong 11.43$).

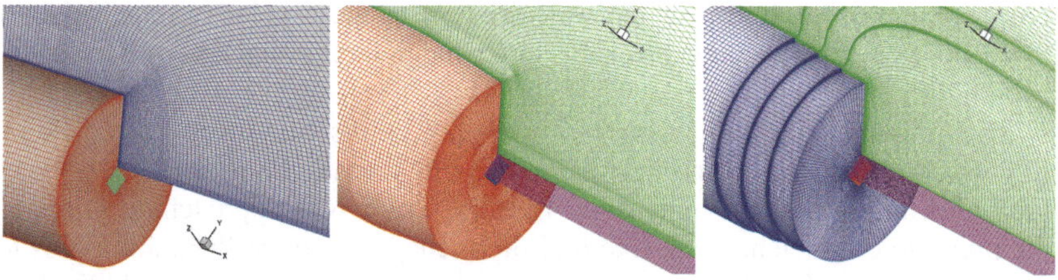

Fig. 5a The three base-configurations and grids (from left to right, Cy., B.T. and M.S.)

Flow conditions of the supersonic axisymmetric base flow are based on the experiments. The free-stream Mach number is set to 2.46, and the Reynolds number based on the cylinder diameter (D* = 63.5 mm) is 2.858×106, and the angle of attack is taken as 0 degrees. Global dimensionless time stepping (Δt=0.005) is implemented to capture the unsteady properties, which is corresponding to 1.37×10-6 second in the physical time. The fundamental turbulence model is SST model with compressible correction.

Figure 5b presents the pressure coefficients on the wall of base and afterbody. The measurements on the base for Cy. and B.T. afterbodies are available from Herrin and Dutton. The numerical results are all obtained using the DES because it can presents more satisfactory results than that by URANS in the Cy. case. At the base, the pressure coefficients between the B.T. and M.S. show little difference from each other, the M.S. afterbody has similar base pressure drag as that of B.T. It is important to note that both B.T. and M.S. afterbodies give a reduction in the level of the pressure coefficient as shown in Figure 5b. Of course, B.T. afterbody can afford a little higher pressure recovery on the base, which means that more base drag reduction can be achieved.

In the streamwise direction, the surface pressure on the B.T. afterbody matches the measurements very well, and the pressure coefficients on the M.S. afterbody are very complex due to the sub-steps starting at x/D=-0.5 before the main base. After each sub-step, the pressure decreases rapidly indicating the existence of sub-base drag. However, the pressure encounters recovery along the sub-step, although it can not resume as the Cy. case.

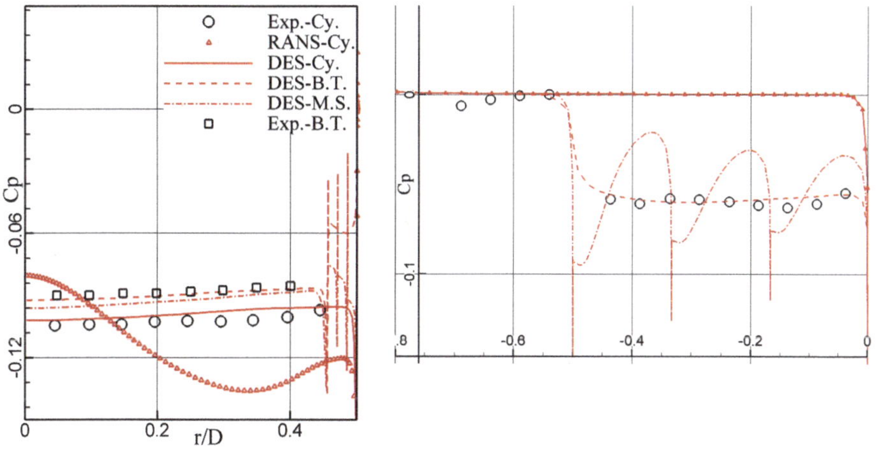

Fig. 5b The pressure coefficients on the base and the afterbodies (by DES and URANS)

Figure 5c presents comparisons on the streamwise and radical velocities and streamwise normal stress at several streamwise positions around the three configurations. The measurement data are based on the Cy. case. From the flow structures behind the base, the velocities around the B.T. and M.S. differ insignificantly from each other, the velocity of Cy. afterbody shows more distinct departure from them. From this figure, the normal stress (uu) of Cy. base is generally higher than those of B.T. and M.S., and the position of the maximum stress is more outward in the radial direction.

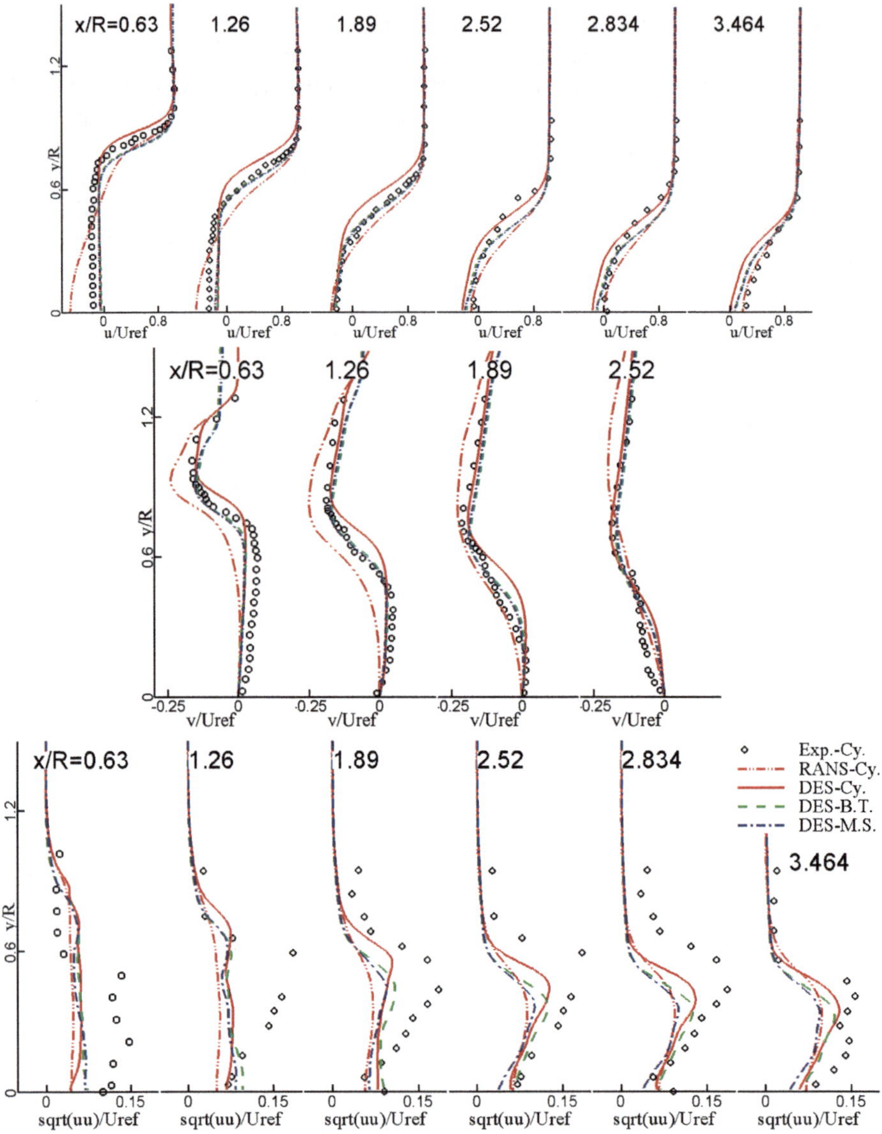

Fig. 5c The comparisons on the streamwise and radial velocity and the normal stress

Figure 5d shows instantaneous iso-contours of Q = -1 around the three configurations by DES. Plenty of small scale vortices are accumulated just downstream of the base. The interaction between the recirculation and the shear layer are obviously visible.

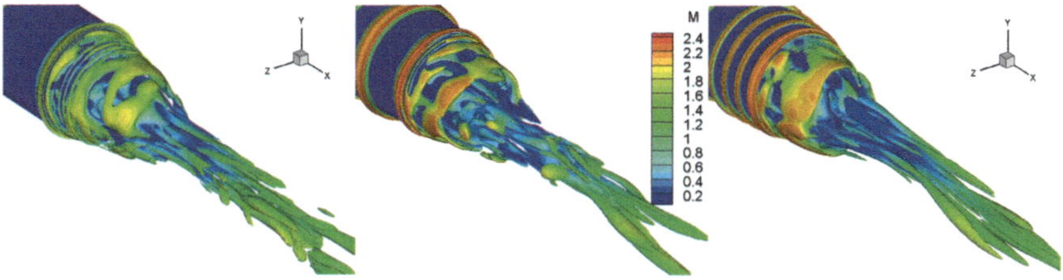

Fig. 5d Instantaneous structures (Q=-1) coloured by Mach number after the base

3 Conclusion

The passive flow control methodologies, such as VG, ring and boat-tail and multistep, are investigated with URANS, DES or DDES to explore the complex interaction of vortices with boundary layer, shock wave, shear layer instability. The hybrid methods show better performances in these cases on overall effects, unsteady flow structures and pressure fluctuations, especially in the flows with massive separations and interactions.

Acknowledgement. This article is supported by EU project (ATAAC) and National Science Foundation of China under Contract 10932005.

References

Durbin, P.A.: On the k–ε stagnation point anomaly. Int. J. of Heat and Fluid Flow 17(1), 89–90 (1996)

Fu, S., Xiao, Z.X., et al.: Simulation of wing-body junction flows with hybrid RANS/LES methods. Int. J. of Heat and Fluid Flow 28, 1379–1390 (2007)

Hermin, J.L., Dutton, J.C.: Supersonic near-wake afterbody boattailing effects on axisymmetric bodies. J. of Spacecraft and Rockets 31(6), 1021–1028 (1994a)

Herrin, J.L., Dutton, J.C.: Supersonic base flow experiments in the near wake of a cylindrical afterbody. AIAA J. 32, 77–83 (1994b)

Janssen, J.R., Dutton, J.C.: Time-series analysis of supersonic base-pressure fluctua. AIAA J. 42, 605–613 (2004)

Jiang, G., Shu, C.W.: Efficient implementation of weighted ENO schemes. Journal of Computational Physics 126, 202–228 (1996)

Kaltenbach, H.J., Fatica, M., Mittal, R., Lund, T.S., Moin, P.: Study of flow in a planar asymmetric diffuser using large-eddy simulation. J. of Fluid Mechanics 390(1), 151–185 (1999)

Lin, J.C.: Review of research on low-profile vortex generators to control boundary-layer separation. Progress in Aerospace Sciences 38, 389–420 (2002)

Menter, F.R.: Two equation eddy viscosity turbulence models for engineering applications. AIAA J. 32, 1598–1605 (1994)

Menter, F.R., Kuntz, M.: A zonal SST-DES formulation. In: DES-WORKSHOP (2003)

Roe, P.L.: Approximate riemann solver, parameter vectors, and difference schemes. Journal of Computational Physics 43(2), 357–372 (1981)

Spalart, P.R., Jou, W.H., Strelets, M., Allmaras, S.: Comments on the feasibility of LES for wings and on the hybrid RANS/LES approach. In: Advances in DNS/LES, 1st AFOSR Int. Conf. on DNS/LES (1997)

Sparlart, P.R., Deck, S., Shur, M.L., Squires, K.D., Strelets, M., Travin, A.: A new version of detached-eddy simulation, resistant to ambiguous grid densities. Thero. Comput. Fluid Dyn. 20, 181–195 (2006)

Strelets, M.: Detached eddy simulation of massively separated flows. AIAA paper 2001-0879, (2001)

van Leer, B.: Towards the ultimate conservative difference scheme V: a second order sequel to Godunov's method. Journal of Computational Physics 32, 101–136 (1979)

Viswanath, P.R.: Drag reduction of afterbodies by controlled separated flows. AIAA J. 39, 73–78 (2001)

Wilcox, D.C.: Reassessment of the scale-determining equation for advanced turbulence models. AIAA Journal 26, 1299–1310 (1988)

Xiao, Z.X., et al.: Study of delayed-detached eddy simulation with weakly nonlinear turbulence model. J. of Aircraft 43, 1377–1385 (2006)

Yao, C.S., Lin, J.C., Allan, B.G.: Flow-field measurement of device-induced embedded streamwise vortex on a flat plate, AIAA 2002-3162, (2002)

Yoon, S., Jameson, A.: Lower-upper symmetric-Gauss-Seidel method for the Euler and Navier-Stokes equations, AIAA Paper 87-0600 (1987)

Advanced DES Methods and Their Application to Aeroacoustics

U. Michel, D. Eschricht, B. Greschner, T. Knacke, C. Mockett, and F. Thiele

Abstract. Detached eddy simulation (DES) is shown to be a suitable method for the simulation of the sound generation of turbulent flows, because it provides access to the resolved turbulent scales at minimal computational cost. The near-wall region is solved efficiently by RANS while LES is applied to all regions containing scales important for the noise generation. In addition to the usual LES resolution requirements, for acoustics the smallest of these scales are defined by the highest frequencies of interest. The sound radiation is generally computed by solving an integral over a data surface surrounding the source region outside the turbulent flow. The grid must be fine enough to resolve the sources and the propagation between the sources and the data surface. Examples for the simulation of noise emission problems are presented for the influence of nozzle serrations on jet mixing noise and three airframe noise sources, which are rod-airfoil interaction noise, the noise of high-lift devices, and airfoil self noise.

1 Introduction

Already in the early design stages, the noise levels of aircraft have to be guaranteed by the airframe and engine manufacturers to the airline customers. This is currently achieved by relying on experience from earlier designs, on semi-empirical prediction schemes, on model test results with new designs, and to an increasing degree on numerical simulations. This paper is concerned with the development of reliable numerical tools for the prediction of the sound emission of turbulent flows of direct relevance to the aircraft industry.

Aircraft noise consists of both engine and airframe noise. Concerning the engine, traditionally the strongest source during take-off has been jet mixing noise. In recent years, a new technique has been developed that may reduce jet mixing noise by a factor in the order of two with respect to sound power (3 dB). This is achieved by adding serrations (chevrons) to the nozzles of the engines. These serrations slightly penetrate into the jet flow, yielding pairs of counterrotating longitudinal vortices along the nozzle circumference. The vortices enhance the mixing in the vicinity of the nozzle and yield a thicker shear layer and a shorter jet with lower turbulence

U. Michel · D. Eschricht · B. Greschner · T. Knacke · C. Mockett · F. Thiele
Institute of Fluid Mechanics and Engineering Acoustics, Technische Universität Berlin, Müller-Breslau-Str. 8, 10623 Berlin
ulf.michel@cfd.tu-berlin.de

S.-H. Peng et al. (Eds.): Progress in Hybrid RANS-LES Modelling, NNFM 111, pp. 59–76.
springerlink.com

levels. The optimum design of the serrations is currently determined with the aid of expensive and time consuming model tests. A trustworthy numerical method could speed up the design and reduce the development costs.

A further contribution to aircraft noise is airframe noise, primarily the noise of the high-lift system and of the landing gear. One of the major airframe noise sources is the slat on the leading edge of the wing. Slats are used on all modern aircraft to ensure that the flow remains attached on the upper surface of the wing at high angles of attack. The slat geometry is optimised with scale models in aerodynamic wind-tunnel tests. Here again, it would be ideal if the optimum design with respect to the noise emission could be investigated with numerical methods.

Detached Eddy Simulation (DES) is a numerical method that promises the required feature of resolving all fluctuations that are important for the noise emission of the flow, while modelling the smaller scales. The method was originally proposed by Spalart et al (1997). Further details can be found in a review article by Spalart (2008) and in the PhD thesis of Mockett (2009). The Institute of Fluid Mechanics and Engineering Acoustics (ISTA) at the Technical University of Berlin has developed numerous implementations of the DES method, several of which are used in the applications described in the following.

The paper consists of three parts. The current status of the DES-method at ISTA is described in section 2. The grid requirements for modelling the noise sources are described in section 2.4.2, those for the sound propagation from the sources up to the integration surface are discussed in section 2.4.3. The application to jet mixing noise is shown in section 3.1 and the application to airframe noise sources in sections 3.2, 3.3, and 3.4.

2 Simulation of Sound Emission with DES

The Navier-Stokes equations do not only describe turbulent flows but also their sound generation and the propagation of the sound waves to any observer position. However, a direct numerical simulation (DNS) of the sound radiation is not possible with available computer resources for Reynolds numbers of technical interest. To reduce the computational effort in the source region, the precise DNS is substituted by less expensive methods such as Large Eddy Simulation (LES) or Detached Eddy Simulation (DES). Both methods rely on modelling the smaller scales while the larger scales are resolved in space and time. The difference between LES and DES lies in the treatment of the turbulent boundary layers. The resolution requirements of LES in the near wall region lead to unfeasibly fine grids for higher Reynolds number flows. DES is a hybrid RANS-LES method that applies RANS to the near-wall region and LES to separated flow regions or, in some applications, to the outer part of a boundary layer, and is aimed at bridging the gap that exists between conventional RANS and LES methods in terms of computational expense and predictive accuracy.

DES is first briefly described in section 2.1. Developments to remedy several shortcomings are then discussed in section 2.2. The numerical scheme used in the computations reported here is described in section 2.3. The special considerations for the description of the sound sources with DES are then discussed in section 2.4.

2.1 DES in Comparison to DNS and LES

The advantage of DES in comparison to DNS and LES with respect to the simulation of the sound sources is that the simulation of the unsteady flow can be restricted to the structures in the flow that are responsible for the aerodynamic noise generation. Boundary layers are solved efficiently in RANS-mode while the sound sources are described with LES-mode. The development of DES is primarily connected with the names Shur, Spalart, Strelets and Travin.

- The original formulation (DES97) was derived by Spalart et al (1997).
- A generalised formulation (for any RANS model) was developed, see e.g. Strelets (2001).
- Spalart et al (2006) derived a version that protects the RANS boundary layer from LES intrusion (Delayed DES or DDES).
- A further development by Travin et al (2006) (Improved DDES or IDDES) made it possible to resolve the turbulence in the outer part of the boundary layer in LES mode, while the region closer to the wall is still computed efficiently with RANS.

A common set of equations is used for RANS and LES. The regions are separated automatically by the choice of the length scale. The RANS models employed for DES at ISTA span a wide range of mathematical complexity, beginning with the one-equation model of Spalart and Allmaras (1992) with the modification of Edwards and Chandra (1996) and strain-adaptive linear Spalart–Allmaras variant (SALSA) of Rung et al (2003). Two-equation models are represented by the k-ω model of Wilcox (1988) and the linear local realisable (LLR) k-ω model of Rung and Thiele (1996). The compact explicit algebraic stress model (CEASM) of Lübcke et al (2002) represents the highest level of modelling complexity considered.

The modification of a RANS model for DES requires the identification of a RANS model length scale L_{RANS}, which is the wall normal distance d_w in the case of the one-equation models considered, or a function of two turbulence parameters (e.g. k and ε, or k and ω) for two-equation models. This RANS length scale is substituted in the computations for DES97 by the compound DES length scale L_{DES}, which is the smaller value of L_{RANS} and the LES length scale L_{LES},

$$L_{\mathrm{DES97}} = \min\left(L_{\mathrm{RANS}}, L_{\mathrm{LES}}\right) . \quad (1)$$

The LES length scale is related to the grid filter length Δ via

$$L_{\mathrm{LES}} = C_{\mathrm{DES}}\Delta . \quad (2)$$

It is standard practise in DES (Spalart et al, 2006) to define the grid filter length as

$$\Delta = \max(\Delta_1, \Delta_2, \Delta_3) . \tag{3}$$

The subscripts 1, 2, and 3 denote the three coordinate directions defined by a curvilinear grid.

Dependent on the result of equation (1) the DES operates in either the original RANS mode, or in a modified LES mode achieved by basing the turbulent length scale on the numerical grid size, Δ, multiplied by a model parameter C_{DES} analogous to the Smagorinsky LES model (Smagorinsky, 1963). Indeed, a model formulation identical to the Smagorinsky model can be derived from a DES model in LES mode, under the assumption of local equilibrium (Travin et al, 2000; Yan et al, 2005; Mockett, 2009). The value of C_{DES} is model and discretization specific, and is obtained by calibration against decaying isotropic turbulence (Haase et al, 2006; Yan et al, 2005; Mockett, 2009). $C_{\text{DES}} = 0.65$ was first calibrated for DES based on the Spalart-Allmaras turbulence model (Shur et al, 1999). Values for other turbulence models and numerical implementations were found by the authors to span the range $C_{\text{DES}} = 0.55 \ldots 0.95$.

2.2 Shortcomings of DES and Remedies

As a result of numerous studies of the original DES method, several key shortcomings were identified, which are related to the switch between RANS and LES mode. A discussion can be found e.g. in Mockett (2009).

2.2.1 Modelled Stress Depletion

For grids with fine wall-tangential resolution, the grid-dependent switch between RANS and LES mode may encroach inside the boundary layer resulting in a depletion of the modelled stresses and subsequent strong under-prediction of the skin friction. In certain cases this may cause a spurious prediction of flow separation known as grid-induced separation (Menter and Kuntz, 2004). This problem has been successfully addressed by a range of shield functions developed to detect and protect the turbulent boundary layer from LES mode incursion. The first of these shield functions were model specific (for the SST model (Menter and Kuntz, 2004)) followed by the CEASM DES implementation (Bunge et al, 2007)), and a generally-applicable shield function was published by Spalart et al (2006) known as "delayed DES" (DDES).

2.2.2 Unphysical Damping of Eddy Viscosity

The near-wall damping terms of some RANS models can become erroneously active in the LES-mode region of DES causing unphysical damping of the eddy viscosity to near zero levels (Spalart et al, 2006). A method for deriving model-specific correction functions, Ψ, was published together with the DDES method described above by Spalart et al (2006). Ψ functions have been derived and implemented for all DES variants applied by ISTA (Mockett, 2009).

2.2.3 Grey Area

A "grey area" (Spalart et al, 1997) of undefined modelling exists between the RANS mode region in a boundary layer and the fully-developed LES region downstream, which is caused by the need to convert modelled RANS turbulence into predominantly resolved LES turbulence. The extent to which this is a problem is highly case specific, and is the reason why DES is better suited to very strongly separated flows than to flows with shallow separation.

The grey area problem is illustrated in figure 1. The ideal situation is shown in figure 1(a). The boundary layer is modelled in RANS mode while the detached shear layer is resolved in LES mode. However, despite being in LES mode, the early detached shear layer is actually neither RANS nor LES due to lack of "LES content", since this needs time to develop from the natural shear layer instability. The actual situation is even worse as shown in figure 1(b), because the convection of upstream eddy viscosity into the LES domain further delays the formation of LES content. A significant improvement of this problem with respect to jet noise is discussed in section 3.1.

(a) ideal DES (b) real DES

Fig. 1 (a) Outline of ideal DES: Boundary layer is in RANS mode, while the large scale turbulence is resolved in detached shear layer. (b) Actual situation in DES. The convection of eddy viscosity from the RANS boundary layer into the LES domain delays the formation of "LES content"

2.2.4 Log-Layer Mismatch

Although not the intended "natural" application of DES, it is nonetheless attractive to attempt to employ the near wall RANS mode as a form of wall-modelling for LES resolution of attached boundary layers. Initial investigations of DES for such applications (Nikitin et al, 2000) revealed a strong "kink" in the mean velocity profiles known as log-layer mismatch (LLM). This problem appears to have been solved by a recently published enhanced version known as "improved DDES" (IDDES)(Shur et al, 2008).

2.3 Numerical Scheme

The flows of the example cases shown later are computed using an in-house finite-volume based code (ELAN), which solves the unsteady Navier–Stokes equations in either compressible or incompressible form. The procedure is implicit and of second order accuracy in space and time (Xue, 1998). All scalar quantities as well as the Cartesian components of tensor quantities are stored in the cell centres of arbitrarily curvilinear, semi-structured grids that can capture complex geometries and allow for local refinement. Diffusive terms are approximated with central schemes, whereas convective terms can be treated with central or upwind-biased limited schemes of higher order. A hybrid blending of both approaches for DES as formulated by Travin et al (2000) is used for all cases shown except for the jet flow simulations. The formulation of the blending function is adopted directly as published, and is designed to deliver upwind differencing in areas of unidirectional flow or relatively coarse grids, whereas central differencing is used for finer grids, higher vorticity and lower strain. As such, in areas where LES is possible and desirable, low-dissipative central differencing is achieved, and more stable upwind differencing is used elsewhere.

The linearised equations are solved sequentially and the pressure is iterated to convergence using a pressure-correction scheme of the SIMPLE type that assures mass conservation as the pressure-correction equation is derived from the continuity equation (Karki and Patankar, 1989). A generalised Rhie and Chow interpolation is used to avoid an odd-even decoupling of pressure, velocity, and Reynolds-stress components (Obi et al, 1991). Transition to turbulence is not treated explicitly by a specialised predictive model, and is either tripped directly at user-defined locations or left to the turbulence model's inherent transition behaviour.

Numerical simulations of the sound emission of compressible flows can be disturbed by reflections from the boundaries of the computational domain. To avoid this, non-reflective boundary conditions are used. The boundary condition of Bogey et al (2007) was implemented. In the jet noise and slat noise simulations a simplified version was used, which does not require knowledge of the origin of the sound waves (Yan et al, 2007b). In addition, damping zones can be set-up on the upstream and downstream ends of the computational domain.

2.4 Aeroacoustic Noise Emission with DES

The sound sources in the unsteady flow have to be simulated with sufficient detail for the frequency range of interest. From an aeroacoustic perspective, DES allows to limit the necessary detail to frequencies below the largest frequency of interest, while the modified RANS model is used to model the unsteady flow that would generate even higher frequencies.

In order to reduce the computational burden, the unsteady flow is simulated only in a flow region of limited extend containing the aerodynamic sound sources and their immediate vicinity. The sound radiation to any observer position outside this domain is then computed by solving integrals over a data surface surrounding the flow region based on acoustic analogies. In special situations one is interested in the contribution of a single element of a complex device to the sound radiation. In such cases, the sound radiation of this element is determined from an integral over the element's surface.

2.4.1 Grid Requirements for Boundary Layer Region

The grid for the boundary layer region is dictated by the usual requirements for RANS models. The first point in the wall-normal direction must be located at a distance appropriate for the turbulent wall boundary condition. In all simulations reported here, full resolution of the viscous sublayer has been conducted, such that the first point is located at $y^+ \leq 1.5$. The tangential wall spacing is limited only by the stability requirements of the numerical scheme with respect to high aspect ratio cells and the capturing of geometrical curvature and discontinuities as well as gradients in the flow field.

2.4.2 Grid Requirements for Aerodynamic Sound Sources

The highest frequency that can be resolved is determined by the grid. The controlling size parameter for the source frequencies is the length scale defined by $L_c = U_c/f$, where U_c is the convection speed of the disturbances in the flow and f is the frequency of the emitted sound. A sufficient resolution may require a grid with approximately $h = 4$ points per L_c for the numerical scheme described in section 2.3. An axial *grid Strouhal number* can be defined,

$$St_s = f\Delta x_s/U_c < 1/h = 0.25\,, \tag{4}$$

where Δx_s is the grid size in the meanflow direction. If the source regions of the various frequencies are known, the condition may be satisfied dependent on the position according to the maximum local source frequency.

The lateral size of the computational domain is important for the simulation of the lateral source interference, which reduces the far-field sound pressure level. For low frequencies a small lateral domain may result in sound pressure levels in the far field that are too high.

2.4.3 Grid Requirements for Sound Wave Propagation

The correct modelling of the sound propagation requires low dispersion and low dissipation. Since the ELAN code is only of second order (Section 2.3), this necessitates a large number of grid points for the simulation of the propagation of the sound waves inside the turbulent flow region. The required number of grid points is roughly $n = 35$ points per wavelength for an amplitude loss of less than 0.5 dB per wavelength. The wave lengths in a flow with uniform mean velocity depend on the wave-normal direction θ_e of the sound propagation. A *grid Helmholtz number* can be defined

$$He_w = f\,\Delta x_w/a = 1 + M\cos\theta_e\ , \tag{5}$$

where Δx_w is the grid size in the wave-normal direction, M is the flow Mach number and θ_e is the wave-normal angle relative to the downstream direction. It can be conclusded that the propagation of sound waves in the upstream directions of a high-speed flow requires much finer grids than in the downstream direction.

2.4.4 Far-Field Radiation with Acoustic Analogy

The far-field sound radiation of the turbulent flow can be calculated by extending the numerical solution for the limited turbulent domain to the far field with linearised approximations. One of the possibilities in the case of the radiation into free space is to use an integral over a surface based on the acoustic analogy. The most commonly used integral is the procedure proposed by Ffowcs Williams and Hawkings (1969) (FW-H), which is based on the density as the wave variable. Problems occur with this approach when vortical structures or density inhomogeneities cross the data surface, which is necessarily the case on the downstream end of the computational domain of heated jets. Modified versions of the acoustic analogy are discussed in Shur et al (2005); Morfey and Wright (2007); Spalart and Shur (2009), in which the local density is absent. The integrals are more tolerant to the truncation of volume sources external to the data surface. These problems are avoided in the jet noise studies shown in section 3.1 by leaving the downstream end of the control volume open and accepting the resulting error in the far field.

The grid requirement on the data surface is very similar to the requirements for the internal grid (see section 2.4.3). Equation (5) has to be satisfied with the Mach number of the external stream. In the case of high subsonic speeds, the grid on the surface needs to be very fine for the propagation in upstream directions.

2.4.5 Required Length of Time Series

In order to be able to compute narrow-band spectra of a stationary random signal with a filter band width $B = \Delta f$, the flow has to be computed with a minimum duration T for a statistically stable result. The statistical error is proportional to $(BT)^{-0.5}$. The narrower the filter bandwidth B is the larger the duration T has to be. A minimum $BT = 25$ would be desirable for a sufficiently stable result. However, this is currently too costly and was not achieved in the computations reported later. The jet noise predictions were based on durations of $T = 0.12$ s. The analysis for $B =$ 50 Hz yields $BT = 6$, which is consequently statistically less stable than desirable.

An initial part of any simulated time series has to be discarded. A minimum guideline for this is $T_1 = L/U_c$, where L is the axial length of the computational domain in the mean flow direction and U_c is the convection velocity of the disturbances. For a jet with $L = 20D$ and $U_c = 0.6U$ we obtain $T_1 = 33D/U$, which equals $T_1 \approx 0.02$ s for the data used in the computations.

Algorithms for the estimation of the random error on statistical estimates from stationary random data, and for the automatic detection of the initial transient to be discarded, have recently been developed at ISTA. The former was applied in the study of Garbaruk et al (2010) in this book, and the methods are to be published by Mockett et al (2010).

3 Noise Predictions

Several examples for the prediction of the noise generation and radiation with DES are now briefly presented.

3.1 Jet Noise

Jet noise is generated by the unsteady part of the flow field and LES is widely used to simulate this problem. However, complex nozzle geometries, like nozzle serrations, struts in the bypass duct of the engine, or forced mixers inside long-cowl nozzles can be treated with less computational effort using DES. DES has been applied by ISTA to a number of industrially-relevant coaxial jets at high Reynolds number. (Yan et al, 2007b,a; Eschricht et al, 2008a,b).

The grey area problem discussed in Section 2.2.3 was found to have a large influence on the development of the initial free shear layer. With the standard DES formulation, the convection of eddy viscosity from the boundary layers into the shear layer strongly damps the development of resolved turbulent structures. With an additional length scale substitution in the eddy viscosity expression of the Wilcox k–ω proposed by Yan et al (2005), the eddy viscosity is quenched more rapidly, resulting in a correspondingly enhanced development of resolved turbulence. A further measure to enhance the development of resolved turbulence is to compute the grid filter length with the cubic root of the grid sizes $\Delta = (\Delta_1 \Delta_2 \Delta_3)^{1/3}$ rather than with equation (3). Arguments for this alternative Δ formulation can be made considering the predominantly 2D nature of the early shear layer, however problems can be anticipated in general complex DES applications. $\Delta = (\Delta_1 \Delta_2 \Delta_3)^{1/3}$ is hence only applied for jet noise simulations at ISTA. The effect of the additional length scale substitution on the simulation of the coaxial flow from a short cowl nozzle is shown in figure 2 (Yan et al, 2007b). In spite of this improvement the development of the large scale content remains delayed.

(a) Standard L_{DES} substitution (b) Dual substitution of L_{DES}

Fig. 2 Density iso-surfaces showing resolved structures. Standard DES (a) and modified DES. Both simulations employ $\Delta = (\Delta_1\Delta_2\Delta_3)^{1/3}$ (b)

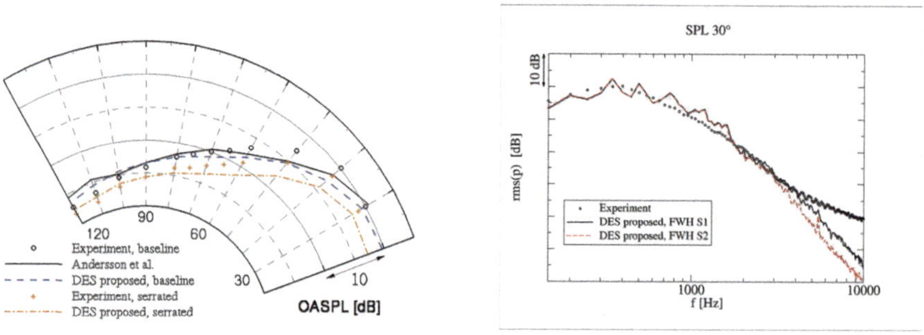

(a) Directivity (b) Frequency spectra of far field pressure

Fig. 3 Directivities of experiment and simulation (a) and frequency spectra of the baseline case for two integration surfaces (b)

The time step used in the aerodynamic simulation is 10^{-6} s. The far-field noise was computed with the FW-H integral for two different control surfaces. The space between the two surfaces was filled with relatively large grid cells. The influence of the control surface on the sound pressure spectra in the far field is shown in figure 3(b) for an emission angle of 30° relative to the jet axis. The spectra based on the simulations show a rapid fall-off for frequencies above 3 to 4 kHz, corresponding to a Strouhal number of $St = fD/U \approx 2$. This is likely caused by insufficient grid resolution in the propagation region according to section 2.4.3. This also explains the different results for the two control surfaces.

It can be concluded from the results achieved so far, e.g. those of figure 3(a) that jet mixing noise can be simulated with an accuracy of 2 dB for the overall sound pressure level and 3 to 4 dB for the spectra if the grid requirements set in section 2.4 are met. The ELAN code is currently applied with grid sizes up to 44M points for internal flow of a coannular serrated nozzle and large scale motion in the jet. One quarter of the grid cells are used in the nozzle. The average grid Strouhal number in the first two nozzle diameters is $St_s \approx 0.02$ and therefore much smaller than needed to resolve the source frequencies. This appears to be necessary to resolve the instability waves in the shear layer.

3.2 Interaction Noise

The sound generation by an airfoil in the wake of a rod was studied by Greschner et al (2008). The flow Mach number is $M \approx 0.2$, the Reynolds number based on the cylinder diameter d is $Re = 48000$. The grid filter is defined according to equation (3) as the maximum of the grid size. The computational grid extends over 180 d in the streamwise direction, 120 d in the cross-stream direction and 3 d in the spanwise direction. The 2-D grid is repeated at 30 spanwise locations, with a constant spacing $\Delta z = 0.1d$ yielding a total grid size of about 2.3 million cells. The first cell size normal to the wall in turbulent units is $y^+ < 1.5$. The time step used in the aerodynamic simulation is 10^{-6} s. The average grid spacing near the leading edge of the airfoil in the flow direction is $\Delta x = 0.5$ mm. This is sufficient for the modelling of the sound sources up to a frequency $f = 0.25\,U_c/\Delta x \approx 30$ kHz by assuming $U_c = 60$ m/s.

The fully three dimensional vortical flow computed with CEASM DES (see section 2.2) is shown in figure 4. A vortex shedding pattern can be observed downstream of the rod, although strong spanwise effects prevent the formation of a regular von Kármán vortex street.

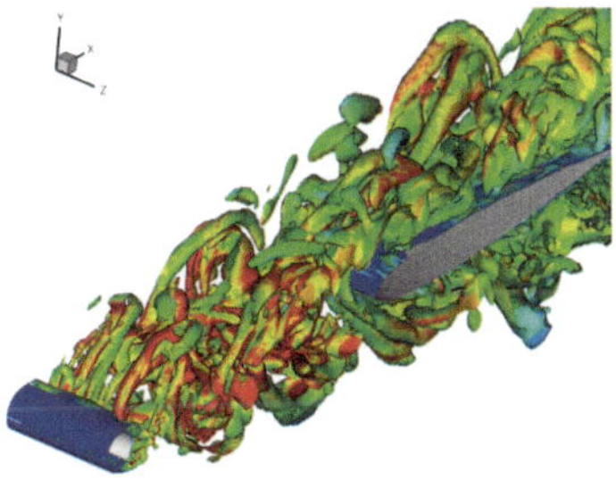

Fig. 4 Flow of rod-airfoil interaction. λ_2 contours are shown (Jeong and Hussain, 1995), coloured with viscosity ratio

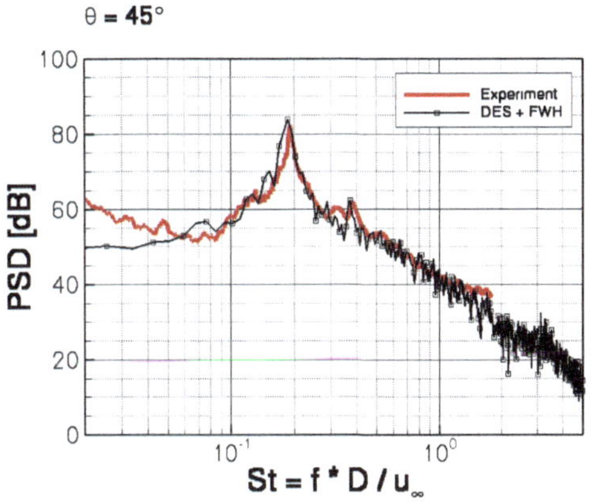

Fig. 5 Far field acoustic power-spectral density (PSD) for an emission angle $\theta = 45°$relative to the flow direction compared to measurements of Jacob et al (2005) ($R = 185\ d$)

The far-field noise is calculated with the FW-H method for permeable surfaces. The observers are located 185 d from the airfoil midpoint and to the airfoil chord in the mid-span plane. The results for 45°(rear arc) are depicted in figure 5. The simulated span of three diameters ($L_S = 3\ d$) is less than the span of the test configuration ($L_{\text{exp}} = 30\ d$), therefore a scaling correction has been applied, as suggested by Kato et al (1993).

The DES predicts the peak of the shedding frequency almost accurately. Its frequency is slightly too small, and the magnitude is under-predicted by 1 to 5 dB for the different observer positions. The OASPL for an observer normal to the flow (90°) is within 1 dB of the experiment and the computation reasonably resolves frequencies up to St = 10. The whole broadband spectrum of the DES/FW-H shows an excellent agreement with the experiment. More information can be found in Greschner et al (2008). By comparing the results with surface integrals over the solid surfaces it was concluded that the volume source terms play a significant role at high frequencies. The difference between simulation and experiment at low frequencies may be related to the short lateral extension of the computational domain for low frequencies.

3.3 Slat Noise

The noise generation of a high-lift system is an example of how airframe noise sources can be simulated with DES. The Reynolds number is 1.7 million based on the clean chord of the wing. Part of the three-dimensional grid of some 25 million points is shown in figure 6. The average grid spacing on the main element in

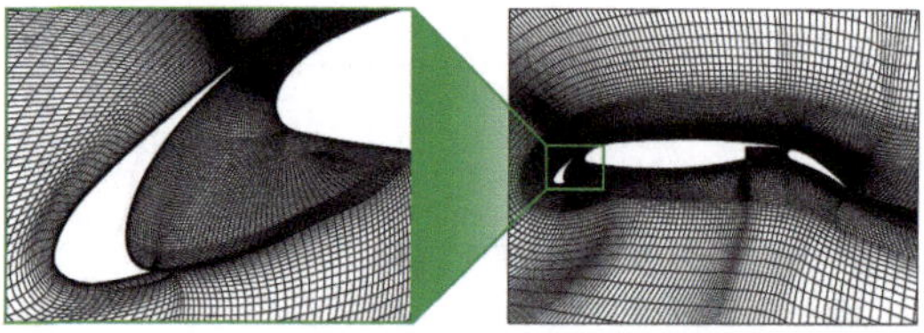

Fig. 6 Grid of three-element high lift configuration

the flow direction is $\Delta x \approx c/500$ with $c = 457$ mm, the spacing in the slat cove is $\Delta x \leq c/1200$, which is further refined near the trailing edge. According to equation (4), this is sufficient for the modelling of the sound sources up to a frequency $f = 0.25 U_c/\Delta x \approx 50$ kHz assuming $U_c \approx 75$ m/s near the slat trailing edge. The boundary layers are resolved with at least 25 points with $y^+ < 1$.

The SALSA-DDES variant (see section 2.1) was used because the grid was fine enough to cause both the invasion of LES mode into the boundary layers and the erroneous activation of the near wall damping terms described in Section 2.2. An impression of the resolved turbulent structures can be obtained from figure 7. The development of the frequency spectra in the separated shear layer as source region is shown in figure 8.

The evolution of the shear layer in the slat cove is shown in figure 9. The far-field noise is predicted by applying the FW-H acoustic analogy to the surface pressures. The directivities of the sound pressure level of the whole wing and of the separate contributions of slat and flap are shown in figure 10. The directivity shows the well

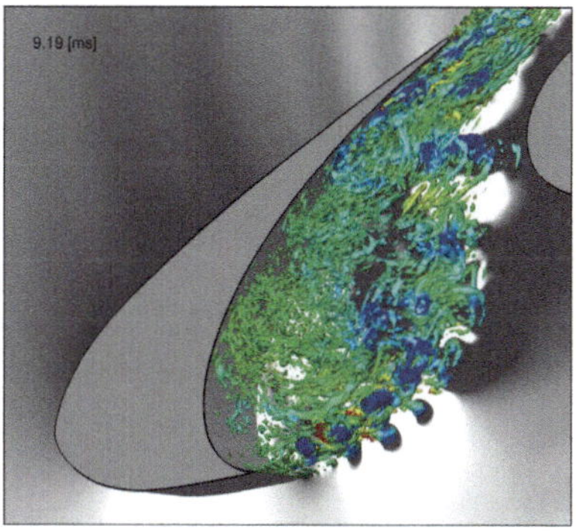

Fig. 7 Separated flow in slat cove: Hydrodynamic and acoustic pressure perturbation contours, vortical structures captured by λ criterion, vortex cores coloured with pressure perturbations

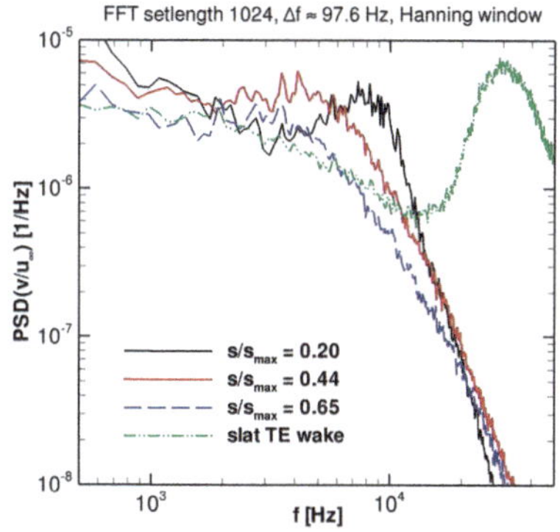

Fig. 8 Early shear layer peak frequency around 8 kHz. Peak moves to lower frequencies in downstream positions as shear layer thickens. Blunt trailing edge shedding frequency 30 kHz (St=0.22)

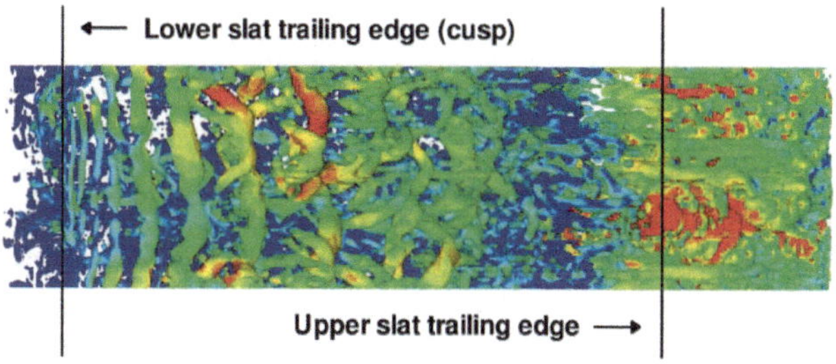

Fig. 9 Slat shear layer evolution. Flow from left to right. Initial 2D state increasingly deformed. Strong acceleration and rapid distortion in the slat gap on the right

known result of slat noise, a dipole like peak radiation in the upper forward and lower rearward directions. The peak frequency of slat noise is found around 3 kHz in this model. The full scale peak frequency of a wing with $c = 3$ m and an airspeed of 80 m/s would be around 750 Hz, which is about the frequency found in flyover experiments. Aerodynamic results are published by Knacke and Thiele (2009). The work with respect to the noise emission has not yet been published.

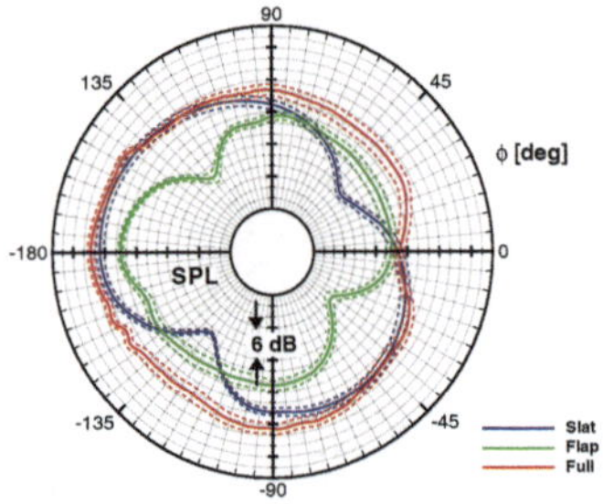

Fig. 10 Directivity of sound pressure level. The slat radiates predominantly into the upper forward arc (135° to 180°) and to the lower rearward direction (-45°to -55°). The flap radiates into the forward arc. The 95% confidence limits of the predictions are indicated

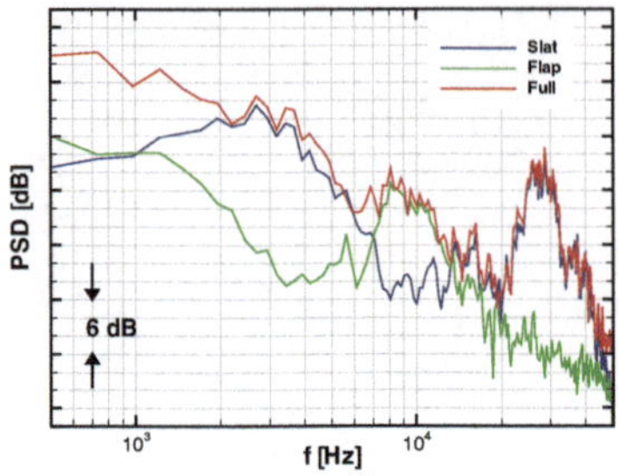

Fig. 11 Frequency spectra of slat, flap and the whole airfoil for a radiation angle of -90° (downward direction)

3.4 Trailing Edge Airfoil Noise Using Wall-Modelled LES

A final example for an application of DES to a flow noise problem is the simulation of the trailing edge noise of a cambered NACA5510 airfoil at 7° angle of attack. Trailing edge noise of the rotor blades and stator vanes is a major contribution to the broadband noise emission of turbofan engines. The IDDES method based on the CEASM model (near wall RANS, outer boundary layer with LES) has been applied (see section 2.2.3) (Greschner et al, 2010). The relatively high Reynolds number of 0.96 million (based on the chord length of $c = 200$ mm) is computed on a grid of 5.3 million cells. The grid in the vicinity of the airfoil is shown in figure 12. The 3D grid is generated by repeating the 2D grid 40 times in the spanwise direction yielding a spanwise extension of $0.05\,c$. The grid size on the suction side is $\Delta x = 0.0025\,c = 0.5$ mm, while it is four times larger on the pressure side. The grid is refined on the leading and especially the trailing edges. Based on a grid Strouhal number $St_s = 0.25$ (see definition (4)) and a convection speed of $U_c = 0.8\,U$ with $U = 70$ m/s, the maximum resolvable frequency on the suction side is $f = 28$ kHz, which increases to the trailing edge due to the reduction of grid size.

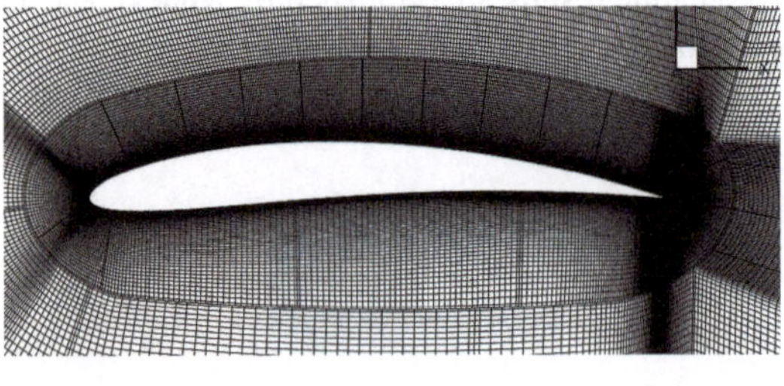

Fig. 12 Chordwise slice through the grid in the vicinity of the airfoil. Every second grid point shown

The initial field is generated with a precursor steady RANS calculation by adding fluctuations in the boundary layer region deduced from isotropic turbulence. Figure 13 shows the resulting unsteady flow field in the boundary layer and the acoustic field in a 3D-view. The vorticity fluctuations in the boundary layer are visualized in term of λ_2 isolines. The sound field is radiated from the region immediately upstream of the trailing edge when these vorticity fluctuations pass the edge. A weaker source is visible near the leading edge, which is due to the strong acceleration of vorticity.

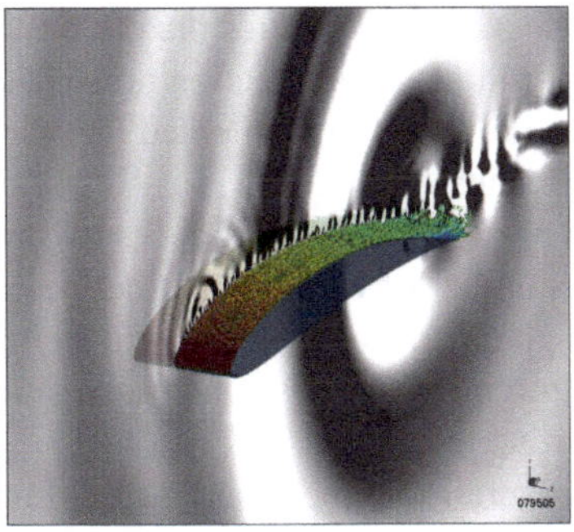

Fig. 13 Vortex structure visualized by a λ_2-isosurface near airfoil and acoustic perturbation field $\partial p/\partial t$ further away from surface. The 3D field has been reproduced 4 times in the spanwise direction to aid visualiation

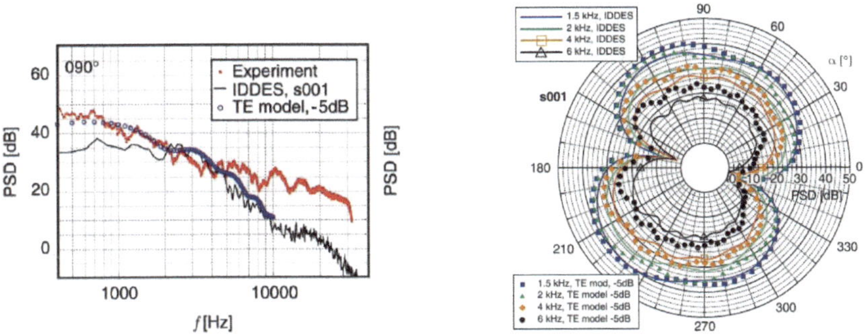

(a) frequency spectra (b) directivities

Fig. 14 (a) Frequency spectra for an emission angle of 90°. (b) Directivities for four frequencies $f = 1.5$ kHz, 2 kHz, 4 kHz, and 6 kHz

The noise emission is calculated by evaluating the surface integral of the pressure fluctuations. One example for the resulting frequency spectra is shown in figure 14(a) for an emission angle of 90° in comparison to a semi-analytical trailing edge model and the experiments. The trailing edge model is based on measured pressure fluctuations near the trailing edge and their lateral coherence. The experimental data are contaminated by background noise at high frequencies. The results of the simulations agree with the semi-empirical model except for the low frequencies, which may be based on highly coherent pressure fluctuations that are dominated by tunnel background noise. The agreement between simulations and the model is equally well for all emission angles (Greschner et al, 2010).

An example for a comparison of directivities between the IDDES and the trailing edge model is shown in figure 14(b). The sound pressure levels of simulation and model agree astonishingly well when the model data are reduced by 5 dB.

4 Conclusions

DES has been shown to be a capable tool for the prediction of far-field noise in the cases of jet mixing noise, rod-airfoil interaction noise, slat noise, and trailing edge noise. Advanced DES features were implemented into the in-house university code in order to achieve these results.

The simulation of the noise emission introduces two new dimensionless parameters. The grid Strouhal number of equation (4) defines the maximum source frequency resolvable with a given grid size and the grid Helmholtz number of equation (5) describes the grid requirement for the sound propagation from the sources to the permeable integration surface.

The simulation of the overall sound pressure level of the far-field noise radiation of a jet was extremely successful. The predictions are within 2 dB of the experimental data. The noise reducing influence of serrations on the nozzle was correctly predicted. The noise emission due to rod-airfoil interaction was also very successful and compares very well with the experimental results, including the peak frequency and the broadband noise. The prediction of slat noise and trailing edge noise was possible by modelling the near wall with RANS and applying LES far from the wall.

Defining the grid is one of the principal challenges facing the user. The grid spacing must be fine enough in the shear layers to resolve the small scales in LES mode. In addition, the grid spacing must be small enough to properly capture the sound propagation inside the flow region up to the integration surface.

Acknowledgements. The authors gratefully acknowledge the support of the European Community within the following 6th Framework Program projects: DESider, CoJeN, PROBAND, and FLOCON. The support of the German Federal Ministry of Economics and Technology in the project FREQUENZ and of Airbus in the project AeroNext+ is appreciated. The computer resources provided by the HLRN (Norddeutscher Verbund für Hoch- und Höchstleistungsrechnen) and by the HLRB (Höchstleistungsrechner in Bayern) are acknowledged with thanks.

References

Bogey, S., Bailly, C., Juvé, D.: Noise investigation of a high subsonic jet using a compressible large eddy simulation. Theoret. Comput. Fluid. Mech. 16, 273–297 (2007)

Bunge, U., Mockett, C., Thiele, F.: Guidelines for implementing detached-eddy simulation using different models. Aerosp. Sc. & Tech. 11, 376–385 (2007)

Edwards, J., Chandra, S.: Comparison of eddy viscosity-transport turbulence models for three-dimensional, shock-separated flowfields. AIAA J. 34, 756–763 (1996)
Eschricht, D., Panek, J.Y., Michel, U., Thiele, F.: Prediction of jet noise from coplanar nozzle. AIAA Paper 2008-2969 (2008a)
Eschricht, D., Panek, L.J.Y., Michel, U., Thiele, F.: Jet noise prediction of a serrated nozzle. AIAA Paper 2008-2971 (2008b)
Ffowcs Williams, J., Hawkings, D.: Sound generation by turbulence and surfaces in arbitrary motion. Phil. Trans. R. Soc. A 264, 321–342 (1969)
Garbaruk, A., Leicher, S., Mockett, C., Spalart, P., Strelets, M., Thiele, F.: Evaluation of time-sample and span-size effects in DES of nominally 2D airfoils beyond stall. In: Peng, S.-H., et al. (eds.) Progress in Hybrid RANS-LES Modelling. NNFM, vol. 111, pp. 86–98. Springer, Heidelberg (2010)
Greschner, B., Thiele, F., Jacob, M.C., Casalino, D.: Prediction of sound generated by a rod-airfoil configuration using EASM DES and the generalised Lighthill/FW-H analogy. Computers and Fluids 37, 402–413 (2008)
Greschner, B., Grilliat, J., Jacob, M.C., Thiele, F.: Maesurement and wall modeled LES simulation of trailing edge noise caused by a turbulent boundary layer. Int. J. Aeroacoustics 9(3), 329–355 (2010)
Haase, W., Aupoix, B., Bunge, U., Schwamborn, D. (eds.): FLOMANIA - A European initiative on flow physics modelling. Notes on Numerical Fluid Mechanics and Multidisciplinary Design, vol. 94. Springer, Heidelberg (2006)
Jacob, M., Boudet, J., Casalino, D., Michard, M.: A rod-airfoil experiment as benchmark for broadband noise modeling. Theor. Comp. Fluid. Dynamics 19, 171–196 (2005)
Jeong, J., Hussain, F.: On the identification of a vortex. J. Fluid Mech. 285, 69–94 (1995)
Karki, K., Patankar, S.: Pressure based calculation procedure for viscous flows at all speeds. AIAA J. 27, 1167–1174 (1989)
Kato, C., Iida, A., Takano, Y., Fujita, H., Ikegawa, M.: Numerical prediction of aerodynamic noise radiated from low Mach turbulent wake. AIAA Paper 1993-0145 (1993)
Knacke, T., Thiele, F.: Time-resolved 3D Simulation of an Aircraft Wing with Deployed High-lift System. In: Proc. 2nd International Conference on Turbulence and Interaction (2009)
Lübcke, H., Rung, T., Thiele, F.: Prediction of the spreading mechanisms of 3D turbulent wall jets with explicit Reynolds-stress closures. Int. J. Heat Fluid Flow 5, 127–145 (2002); In: Schmidt, F.W., Launder, B.E. (eds.) The Seventh International Symposium on Engineering Turbulence Modelling and Measurements, ETMM7, Limassol, Cyprus 04-08 (June 2008)
Menter, F., Kuntz, M.: Adaption of eddy-viscosity turbulence models to unsteady separated flow behind vehicles. In: McCallen, R., Brownad, F., Ross, J. (eds.) The Aerodynamics of Heavy Vehicles: Trucks, Buses, and Trains. LNACM, vol. 19, pp. 339–352. Springer, Heidelberg (2004)
Mockett, C.: A comprehensive study of detached-eddy simulation. PhD thesis, Technische Université Berlin, Fakultät V - Verkehrs- und Maschinensysteme (2009), `http://opus.kobv.de/tuberlin/volltexte/2009/2326/`
Mockett, C., Knacke, T., Thiele, F.: Detection of initial transient and estimation of statistical error in time-resolved turbulent flow data. In: Proc. 8th International ERCOFTAC Symposium on Engineering Turbulence Modelling and Measurement (2010)
Morfey, C., Wright, M.: Extensions of Lighthill's acoustic analogy with application to computational aeroacoustics. Proc. Roy. Soc. A 463, 2101–2127 (2007)
Nikitin, N., Nicoud, F., Wasistho, B., Squires, K., Spalart, P.: An approach to wall modeling in large-eddy simulations. Phys. Fluids 12, 1629–1632 (2000)

Obi, S., Peric, M., Scheurer, M.: Second moment calculation procedure for turbulent flows with collocated variable arrangement. AIAA J. 29, 585–590 (1991)

Rung, T., Thiele, F.: Computational modelling of complex boundary-layer flows. In: Proc. 9th International Symposium on Transport Phenomena in Thermal-Fluid Engineering, pp. 321–326 (1996)

Rung, T., Bunge, U., Schatz, M., Thiele, F.: Restatement of the Spalart–Allmaras eddy-viscosity model in strain-adaptive formulation. AIAA J. 41, 1396–1399 (2003)

Shur, M.L., Spalart, P.R., Strelets, M., Travin, A.: Detached eddy simulation of an airfoil at high angle of attack. Engineering Turbulence Modelling and Experiments 4, 669–678 (1999)

Shur, M.L., Spalart, P.R., Strelets, M.K.: Noise Prediction for Increasingly Complex Jets. Part I: Methods and Tests. International Journal of Aeroacoustics 4, 213–246 (2005)

Shur, M.L., Spalart, P.R., Strelets, M., Travin, A.: A hybrid RANS-LES model with delayed DES and wall-modelled LES capabilities. Int. J. Heat Mass Transfer 29, 1638–1649 (2008)

Smagorinsky, J.: General circulation experimentswith the primitive equations: I. The basic experiment. Monthly Weather Review 91, 99–164 (1963)

Spalart, P., Allmaras, S.: A one-equation turbulence model for aerodynamic flows. AIAA-1992-0439 (1992)

Spalart, P., Deck, S., Shur, M., Squires, K., Strelets, M., Travin, A.: A new version of detached-eddy simulation, resistant to ambiguous grid densities. Theor. Comp. Fluid Dynamics 20, 181–195 (2006)

Spalart, P.R.: Detached-Eddy Simulation. Ann. Rev. Fluid Mech. 41, 181–202 (2008)

Spalart, P.R., Shur, M.L.: Variants of the Ffowcs Williams – Hawkings equation and their coupling with simulations of hot jets. Int. J. Aeroacoustics 8, 477–492 (2009)

Spalart, P.R., Jou, W.H., Strelets, M., Allmaras, S.R.: Comments on the Feasibility of LES for Wings, and on a Hybrid RANS/LES Approach (1997)

Strelets, M.: Detached eddy simulation of massively separated flows. AIAA-2001-0879 (2001)

Travin, A., Shur, M., Strelets, M., Spalart, P.: Physical and numerical upgrades in the detached-eddy simulation of complex turbulent flows. In: Friedrich, R., Rodi, W. (eds.) Proc. 412th Euromech Coll. LES and Complex Transitional and Turbulent Flows, Munich, Germany. Fluid Mechanics and it's Application, vol. 65, pp. 239–254 (2000)

Travin, A., Shur, M., Spalart, P., Strelets, M.: Improvement of delayed detached-eddy simulation for LES with wall modelling. In: Proc. Eur. Conf. Comp. Fl. Dyn. ECCOMAS CFD, Egmond aan Zee, The Netherlands (2006)

Wilcox, D.: Reassessment of the scale-determining equation for advanced turbulence models. AIAA J. 26, 1299–1310 (1988)

Xue, L.: Entwicklung eines effizienten parallelen Lösungsalgorithmus zur dreidimensionalen Simulation komplexer turbulenter Strömungen. PhD thesis, TU Berlin (1998)

Yan, J., Mockett, C., Thiele, F.: Investigation of alternative length scale substitutions in detached-eddy simulation. Flow, Turb. & Comb. 74, 85–102 (2005)

Yan, J., Panek, L., Thiele, F.: Simulation of jet noise from a long-cowl nozzle with serrations. AIAA Paper 2007-3635 (2007a)

Yan, J., Tawackolian, K., Michel, U., Thiele, F.: Computation of jet noise using a hybrid approach. AIAA Paper 2007-3621 (2007b)

Detached-Eddy Simulation of Supersonic Separated Flows

V. Togiti, M. Breuer, and J. Longo

Abstract. In the present work, the supersonic flow past an axisymmetric cylinder and over a three-dimensional cavity are investigated using detached-eddy simulation to evaluate the predictive capabilities of this simulation technique. The numerical predictions are compared with experimental data of Herrin and Dutton (1994) for the base flow and of Zhuang (2007) for the cavity flow. In the case of base flow, the mean velocity field and pressure along the base are well reproduced. In the cavity flow case, the predictions follow the trends of the measurement concerning the mean velocity field. Pertaining to the sound pressure level along the cavity floor, slightly higher pressure levels are predicted and the reasons are discussed.

1 Introduction

In aerospace industry, base flows and cavity flows are of major importance, as these kinds of flows often occur in practice. E.g. base flows are found behind projectiles, re-entry capsules, aerospike nozzles and re-usable launch vehicles. In this case, the flow is separated due to sudden changes in the geometry. Thereby, the pressure behind the base is lower than in the free stream which causes drag affecting the aerodynamic performance of the vehicles. At times, the unsteady pressure fluctuations behind the base derogate the aerodynamic stability of the vehicle. Cavity flows represent the kind of flow phenomenon occurring in weapon bays and landing-gear wells of aircrafts. Due to the existence and dynamics of free shear layers in cavity flows, high levels of noise are produced. Such high sound pressure levels can be harmful to the structural life of weapon systems inside the weapon bay and landing-gear deployment.

Numerical simulations of such complex flows using direct numerical simulation (DNS) are computationally too expensive. Turbulence models are required to solve these flows numerically. Detached-eddy simulation (DES) proposed by Spalart et al. (1997) is a hybrid model which combines Reynolds-Averaged Navier-Stokes

V. Togiti · J. Longo
Institute of Aerodynamics and Flow Technology, Branch: Spacecraft, German Aerospace Center (DLR), Braunschweig, Germany
vamshi.togiti@dlr.de

M. Breuer
Department of Fluid Mechanics (PfS), Institute of Mechanics, Helmut-Schmidt-University Hamburg, Germany

S.-H. Peng et al. (Eds.): Progress in Hybrid RANS-LES Modelling, NNFM 111, pp. 77–86.
springerlink.com

(RANS) and large-eddy simulation (LES) in such a way that the method works as a RANS model close to solid walls and as a LES in separated flow regions. The advantage of this technique is that flows involving thin boundary layers and large separation regions can be computed with moderate computational effort.

The main objective of the present work is to investigate a supersonic base flow and a three-dimensional cavity flow using DES and to evaluate the predictive capabilities of DES by comparing numerical predictions with available experimental data. These include boundary layer profiles, pressure distributions along the base, centerline velocity profiles, and distributions of the Mach number, resolved Reynolds stresses and turbulent kinetic energy for the base flow. For the cavity flow, unsteady pressure spectra, overall sound pressure levels, mean velocity fields and resolved Reynolds stresses on the symmetry plane are taken into account.

2 Numerical Methodology

In the current investigation the DLR-TAU code is used, which is an unstructured compressible flow solver. Details about the finite-volume flow solver can be found in Mack and Hannemann (2002). In the present study the convective fluxes are approximated by the advection upstream splitting method AUSM+ -UP, and the diffusion fluxes are evaluated by a central scheme. The time advancement is done using an implicit dual time-stepping scheme of second-order accuracy with an implicit lower-upper symmetric Gauß–Seidel algorithm, LUSGS.

3 Details of the Test Cases

In the current numerical studies, the investigated base flow case and three-dimensional cavity flow configuration are based on the experimental work carried out by Herrin and Dutton (1994) and Zhuang (2007), respectively. In the base flow experiments, the free-stream Mach number is set to 2.46 and the free-stream pressure and temperature are 31,415 Pa and 145 K, respectively. The Reynolds number per unit meter is 45 x 10^6. In these studies a cylindrical afterbody with a radius R of 31.75 mm is used. In the experiments of Zhuang, the cavity has an aspect ratio of L/D = 5.2 and L/W = 5.8, the Reynolds number per meter is 23 x 10^6 and the Mach number is 2.

In the case of the supersonic base flow investigation, the computational domain covers a region of 4R upstream of the base, where a steady turbulent boundary layer profile is set. This distance was chosen to match the experimental boundary layer thickness of 3.2 mm close to the cylinder base. The outflow boundary is placed 12R downstream of the base, where supersonic outflow conditions are prescribed. In the radial direction, the outside boundary is a conical frustum, with radius 4R at the inflow plane and 8R at the outflow plane, where the farfield boundary condition is imposed.

In the supersonic cavity flow case, the computational domain has an inflow boundary at 3D upstream of the cavity where a steady inflow boundary layer profile is prescribed in a way that the boundary layer thickness at the leading-edge of the cavity matches with the boundary layer thickness apparent in the experiment. The upper and outflow boundaries are placed at about 2D above the cavity and 6D downstream of the cavity, respectively. Supersonic outflow conditions are applied at the outflow plane and no-slip adiabatic wall conditions are prescribed at the upper boundary. In spanwise direction, the walls of the test-section are not resolved and Euler wall conditions (i.e. slip wall) are used.

Owing to higher accuracy the computational grids used in the present investigation are structured grids which consist of about 3.5 and 4 million nodes for the base flow and the cavity flow, respectively. These grids are refined to study the influence of grid resolution and the refined grids have about 7.5 and 6 million nodes for the base flow and cavity flow, respectively. Since predictions are not much improved by grid refinement, predictions on coarse grids are discussed. The computational grids on the symmetry plane for the axisymmetric cylinder and for the cavity are shown in Figs. 1(a) and (b), respectively.

In the current study, for the base flow detached-eddy simulation proposed by Spalart et al. (1997) is applied and for the cavity flow delayed detached-eddy simulation proposed by Spalart et al. (2006) is applied to ensure the boundary layer to be treated in RANS mode, as the grid has small cells in the near-wall region.

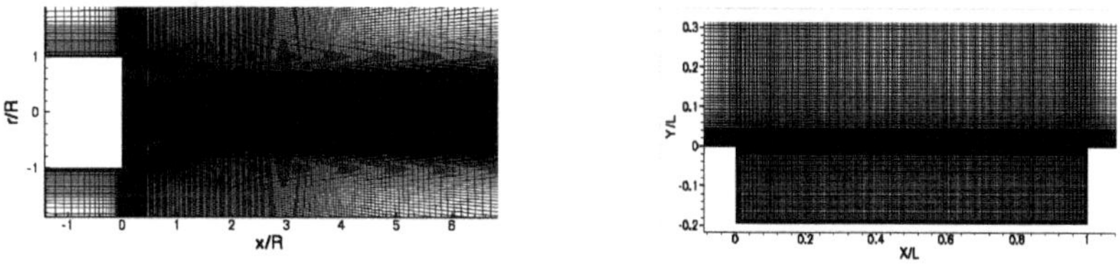

(a) Base flow (b) Cavity flow

Fig. 1 Computational grids on the symmetry plane

4 Results

4.1 Supersonic Base Flow

In the experiment, the oncoming flow is turbulent which need to be resolved correctly, as it has an influence on the free shear layer growth downstream of the base. For that purpose, the time-averaged turbulent boundary layer profile at a radial cross-section 1 mm upstream of the cylinder base is compared with the experiments and the law of the wall (see Fig. 2). A very good agreement is observed.

In Fig. 3, the mean pressure contours along with the contours of instantaneous vorticity magnitude are depicted. The mean pressure contour lines render the expansion fan at the base corner where the flow separates and further downstream of the base the re-compression during which the free shear layer tries to realign with the flow axis. The contours of instantaneous vorticity magnitude on the symmetry plane and a plane extracted at X/D = 1.6 downstream of the base represent the resolved large-scale structures which indicate the unsteadiness of the flow.

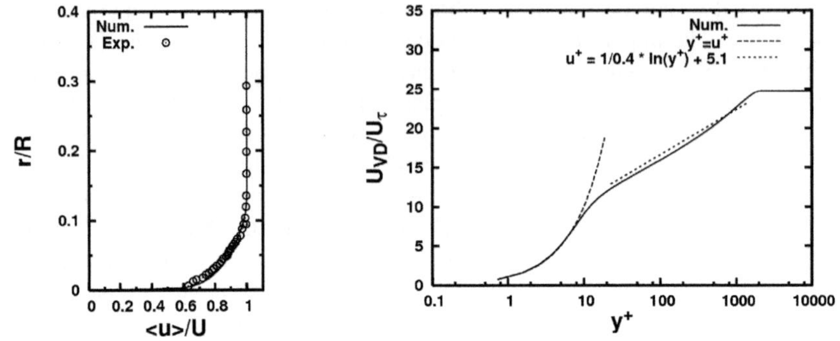

(a) Velocity profile (b) Log law

Fig. 2 Time-averaged boundary layer profile at a radial cross-section 1 mm before the base corner

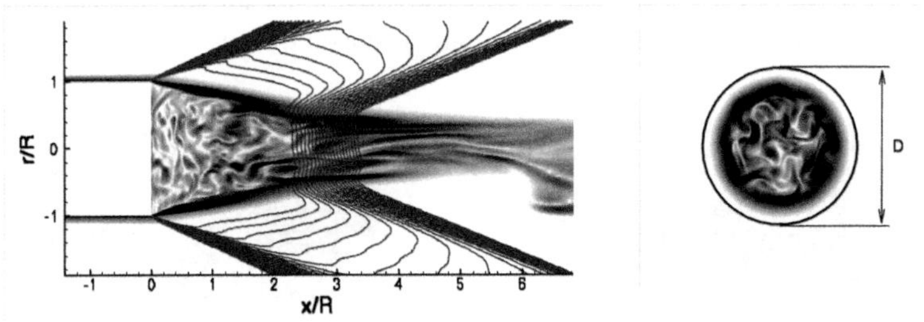

(a) (b)

Fig. 3 Instantaneous vorticity magnitude contours along with the mean pressure contour lines; (a) on the symmetry plane, (b) downstream of the base at x/R = 1.6

The time-averaged streamwise and radial velocity profiles produced by DES are depicted in Fig. 4 and compared with the experimental data and RANS predictions. These profiles show that the DES predictions not only follow the trends of the measurements but also agree rather well with the experiments. Contrarily, RANS predictions poorly agree with the measurements.

The mean streamwise velocity profile along the wake axis behind the base predicted by DES and RANS are shown in Fig. 5(a) and compared with the experiments. The profiles illustrate the length of the recirculation which is slightly overpredicted by DES. Based on the RANS approach the length of the recirculation is underpredicted. The peak reverse velocity inside the recirculation is overpredicted, as the shorter recirculation is predicted by RANS, while the peak velocity delivered by DES is in good agreement with the measurement. The shorter recirculation is possibly due to higher turbulent viscosity behind the base predicted by RANS, which forces the free shear layer to deflect towards the axis of the cylinder earlier.

The vital parameter for aerospace industry regarding blunt bodies is the pressure distribution behind the body, as it has large impact on the total drag. The mean pressure coefficient along the base is depicted in Fig. 5(*b*). A flat pressure coefficient typical for large separation regions is apparent from the experimental data. The DES prediction also delivers a flat pressure distribution as expected. The predicted level slightly deviates from the experimental one. The RANS prediction, however, predicts a much lower pressure coefficient and the distribution along the base differs considerably from the experimental data. This lower pressure coefficient is clearly linked to the distribution of the centerline velocity which shows a higher reverse velocity and a shorter recirculation length as mentioned above leading to a lower pressure level.

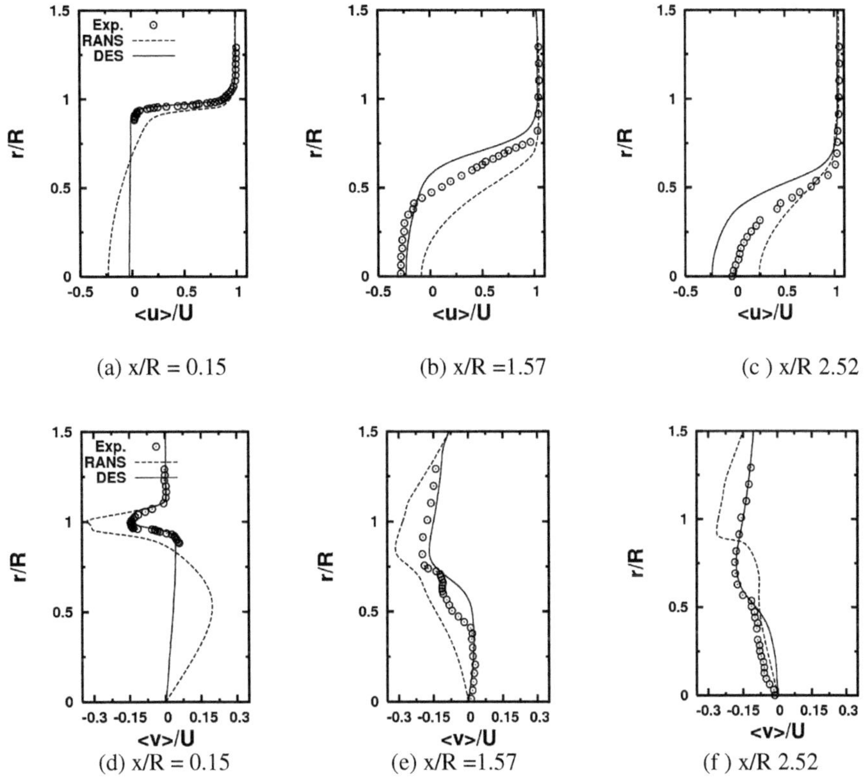

(a) x/R = 0.15 (b) x/R =1.57 (c) x/R 2.52

(d) x/R = 0.15 (e) x/R =1.57 (f) x/R 2.52

Fig. 4 Time-averaged streamwise [(a) - (c)] and radial [(d) - (f)] velocity profiles at different locations behind the base

In Fig. 6, profiles of the resolved streamwise turbulent intensity at three different locations downstream of the base are depicted. These profiles show that the experimental trend is well reproduced. However, at x/R = 0.15 and 1.52 the peak levels are not predicted correctly by DES. The reason is that at this location a major part of the turbulence is modeled. Thus, the resolved part is low compared to the experimentally determined streamwise turbulent intensity. Near the free stagnation point (x/R = 2.57) the level predicted by DES is in good agreement with the measurements but the location of the peak is not.

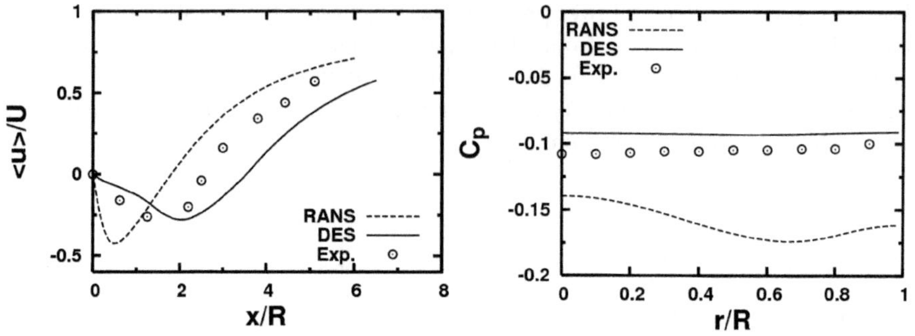

(a) (b)

Fig. 5 (a) Time-averaged streamwise velocity profile along the wake axis, (b) mean pressure coefficient along the base

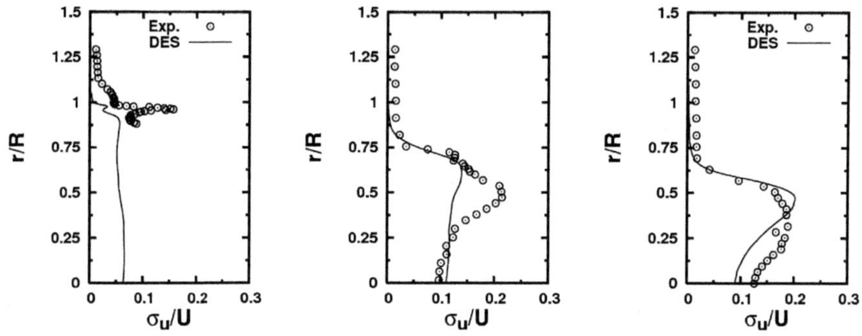

(a) x/R =0.15 (b) x/R =1.52 (c) x/R =2.57

Fig. 6 Resolved streamwise turbulent intensity at different locations downsteam of the base

4.2 Supersonic Cavity Flow

In the current investigation, the experimental conditions of Zhuang (2007) are reproduced. In the experiment, the boundary layer thickness at the leading-edge of the cavity is 3.2 mm and the incompressible momentum thickness is 0.4 mm. In the current study, the boundary layer thickness matches with the boundary layer thickness apparent in the experiment but the momentum thickness is not. An additional computation is carried out to investigate the influence of the momentum thickness at the cavity leading-edge by introducing a thicker boundary layer.

Contours of the instantaneous vorticity magnitude are shown in Fig. 7 for the two inflow conditions investigated applying a thin and thick turbulent boundary layer at the inlet. In all cases large-scale structures in the free shear layer and the subsequent shear layer breakup are visible. In the case of the thick turbulent boundary layer at the inlet, in which the momentum thickness of the boundary layer is larger than for the other turbulent boundary layer case, an earlier shear layer breakup can be noticed (see Fig. 7(b)).

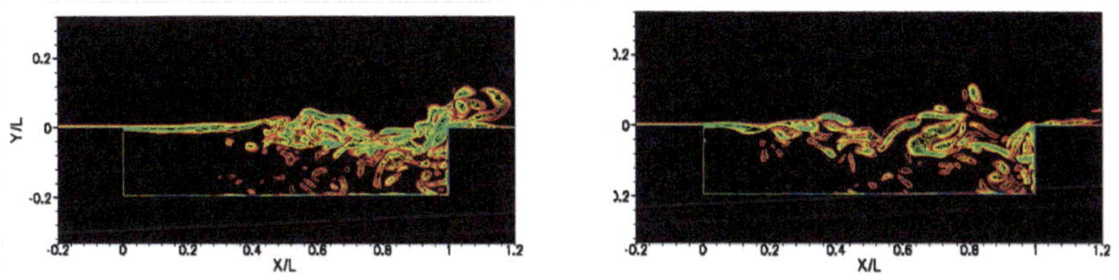

(a) (b)

Fig. 7 Instantaneous vorticity magnitude contours on the symmetry plane, (a) thin boundary layer at the inlet, (b) thick boundary layer at the inlet

Time-averaged streamwise velocity contours $<u>$ on the symmetry plane are depicted in Fig. 8 for the two different cases and compared with the experiment. The contours show a recirculation inside the cavity with very high reverse velocities which are as high as 40-45% of the free-stream velocity. In all cases the contours seem to follow the trends of the experiment.

Distributions of the pressure coefficient along the cavity floor are depicted in Fig. 9(a) and compared with the measurements of Kaufman et al. (1983). The general trend in the experimental data is correctly predicted by all numerical predictions. In both cases, almost the same level of Cp is observed and no big influence of the boundary layer thickness on the pressure distribution along the cavity floor is noticed.

For the aeronautical industry the major concern concerning cavity flows is the sound generated due to flow interaction. In order to characterize the sound generated by the cavity, the overall sound pressure level (OASPL) along the cavity floor for the different inflow boundary layers is depicted in Fig. 9(b). For comparison the experimental data are included. The trend observed in the experiment is reproduced by the numerics but the levels deviate. In the case of the thick turbulent boundary layer a lower level of OASPL can be observed which is due to earlier an shear layer breakup that destabilizes the free shear layer and thus smaller vortical structures are evolving.

The spectra of the unsteady pressure monitored along the base of the cavity near the trailing edge at X/L = 0.87 are displayed in Fig. 10 for the thin and thick boundary layers and compared with the experimentally determined first four modes shown by straight lines. In all cases, the pressure spectra illustrate a broadband content with a wide range of frequencies and also exhibit peaks at the modes experimentally determined. In the case of the thin boundary layer, the experimentally determined dominant mode, i.e., mode 4, is correctly predicted. In the thick turbulent boundary layer case low amplitudes are predicted compared to the thin boundary layer case and mode 2 turns out to be the dominant mode.

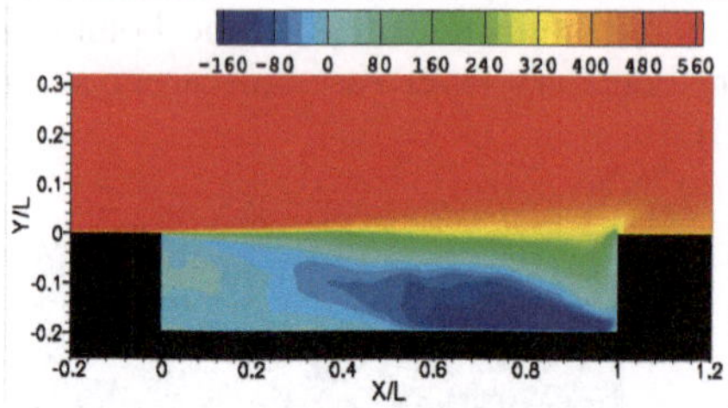

(a) Thin boundary layer

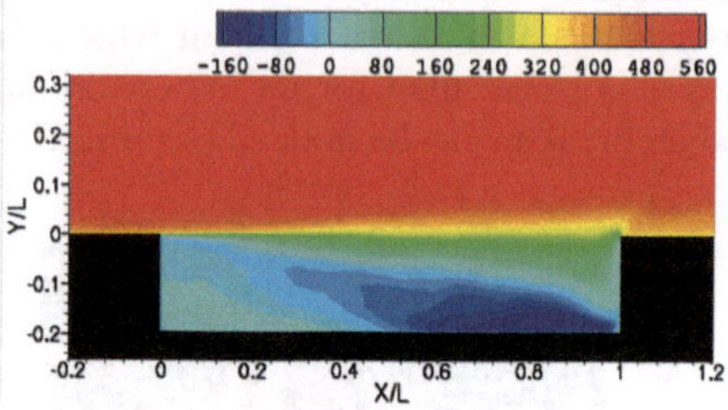

(b) Thick boundary layer

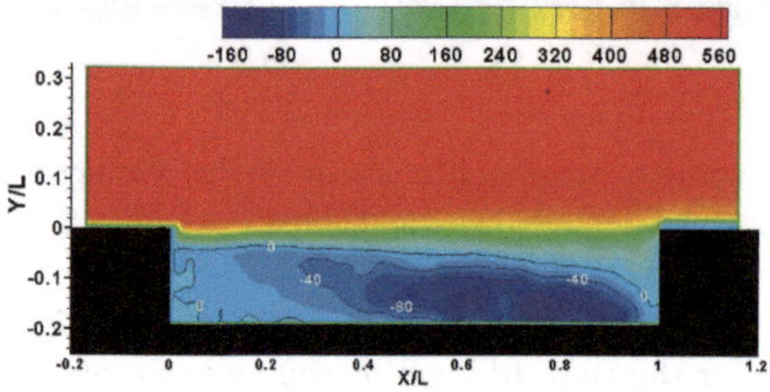

(c) Experiment

Fig. 8 Time-averaged streamwise velocity contours in the cavity

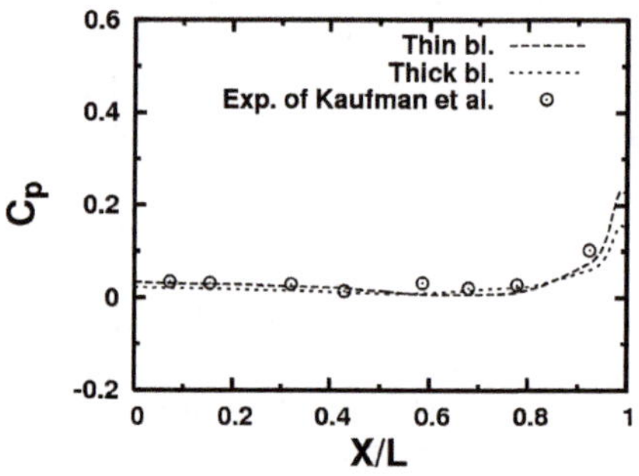

(a) Pressure coefficient

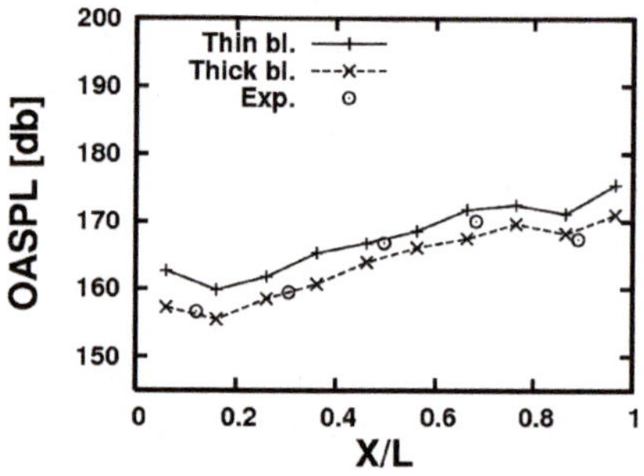

(b) OASPL

Fig. 9 Distribution of the pressure coefficient and overall sound pressure level (OASPL) along the cavity floor

The profiles of resolved streamwise turbulent intensity at three different locations inside the cavity are shown in Fig. 11. The profiles represent a clear evidence that the predicted stress levels are higher in the cases of thin boundary layers, whereas in the case of the thick turbulent boundary layer good agreement with the experiment is found. Here it should be noticed that in the latter case the boundary layer possesses the same momentum thickness near the leading edge of the cavity as in the experiment, but the boundary layer thickness is larger. The reason for the higher resolved stresses compared to the experiment are the large-scale structures present in the free shear layer (see Fig. 7). This observation is also consistent with the increased overall sound pressure level discussed before (see Fig. 9(b)).

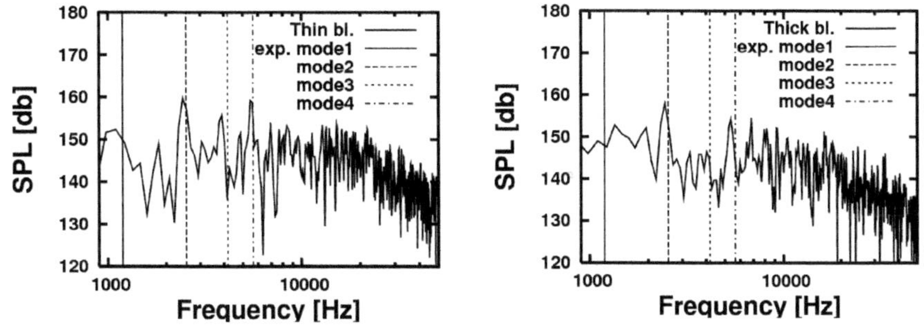

(a) Thin boundary layer (b) Thick boundary layer

Fig. 10 Pressure spectra at X/L = 0.87 for thin and thick boundary layers

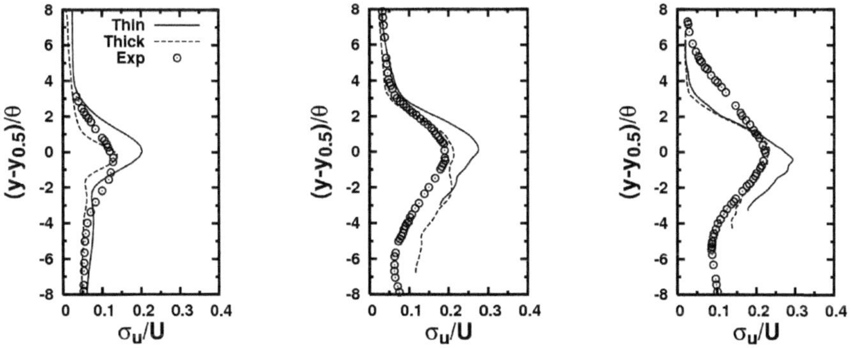

(a) X/L = 0.1 (b) X/L = 0.5 (c) X/L = 0.9

Fig. 11 Resolved streamwise turbulent intensity at difference locations inside the cavity

5 Summary

To evaluate the predictive capabilities of detached-eddy simulation, supersonic flows past an axisymmetric cylinder and over a three-dimensional cavity are investigated in the current study. In the base flow case, the distributions of the mean Mach number, streamwise and normal velocity components and the mean pressure coefficient along the cylinder base are in good agreement with the experiments. However, the length of the recirculation is slightly overpredicted and the second-order statistics, i.e., resolved Reynolds stresses and turbulent kinetic energy, are slightly underpredicted in the shear layer. The reason for this discrepancy is thought to be insufficient grid resolution in the shear layer.

In the case of the cavity flow, two boundary layer profiles are introduced at the inlet of the cavity to match the experimental visual boundary layer thickness and the momentum thickness at the leading edge of the cavity. In all cases, the numerical predictions follow the trends of the experiment concerning the mean velocity field, pressure distribution along the cavity floor and the resolved Reynolds stresses. The pressure spectra show a broadband content with a wide range of frequencies. Experimentally determined frequency spectra of different modes match with the present numerical results and the dominant mode in the cavity, i.e., mode 4, is predicted correctly with the thin boundary layer, whereas in the case of thick boundary layer mode 2 turns out to be the dominant case.

In the case of the thin boundary layer the overall sound pressure level along the cavity floor at the base of the leading and trailing-edge is 5 dB higher than in the experiments. However, good agreement is observed with the thick boundary layer for which the momentum thickness at the leading edge of the cavity matches the measurement but the boundary layer thickness does not.

For better agreement of predictions with the measurements, further research will be focused on introducing turbulent content at the inflow plane of the domain. In this case, earlier shear layer breakup is anticipated which would eventually improve predictions.

References

Spalart, P.R., Jou, W.-H., Strelets, M., Allmaras, S.R.: Comments on the Feasibility of LES for Wings, and on a Hybrid RANS/LES Approach. In: Advances in DNS/LES, 1st AFSOR Int. Conf. on DNS/LES, Greyden Press, Col., OH (1997)

Spalart, P.R., Deck, S., Shur, M.L., Squires, K.D., Strelets, M.K., Travin, A.: A New Version of Detached-Eddy Simulation, Resistant to Ambiguos Grid Densities. J. Theor. Compt. Fluid Dynamics 20, 181–195 (2006)

Mack, A., Hannemann, V.: Validation of the Unstructured DLR TAU-Code for Hypersonic Flows, AIAA Paper, 2002-3111 (2002)

Herrin, J.L., Dutton, J.C.: Supersonic Base Flow Experiments in the Near Wake of a Cylindrical Afterbody. AIAA Journal 32(1) (1994)

Zhuang, N.: Experimental Investigation of Supersonic Cavity Flows and Their Control. PhD Thesis, Florida State University (2007)

Kaufman II, L.G., Maciulatis, A., Clark, R.L.: Mach 0:6 to 3:0 Flows Over Rectangular Cavities. AFWAL-TR-82-3112.

Evaluation of Time Sample and Span Size Effects in DES of Nominally 2D Airfoils beyond Stall

A. Garbaruk, S. Leicher, C. Mockett, P. Spalart, M. Strelets, and F. Thiele

Abstract. A detailed investigation of detached-eddy simulation (DES) applied to the flow around nominally 2D airfoils in deep stall has been conducted. The sensitivity of the flow to a wide range of computational parameters has been assessed. The principal parameters of influence were found to be the length of the time sample computed for Reynolds averaging and the size of the periodic computational domain in the spanwise direction. In contrast, wind tunnel wall effects, the choice of underlying RANS model and the boundary layer transition treatment counted among the parameters for which a low to negligible level of sensitivity was observed. The magnitude of the error introduced by short time samples has been estimated using a statistical analysis technique. This combined with very long time samples (in excess of 2000 non-dimensional units) allows a reliable description of the span size effect independent of the time sample error. The study significantly enhances the body of knowledge concerning this standard validation flow for hybrid RANS-LES methods. The original DES study of Shur et al. (1999), which was conducted for such a flow, is reassessed in light of the new information and its principal conclusions are upheld. The strong level of scatter observed between different experimental sources represents the chief remaining uncertainty.

1 Introduction

The flow over an airfoil in deep stall is one of the most natural applications of detached-eddy simulation (DES) (Spalart et al., 1997, Spalart, 2009). Not

---

A. Garbaruk · M. Strelets
New Technologies and Services, St.-Petersburg, Russia
strelets@mail.rcom.ru

S. Leicher
EADS Military Air Systems, Munich, Germany

Mockett · F. Thiele
Technische Universität Berlin & CFD Software GmbH, Germany
mockett@cfd.tu-berlin.de

P. Spalart
Boeing Commercial Airplanes, Seattle, USA
philippe.r.spalart@boeing.com

S.-H. Peng et al. (Eds.): Progress in Hybrid RANS-LES Modelling, NNFM 111, pp. 87–99.
springerlink.com

surprisingly, precisely such a flow (the NACA0012 airfoil beyond stall) was the first real 3D DES study (Shur et al., 1999), and its success largely ensured the rapid spread of the DES methodology. It has also become a standard test case for the validation of a new DES implementation. However, owing to limited computer power, in this first DES application and in some subsequent studies, simulations were carried out over relatively short time samples and narrow spanwise domains. Furthermore, the experimental data, quoted from the textbook of Hoerner (1965), only involved the mean lift and drag coefficients and no details were given concerning important wind tunnel details such as aspect and blockage ratios. DES was found to perform far better than URANS, which was a very welcome outcome, however it is now desirable to judge accuracy to better than 10%, more in line with industrial needs. Furthermore, it is necessary to identify and quantify as far as possible the inherent uncertainties in the test case, including the variability seen between equivalent experiments and the sensitivity to computational parameters.

This motivated a further, broader and more computer-intensive, evaluation of DES in predicting stalled airfoil flows, which was carried out within and following the DESider EU project, Haase et al. (2009), and is the subject of this paper. A comprehensive range of parameter studies has been carried out to identify the relative importance of each. The comparison of results also encompasses further levels of analysis, including unsteady force fluctuations and spectra, thereby allowing a deeper evaluation of the predictive capability of DES. In addition to a reassessment of the validity of the conclusions of the original study, this work is therefore also intended to raise the value of the flow as an entry level test case for all hybrid RANS-LES methods.

The paper begins with a description of the flow case and a summary of the available experimental benchmark data, before a concise account of the computational methodologies applied. The analysis of the results begins with a very brief mention of the numerous parameter variations that were found to be of secondary or negligible importance. The bulk of the discussion is thereby dedicated to the two principal sources of test case uncertainty identified, namely the length of the time sample computed for statistical analysis and the extent of the computational domain in the spanwise direction. Following this, the effect of including wind tunnel wall effects in the simulation is assessed. Observe how both the experimental and the numerical studies are aimed at an "ideal" flow, namely the 3D flow over a 2D geometry without any confinement spanwise. Before the conclusion, the implications of the findings upon the interpretation of the original results are discussed.

2 Experimental Benchmark Data

The experimental benchmark data employed in the original DES study of Shur et al. (1999) is not considered ideal for the reasons outlined in the introduction. This, together with a desire to compare unsteady flow characteristics motivated a search for alternative experimental sources. Data for airfoils at such high angles of attack are unfortunately rather rare, as a consequence of which the inclusion of data for different profiles and Reynolds numbers was necessary. In total, four sets of experimental data were identified, the salient features of which are summarised in Table 1.

The experimental review has two key outcomes. Firstly the effects of Reynolds number, boundary layer tripping and profile thickness (Sheldahl & Klimas, 1981) as well as free-stream turbulence intensity (Swalwell, 2005) were found to be negligible in the range $25° < \alpha < 180°$. The second outcome is highly frustrating for the purposes of CFD validation: A tremendous degree of scatter is apparent between facilities, e.g. 12% for C_l at $\alpha = 60°$. In the absence of any information about the relative fidelity of the experiments, no strong conclusions can be drawn from comparison of mean force coefficients between DES results and experimental data.

Table 1 Summary of relevant experimental data for symmetric NACA profiles. The mean and standard deviation values quoted for $\alpha = 60°$

Reference	Profile	$Re_c (\times 10^5)$	α_{max}	$\overline{C}_l$	$\overline{C}_d$	$\sigma[C_l]$	$\sigma[C_d]$	Spec	$\overline{C}_p(x)$
Hoerner (1965)	0012	1.0	180°	0.92	1.65	—	—	—	—
Sheldahl & Klimas (1981)	0021	3.6	180°	0.875	1.470	—	—	—	—
Swalwell et al. (2004)	0021	2.7	90°	0.93	1.55	0.104	0.151	Yes	Yes
Raghunathan et al. (1988)	0021	2.6	90°	0.82	—	—	—	—	—

Re_c Reynolds number based on chord length and free-stream velocity, α_{max} maximum angle of attack measured, $\overline{C}_l$, $\overline{C}_d$ time-averaged lift & drag coefficient at $\alpha = 60°$, $\sigma[C_l]$,$\sigma[C_d]$ standard deviation of lift and drag coefficient, *Spec* force spectra available? $\overline{C}_p(x)$ surface pressure distribution available?

3 Computational Methodology

Table 1 reveals only one experiment with time-resolved data, namely that of Swalwell et al. (2004). For this reason, the test case was adopted within later studies, beginning with the EU DESider project (Haase et al., 2009). The corresponding NACA0021 airfoil at $Re_c = 2.7 \times 10^5$ was simulated and the decision was made to focus upon a single angle of attack, namely $\alpha = 60°$. All reported results were obtained from fully-turbulent computations and the force coefficients were integrated over a single spanwise position (as in the experiment). Results computed by three partners are drawn upon, referred to by the following acronyms: EADS Military Air Systems (EADS), New Technologies and Services (NTS) and Technische Universität Berlin (TUB). The NTS and TUB simulations assumed infinite far-field conditions and periodicity in the spanwise (z) direction. The EADS computation by contrast included free-slip wall boundary conditions in the vertical and lateral directions, at locations corresponding to the wind tunnel geometry. In all cases a sufficient number of time steps were excluded before Reynolds averaging to remove initial transient. A summary of all computations is given in Table 2.

The computational grids employed are depicted in Fig. 1. The structured, O-topology grid generated by NTS was used in the simulations of NTS and TUB. EADS generated their own unstructured hybrid mesh in which the boundary layer region is resolved using prismatic cells. In both cases the grids have been uniformly extruded in the spanwise direction.

Due to space constraints only a summary description of the flow solvers will be given. EADS employed the unstructured DLR Tau code (Schwamborn et al., 1998) whereas NTS (Strelets, 2001) and TUB (Xue, 1998) used in-house multi-block structured codes. The convective terms in Tau are approximated using a central differencing scheme with matrix dissipation of the Jameson type for stabilisation. The NTS and TUB solvers both employ a hybrid convection scheme conceived for DES (Travin et al., 2000), whereby the localised blending is between 4th order central and 5th order upwind schemes in the NTS code and between 2nd order central and upwind-biased TVD schemes in the TUB code.

Table 2 Summary of simulations conducted

Partner	Model	L_z/c	Δ_z/c	Δ_{xy}/c	N_{xyz}	$\Delta t \times \lvert U_\infty \rvert / c$	$T_{avg} \times \lvert U_\infty \rvert / c$
EADS	SA-DES	7.2	0.025	0.014	19.2M	0.07	430
NTS	SA-DES	1.0	0.03	0.037	0.5M	0.03	2100
		1.4			0.7M		2100
		2.0			1.0M		2000
		2.8			1.4M		2000
		4.0			2.0M		2500
TUB	SALSA-DES	1.0	0.03	0.037	0.5M	0.025	1000
	LLR-DES	1.0			0.5M		1200
	CEASM-DES	1.0			0.5M		1600
		3.24			1.6M		700
	SA-DES	4.0			2.0M	0.03	2300

SA Spalart & Allmaras (1992) model, *SALSA* strain-adaptive linear SA model of Rung et al. (2003), *LLR* local linear realisable model of Rung & Thiele (1996), *CEASM* compact explicit algebraic stress model of Lübcke (2001), L_z spanwise domain size, Δ_z spanwise grid cell size, Δ_{xy} typical *x*/*y* cell size in focus region, N_{xyz} number of grid cells, Δt time step size, T_{avg} physical time sample for Reynolds averaging.

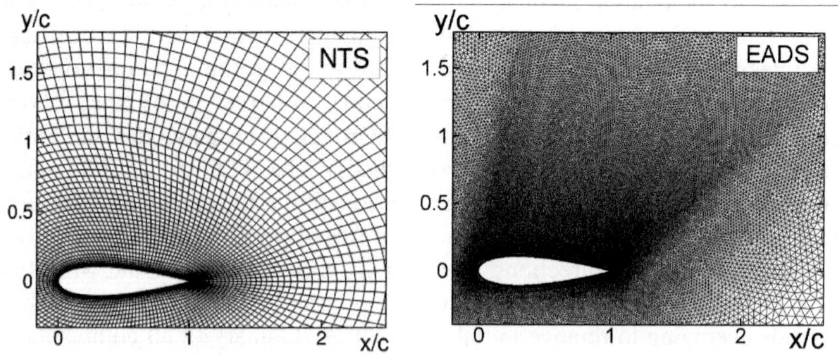

Fig. 1 The computational grids employed (the spanwise-normal plane is shown)

4 Presentation and Discussion of Results

4.1 Summary of Related Findings

The detailed discussion of the results in the following sections concern the parameters to which the flow proved highly sensitive. However, a great deal of other parameter variation studies have been carried out, the results of which have in many cases remained unpublished because of the weak sensitivity found. It is nonetheless important to summarise these findings.

A minor sensitivity of the integral forces and negligible sensitivity of the spectra to the underlying RANS model has been observed in studies by TUB (Mockett, 2009) and NTS (Haase et al., 2009). In comprehensive studies by NTS, the sensitivity to the boundary layer transition treatment, mild variations of the time step size and the overall grid resolution as well as the resolution of the leading edge curvature and the time step size has been found to be negligible. Many of these results indicate the weak overall importance of the attached boundary layer prediction, and can be related to the weak effects of Reynolds number and free stream turbulence reported in the experiments.

4.2 Effect of Time Sample

The accuracy of statistical estimates naturally depends upon the sample length available. Usual practice in such unsteady simulations is to compute "long" samples and to assume that the statistical error is negligible. What time sample length is considered "long" is unfortunately often a strong function of the computational resources and time available. In more thorough analyses, e.g. Haase et al. (2009), running averages are inspected to give a qualitative feel for the level of variability present in the statistics. However, even after long averaging durations these trends can still give rise to surprises in the form of strong and sudden drift in the mean value. Most importantly, no quantitative error estimate is obtained from this technique. The time signals of the integral forces from this flow case unfortunately exhibit very long wavelength random modulation and a multi-modal character, as seen in both experiment (Swalwell et al., 2004) and DES (Haase et al., 2009, Mockett, 2009). As a result it is expected that very long time samples are necessary.

To obtain quantitative estimates of the statistical error arising due to the finite time sample available, a technique developed recently at TUB is applied. A full description of this technique is outside the scope of this paper and would indeed necessitate a dedicated paper. The method employs a standard definition of the random error occurring in statistically stationary data (Bendat & Piersol, 1986), which is computed by subdividing the available data into windows. Employing analytically-derived relationships for bandwidth-limited white noise, case-specific expressions of the statistical error as a function of sample size, T_{avg}, are returned. The sole input is the time series in question and no further user parameters are required. The following expressions have been derived for the normalised error, ε, in the estimated mean, $\hat{\mu}$, and standard deviation, $\hat{\sigma}$, of the lift and drag coefficient:

$$\varepsilon\left[\hat{\mu}_{C_l}\right]=\frac{1.83}{\sqrt{T_{avg}}}\left(\frac{\sigma_{C_l}}{\mu_{C_l}}\right)\times 100\%\,,\quad \varepsilon\left[\hat{\mu}_{C_d}\right]=\frac{2.36}{\sqrt{T_{avg}}}\left(\frac{\sigma_{C_d}}{\mu_{C_d}}\right)\times 100\%\,,$$

$$\varepsilon\left[\hat{\sigma}_{C_l}\right]=\frac{1.29}{\sqrt{T_{avg}}}\times 100\%\,,\quad \varepsilon\left[\hat{\sigma}_{C_d}\right]=\frac{1.67}{\sqrt{T_{avg}}}\times 100\%\,, \tag{1}$$

where μ_x and σ_x are the mean and standard deviation of the quantity x computed from the entire signal, respectively. T_{avg} is given in dimensionless quantities, as specified in Table 2. These expressions, the constants for which are computed from the experimental time series (of length T = 8860 c/U_∞ and derived from a single section), have been found to hold for the DES data of varying span size when the respective computed values of μ_x and σ_x are used. The error trends for the wide domain DES furthermore agreed excellently with those from the experiment. Notably, the difference in the error trend between sectional force data and that integrated over the entire span was seen to be negligible.

The expressions of Eq. (1) therefore allow statistical error bars to be placed on the mean and standard deviations of the DES forces, which is indeed done in the following section. Such analysis furthermore enables the engineer to estimate the number of time steps required to achieve the desired level of statistical accuracy for a particular flow case. For the DES with L_z = 4c for example, Eq. (1) stipulates that a time sample of roughly 530 c/U_∞ is required to achieve 1% accuracy in the mean forces.

4.3 Span-Size Effects

The spanwise distance between the periodic boundaries to the computational domain has been varied between L_z = 1c and 4c in the calculations of NTS and TUB, allowing the influence of this user parameter to be studied. Figures 2 and 3 draw upon all DES results employing a periodic spanwise domain and all applicable experimental data to provide an overview. The magnitude of the statistical error obtained from Eq. (1) is portrayed in the figures by error brackets.

The effect of a narrow span size is to increase all quantities depicted. The effect is indeed pronounced, far outweighing the statistical error and the model/code dependency. When long time samples are computed, L_z hence constitutes the most influential source of error. It is to be expected that this finding would apply to nominally 2D bluff bodies in general. Furthermore, since the ratio σ/μ increases, it can be seen from (1) that the statistical error of the mean forces is greater for shorter L_z and a given T_{avg}. This effect is so strong that L_z = 4c becomes computationally less expensive than L_z = 1c to a desired statistical accuracy.

The plots of Figs. 2 and 3 suggest that the L_z influence becomes marginal at values of around 4c for this test case, although further computations at still larger L_z values would strictly be required to demonstrate this - with exception of the L_z = 1c data, the results appear to scale according to L_z^{-2}. An indication that the required L_z is strongly case-specific is given by comparison with the results of Guenot (2004) for the lower $\alpha = 45°$ case. In that case, the L_z effect was seen to settle at lower values of $L_z \approx 2c$. It is believed that the wider wake is responsible for the larger L_z required for our $\alpha = 60°$ case.

The mean surface pressure coefficient, C_p, plotted in Fig. 4, corroborates the trend shown by Fig. 2 and shows that the L_z effect is manifested in a strong variation of the base pressure on the downstream (upper) surface of the airfoil. The paradox that the agreement with experiment degrades with increasing L_z was initially a source of concern. However, the observation of a strong scatter between the experiments (particularly visible in Fig. 2) reduces this. Furthermore, the conclusion to be reached from Fig. 2 is that increasing L_z brings the mean C_l inside the experimental range and in excellent agreement with the Sheldahl & Klimas data. For the mean C_d however, the agreement is less positive. This could in part be because the C_d value from the Raghunathan et al. experiment, expected to be low, is unknown.

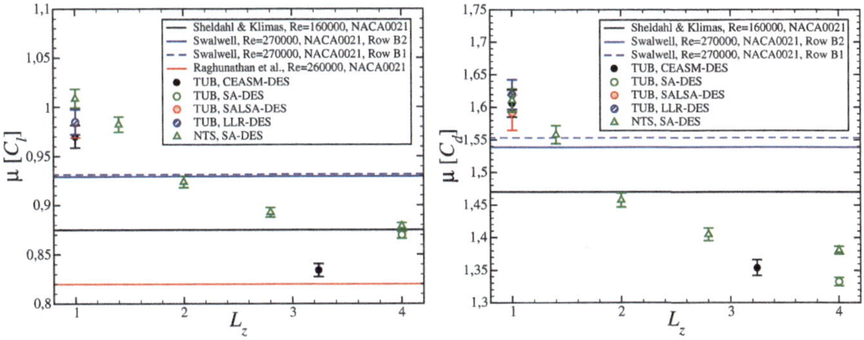

Fig. 2 Trends of mean lift and drag coefficient against L_z for the NACA0021 at $\alpha = 60°$ and Re = 270000 (unless otherwise stated)

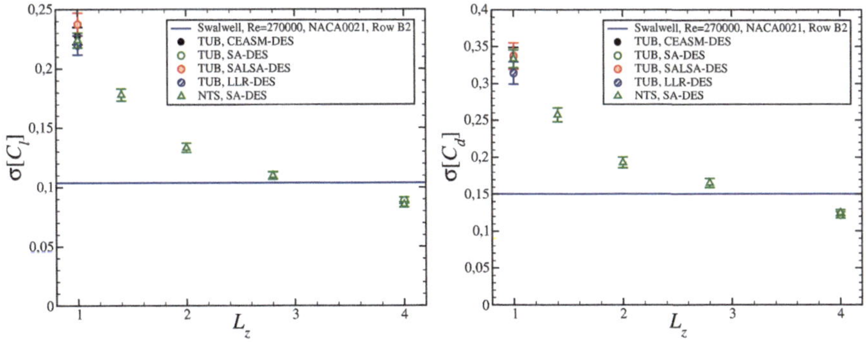

Fig. 3 Trend of standard deviation of the lift and drag coefficients with L_z for the NACA0021 at $\alpha = 60°$ and Re = 270000

Despite the poor agreement of the mean quantities with Swalwell's experimental data, the story for the unsteady quantities is very different: A strong improvement in the standard deviation of the integral forces is achieved at longer L_z values (Fig. 3). More detail is offered by the power spectral density (PSD) of C_l, plotted in Fig. 5. Particularly regarding the lowest frequencies, the agreement of the broadband energy is strongly improved at $L_z = 4c$. The sharp spectral peak at St $\approx$ 0.2 represents the vortex shedding frequency and the strong disagreement in its energy is troublesome, however. Although not clearly visible in the plot of Fig. 5, the computed peak energy ranges between 30% (TUB) and 50% (NTS) of the experimental value.

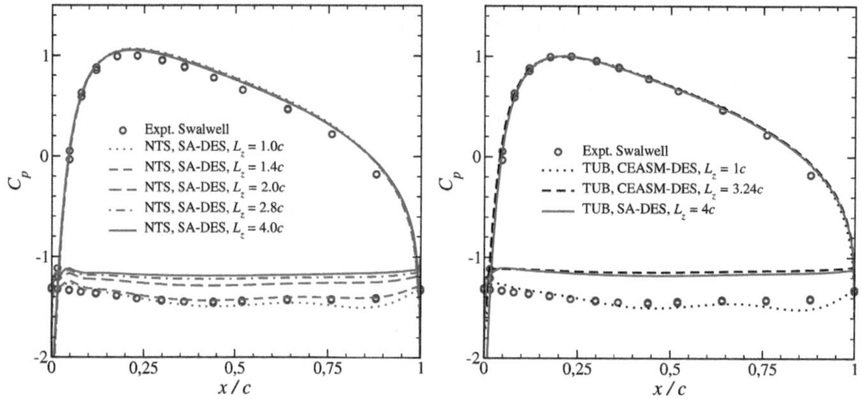

Fig. 4 The effect of L_z on the mean C_p distribution. Comparison of NTS (*left*) and TUB DES (*right*) with experiment

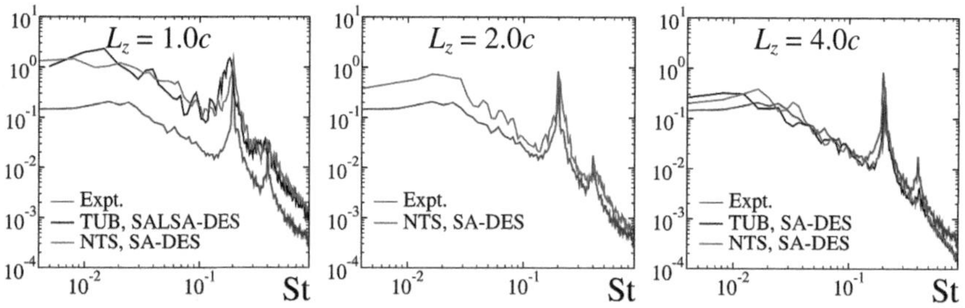

Fig. 5 PSD of C_l comparing NTS and TUB DES for varying L_z with experiment

A natural statistical quantity to examine in such an investigation is the two-point correlation of velocity in the spanwise direction, $R_{ii}(z)$, where i represents the u, v and w components. These are plotted in Fig. 6 at four probe positions in the near wake for TUB computations at $L_z = 1c$ and $L_z = 4c$. For the u and v components, the correlations settle to lower values in the wider domain, however they clearly remain above zero. Such long spanwise correlation lengths naturally arise due to the strong spanwise coherence of the large-scale vortex shedding. For the w component a different behaviour is seen: In the narrow domain the correlation consistently reaches a negative value, implying an anti-phase behaviour. For the longer domain however, the w correlation settles around zero.

It is therefore suggested that $R_{ww}(z)$ could provide a generic test for sufficient span size in practice. This would however require more testing to confirm.

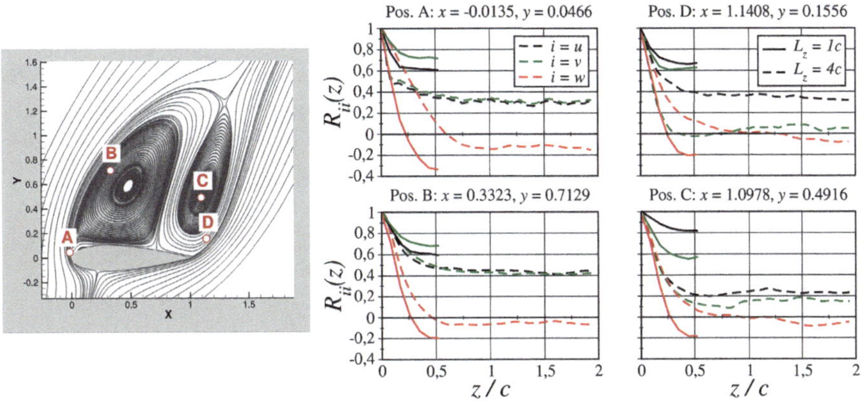

Fig. 6 Two-point spanwise correlations of u, v and w compared for $L_z = 1c$ (TUB CEASM-DES) and $L_z = 4c$ (TUB SA-DES) at various probe locations

Finally, further insight into the span size effect can be obtained from instantaneous visualisations of the resolved vortex structures (Fig. 7). The streamwise rib vortices and spanwise von Kármán vortices maintain a high degree of orthogonality in the narrow spanwise domain. In contrast, these exhibit less orthogonality with spanwise dislocations and higher degrees of slant apparent in the wider domain.

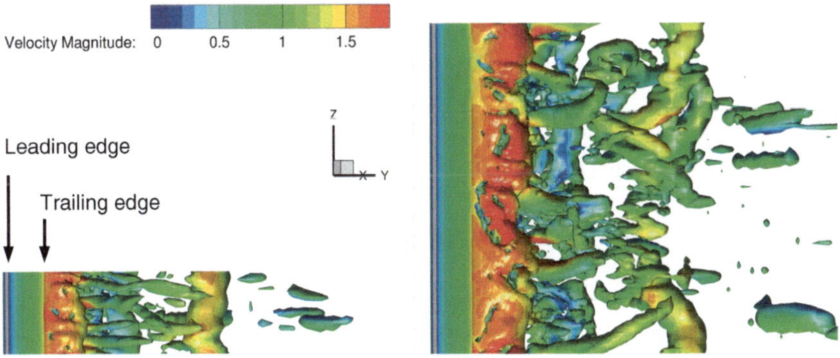

Fig. 7 Instantaneous snapshots of resolved vortical structures compared between $L_z = 1c$ (TUB CEASM-DES) and $L_z = 4c$ (TUB SA-DES) calculations. Isosurfaces of $\lambda_2 = -4$ (Jeong & Hussain, 1995) shaded with velocity magnitude, view from below (facing the trailing edge)

4.4 Effect of Wind Tunnel Walls

It was suspected that the surrounding wind tunnel walls, neglected in the spanwise-periodic computations, may contribute to the discrepancy with the Swalwell experiments. This was the motivation for the EADS simulation including the wind tunnel walls with free-slip boundary conditions. The mean surface C_p and PSD of C_l plotted in Fig. 8 indicate a negligible impact of the tunnel walls. The time-averaged force coefficients give a similar message: The $C_l = 0.89 \pm 0.01$ and $C_d = 1.43 \pm 0.02$ from the computation with wind tunnel walls do not differ significantly from the NTS periodic computation ($C_l = 0.88 \pm 0.003$ and $C_d = 1.38 \pm 0.006$). Although the comparison is unfortunately not direct (employing different L_z, solvers, grid resolution and time step), the experience of various related investigations implies that these factors have a negligible influence for this test case.

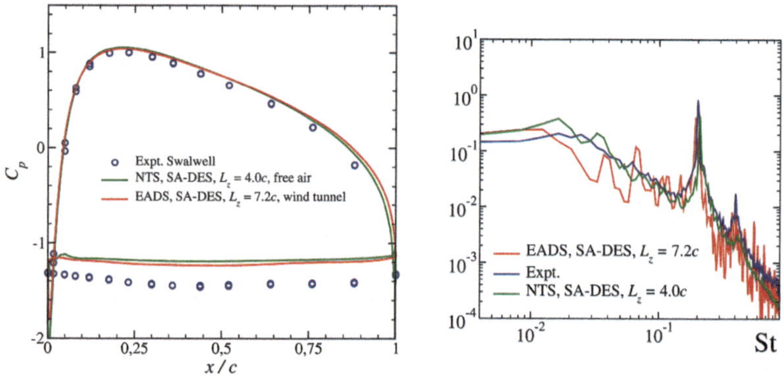

Fig. 8 Comparison of mean surface C_p (*left*) and PSD of C_l (*right*) between the spanwise-periodic NTS computation (L_z = 4c) and the EADS computation with tunnel walls ($L_z = 7.2c$) and with experiment

5 The Original DES Study Revisited

In light of all that has since been learned about this test case, it is interesting to revisit the pioneering DES study of Shur et al. (1999): To what extent are the original conclusions regarding the DES predictive capability for such flows still valid? It is now clear that the employed span size of one chord significantly distorts the mean force coefficients. Likewise, the simulated time samples (quoted in the range $100 < T_{avg} \times |U_\infty| / c < 200$) can be expected to add an appreciable error bracket to the statistical quantities. How can we therefore interpret the high level of agreement that was seen with the experimental data of Hoerner (1965)?

A single data point from the Shur et al. results, namely at the $\alpha = 60°$ value that formed the focus of subsequent investigations, has been analysed in Table 3. Assuming the same statistical error trend as that determined for the NACA0021 profile, corresponding error bands have been added to the original mean force coefficients. To estimate the influence of the narrow span size, the original values have furthermore been re-scaled using the difference in force coefficient observed between $L_z = 1c$ and $L_z = 4c$ for the NACA0021. Although the original results agreed well with the experimental values quoted by Hoerner, the values re-scaled to $L_z = 4c$ agree excellently with the alternative experimental data provided by Sheldahl & Klimas. The message is therefore that the latest findings serve to reinforce rather than undermine the original conclusions of Shur et al.

Table 3 Comparison of the original and span size rescaled NACA0012 DES results of Shur et al. (1999) with the Sheldahl & Klimas (1981) (S&K) experimental data

Case:	Mean C_L:	C_L deviation rel. to S&K	Mean C_D:	C_D deviation rel. to S&K
DES, Shur et al. (1999), Re = 100000, $L_z = 1c$	1.00 ± 0.02	+14 %	1.69 ± 0.04	+15 %
Above result rescaled to $L_z = 4c$	0.87 ± 0.02	–0.6 %	1.46 ± 0.03	–0.7 %
Exp, Sheldahl & Klimas (1981), Re = 160000	0.875	—	1.470	—

6 Conclusions

This study has drawn upon a large body of simulations involving numerous parameter variations and has shed considerable light on the dominant sources of uncertainty for DES of nominally 2D airfoils in deep stall. Two key factors have been found to affect the results, namely the length of the time sample simulated and the spanwise domain size. Thanks to the statistical error analysis method, the former effect could be quantified. It emerges that around 530 non-dimensional time units must be computed to achieve at least 1% accuracy in mean lift and drag for $L_z = 4c$. The strong influence that a narrow span size has on the mean and fluctuating forces has been documented. The computation of very long time samples has furthermore served to separate the span size influence from the statistical error. A key result of this study concerns the interaction of these dominant parameters of influence: A further effect of a narrow domain size is a strong increase in the time sample requirements to reach the same level of statistical accuracy in the mean force coefficients. For $L_z = 1c$, time samples in the region of 2400 non-dimensional units are required to achieve the same 1% accuracy in the mean forces. So strong is the effect that once initial transient has been removed, it is in fact computationally *less* expensive to simulate the longer spanwise domain size of $L_z = 4c$ when the mean forces are of primary interest[1].

Standing in contrast to the time sample and span size, the variation in the results due to the choice of underlying RANS model, conducted primarily for $L_z = 1c$, is weak and of a similar magnitude to the statistical error. This is in accordance with the experimental observation that the flow is robust to variations in the Reynolds number and the free stream turbulence intensity. This weak model sensitivity for DES when applied to such flows represents a key advantage over

[1] Note that this does not apply to the standard deviations, the statistical error upon which is unaffected by the span size (Eq. (1)).

URANS (for which significant model sensitivity has been established). A further encouraging result has been the demonstration of excellent agreement between the NTS and TUB computations in a direct comparison of the different codes. Finally, an indication has furthermore been given that the influence of wind tunnel walls is also weak, although a more direct comparison would be desirable.

The best agreement of the mean forces is with the Sheldahl & Klimas data and the comparison of force spectra with the data of Swalwell is encouraging in many respects. However, a fundamental obstacle remains due to the strong experimental scatter. A more detailed and critical review of the experiments with the goal of reducing this would hence be a valuable future contribution. As things stand, the primary suitability of the flow case is for the cross-verification of CFD methods.

Finally, rather than challenging the conclusions of the pioneering DES study of Shur et al. (1999), which established the significant advantages of DES over URANS, the current work has served to reinforce these.

Acknowledgements. This work has been partially funded by the EU projects DESider (Detached-Eddy Simulation for Industrial Aerodynamics) and ATAAC (Advanced Turbulence Simulation for Aerodynamic Application Challenges). DESider was funded in the 6th Framework Programme under Contract No. AST3-CT-2003-502842. ATAAC is funded in the 7th Framework Programme under Contract No. ACP8-GA-2009-233710-ATAAC. Some of the TUB computations were conducted on the IBM pSeries 690 at the Norddeutschen Verbund für Hoch- und Höchstleistungsrechnen (HLRN).

The authors are grateful to Dr. K. Swalwell (formerly of Monash University) for the provision of the experimental data in electronic form and to S. Julien (Université Laval, Quebec) and T. Knacke (Technische Universität Berlin) for advice concerning the statistical error analysis.

References

Bendat, J.S., Piersol, A.G.: Random data, analysis and measurement procedures. John Wiley & Sons, Inc., Chichester (1986)

Guenot, D.: Simulation des effets instationnaires à grande échelle dans les écoulements décollés. Ph.D. thesis, SUPAERO (2004)

Haase, W., Braza, M., Revell, A. (eds.): DESider – A European effort on hybrid RANS-LES modelling. Notes Numer. Fluid Mech. Multidiscip. Design, vol. 103. Springer, Heidelberg (2009)

Hoerner, S.: Fluid-dynamic drag. Hoerner Fluid Dynamics, New Jersey (1965)

Jeong, J., Hussain, F.: On the identification of a vortex. J. Fluid Mech. 285, 69–94 (1995)

Lübcke, H.: Entwicklung expliziter Darstellungen zweiter statistischer Momente zur numerischen Simulation turbulenter Strömungen. Ph.D thesis, Technische Universität Berlin (2001)

Mockett, C.: A comprehensive study of detached-eddy simulation. Ph.D. thesis, Technische Universität Berlin (2009)

Raghunathan, S., Harrison, J.R., Hawkins, B.D.: Thick airfoil at low Reynolds number and high incedence. J. Aircr. 25(7), 669–671 (1988)

Rung, T., Thiele, F.: Computational modelling of complex boundary-layer flows. In: Proc 9th Int. Symp. Transp. Phenom. Therm-Fluid Eng., Singapore (1996)

Rung, T., Bunge, U., Schatz, M., Thiele, F.: Restatement of the Spalart–Allmaras eddy-viscosity model in strain-adaptive formulation. AIAA J. 41(7), 1396–1399 (2003)

Schwamborn, D., Gerhold, T., Hannemann, V.: On the validation of the DLR Tau code. In: Nitsche, W., Heinamann, H.J., Hilbig, R. (eds.) New Results Numer. Exp. Fluid Mech. II. Springer, Heidelberg (1998)

Sheldahl, R.E., Klimas, P.C.: Aerodynamic characteristics of seven symmetrical airfoil sections through 180-degree angle of attack for use in aerodynamic analysis of vertical axis wind turbines. Sandia National Laboratories SAND80-2114 (1981)

Shur, M., Spalart, P.R., Strelets, M., Travin, A.: Detached-eddy simulation of an airfoil at high angle of attack. In: Rodi, W., Laurence, D. (eds.) Eng. Turbul. Model. Exp., vol. 4, pp. 669–678. Elsevier, Amsterdam (1999)

Spalart, P.R.: Detached-eddy simulation. Annu. Rev. Fluid Mech. 41, 181–202 (2009)

Spalart, P.R., Allmaras, S.R.: A one-equation turbulence model for aerodynamic flows. In: Proc 30th AIAA Aerosp. Sci. Meet. Exhib., Reno (1992)

Spalart, P.R., Jou, W., Strelets, M., Allmaras, S.R.: Comments on the feasibility of LES for wings, and on a hybrid RANS/LES approach. Adv DNS/LES 1 (1997)

Strelets, M.: Detached eddy simulation of massively separated flows. In: Proc 30th AIAA Aerosp. Sci. Meet. Exhib., Reno (2001)

Swalwell, K.: The effect of turbulence on stall of horizontal axis wind turbines. Ph.D thesis, Monash University (2005)

Swalwell, K., Sheridan, J., Melbourne, W.: Frequency analysis of surface pressure on an airfoil after stall. In: Proc 21st AIAA Appl. Aerodyn. Conf. (2004)

Travin, A., Shur, M., Strelets, M., Spalart, P.R.: Physical and numerical upgrades in the detached-eddy simulation of complex turbulent flows. In: Friedrich, R., Rodi, W. (eds.) Advances in LES of Complex Flows. Kluwer Acad., New York (2000)

Xue, L.: Entwicklung eines effizienten parallelen Lösungsalgorithmus zur dreidimensionalen Simulation komplexer turbulenter Strömungen. Ph.D thesis, Technische Universität Berlin (1998)

Initial RANS and DDES of a Rudimentary Landing Gear

P. Spalart, M. Shur, M. Strelets, and A. Travin

Abstract. Results of RANS and Delayed Detached Eddy Simulation (DDES) are presented for a new research model of landing gear. The design is as simple as that of Lazos but a primary motivation for it is a weak and well-understood Reynolds-number dependence. The simplicity should allow essentially grid-converged RANS-LES solutions now or within a few years, but it preserves some physics, which challenge CFD, in a manner meaningful to noise-prediction capability for landing gear. The initial simulations presented in this work parallel an experimental study launched recently at NAL (India). A major goal of the simulations is to verify the potential of the RLG geometry in terms of CFD validation and to attract the attention of the turbulence and airframe noise communities to the experimental database, which will be available in the near future. The principal early challenge to CFD is to predict the wall-pressure fluctuations. Far-field noise will be addressed later.

1 Introduction

The landing-gear area is one of the many fronts on which industry is fighting the "war on noise." Unsteady CFD methods have matured very nicely for jet noise, although coming far short of covering the audible range. In contrast, airframe-noise CFD prediction remains very incomplete, and has few "success stories." One reason is the absence of experimental data which isolate the noise due to one region of the flow. Other reasons are the higher geometric complexity, and difficult integration with Ffowcs-Williams-Hawkings (FWH) and similar post-processing utilities. Today it would be difficult to distinguish the errors caused by the simulation itself and those caused by the acoustics component. This motivated the experimental study of simplified landing gear carried out by Lazos (2004). Its complexity was very appropriate since there is no doubt that a full or even "partially-dressed" landing-gear geometry is out of reach for grid convergence, and will be for years. Even one stripped of items such as brake lines, but with links and other items with non simple shapes, appears to demand too many grid

P. Spalart
Boeing Commercial Airplanes, USA
philippe.r.spalart@boeing.com

M. Shur · M. Strelets · A. Travin
New Technologies and Services, St.-Petersburg, Russia

S.-H. Peng et al. (Eds.): Progress in Hybrid RANS-LES Modelling, NNFM 111, pp. 101–110.
springerlink.com

points, without adding much understanding. On the other hand, proposals such as a single wheel or tandem wheels lack shape variety and interference. The Lazos study turned out helpful in validating DES (see, e.g., a paper of Hedges et al. (2002) and results of DESider Project, Haase et al. (2009)) but lacked the unsteady pressure data which are felt to be the key product of the flow for noise purposes, at low Mach numbers. Other than that, it was in an unfortunate Reynolds-number range for the (circular) axles and post presumably suffering a tangible Reynolds-number dependence which is very undesirable since moving to acoustic facilities often imposes size or speed changes.

Based on these considerations and the current desire to address airframe noise numerically in an ordered manner, a new experimental project funded by Boeing has been recently launched at NAL (India). The priorities in the study design have been to make possible essentially grid-converged RANS-LES solutions now or within a few years, while containing major physics challenging CFD in a manner meaningful to its noise-prediction capability for landing gear. Another focus has been weak Reynolds-number dependence. This led to revising the Lazos model, primarily in order to eliminate Reynolds-number dependence and increase the blockage between the wheels. The principal new features of this, Rudimentary Landing Gear (RLG), are in having larger axles with square cross-sections and tripping the wheel boundary layers. The trips were designed by CFD in terms of location and size, and found effective. As a result, CFD can be run with fully turbulent boundary layers; this greatly simplifies setting up the hybrid RANS-LES methods and, again, reduces Reynolds-number sensitivity. It is also realistic of full-size systems, with their rubber tires.

The goals and policies adopted here are similar to those of the LAGOON effort, which however has a more complex geometry and has limited distribution (Manoha et al. 2009).

An objective of this paper is to make the turbulence and airframe noise communities aware of the experimental study of the RLG, which has the potential to become the focus of CFD and CAA workshops, and of other experimental campaigns including noise measurements. It presents results of RANS computations aimed at designing of the trips used in the experiment and DDDES predictions of unsteady and mean flow characteristics which can be considered as a "blind test" of the DDES capabilities. Correspondingly, we first briefly outline the RLG geometry and experimental set-up (Section 2), then discuss S-A RANS results of a single wheel with different tripping systems (Section 3), and after that present the numerical set-up used for S-A based DDES (see Spalart et al. (2006)) of the full geometry and results of the simulations (Section 4).

2 RLG Geometry and Experimental Set-Up

A general view of the RLG model installed in the wind tunnel is presented in Fig. 1.

The model can be defined with the few numbers presented in Table 1 (all lengths are normalized with the wheel diameter, D=0.4064 m) and shown also in Fig. 1.

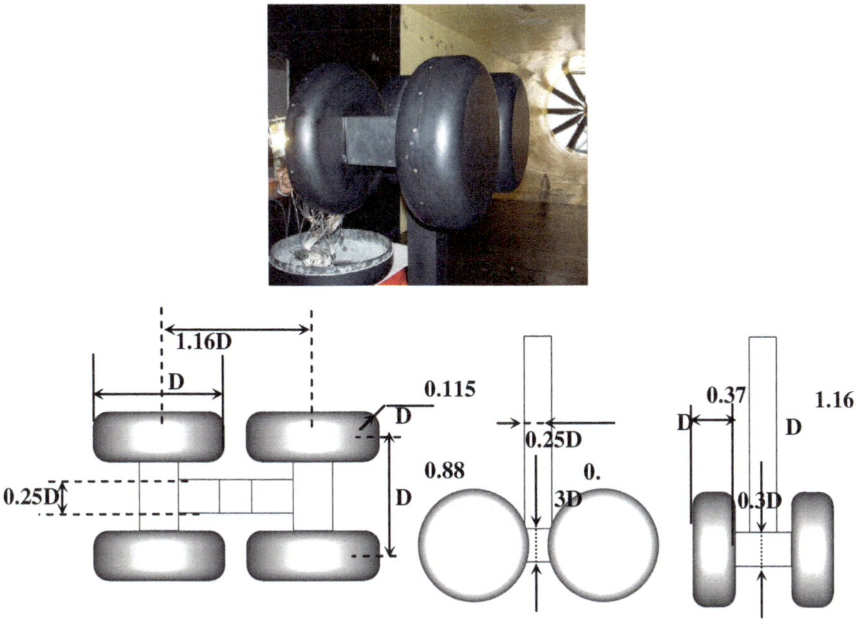

Fig. 1 General view of the RLG installed in the wind tunnel and its dimensions

Table 1 RLG dimensions

Wheel width	Tire shoulder radius	Wheel-base	Track	Beam cross-section	Axles cross-section	Post cross-section
0.37	0.115	1.16	0.88	0.3×0.25	0.3×0.3	0.25×0.25

The model is placed centrally in the tunnel, which is a square with sides of 3.69*D* (the wheel centres are in the central horizontal plane of the tunnel, y=1.845 *D*). The Reynolds number based on the wheel diameter and reference velocity ($U_0 = 40$ m/s) is equal to 10^6.

The 2009-2010 campaigns will produce average and unsteady wall-pressure fields. Pressure fluctuations will be measured over the wheel surfaces and at a few points on the axles and post, and their spectra are expected to be a challenging and meaningful test.

Oil flow is taken in order to determine separation locations and in particular confirm that laminar separation is not taking place on the wheels. The square axles are predictable in terms of separation, but not reattachment, providing additional comparison material.

Future campaigns will include flow measurements, by LDV and/or PIV.

3 RANS-Based Design of Tripping Systems for RLG Wheels

As mentioned in the Introduction, during the RLG model design, S-A RANS computations of the flow past a single wheel were carried out for different geometries of tripping lines on the wheel surface. As a result, an optimal tripping system with two pairs of trip-lines on the tire shoulder and on the tread shown in Fig.2 was proposed which is relatively simple, on one hand, and as shown in Fig.3 provides for a reliable transition to turbulence in the attached boundary layer, on the other hand. The shoulder trip ensures transition before separation on the wheel sides, and the tread trip ensures transition before separation on the top and bottom sides. There is no early separation, compared with a Fully Turbulent run (that is, with sufficient inflow eddy viscosity to cause immediate transition). The calculations also provided the boundary-layer thickness, which was used to size the trip dots.

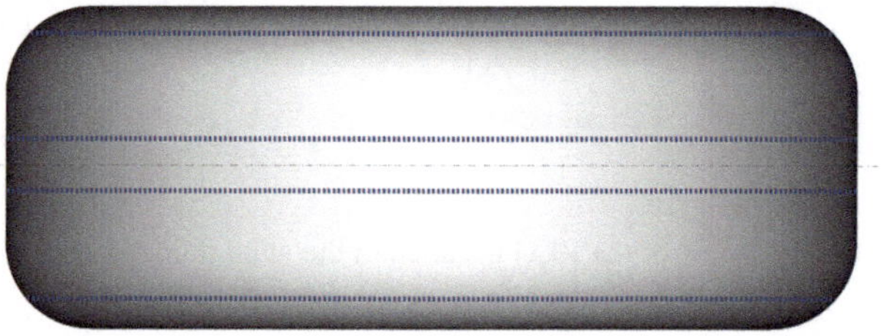

Fig. 2 Schematic of trip lines on wheel surface

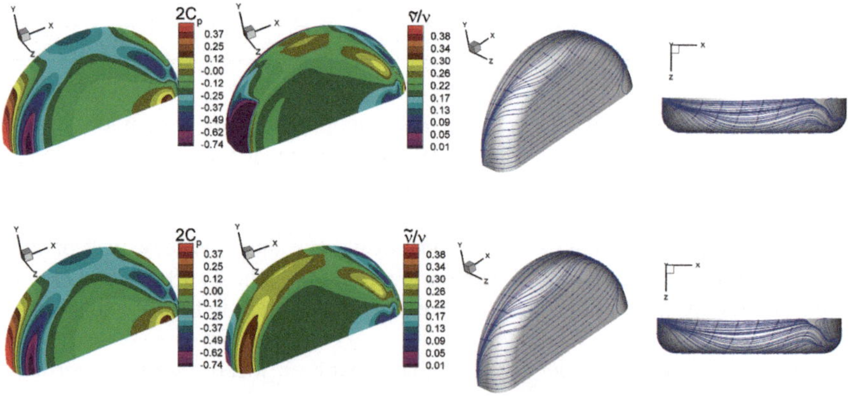

Fig. 3 Comparison of wheel surface pressure, and eddy viscosity and streamlines at the nearest to wall grid surface computed with boundary layer tripping (upper row) and within the Fully Turbulent (FT) approach (lower row)

4 Initial Results of DDES of the Full RLG Geometry

Figure 4 displays the computational domain and 11-block structured grid of about 7.6 million cells total used in the simulation.

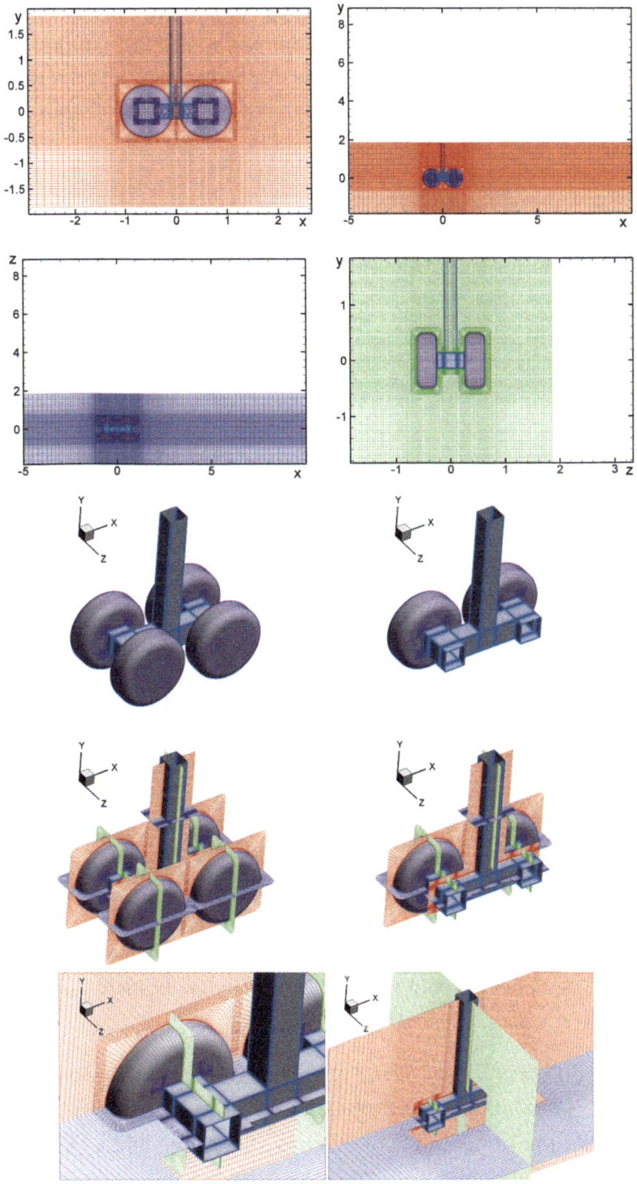

Fig. 4 Computational domain and some details of structured multi-block grid

The grid is clustered near the RLG surface so that the value of y^+ nearest to the wall is about 1.0, and is refined in the focus region of DES so that the target size of the grid there is about 0.02D. The test section walls in the simulation are treated as inviscid. The time step in the simulation is 0.005 (D/U_∞).

Figure 5 displays different views of vorticity snapshots from the simulation, which reveal rather fine-grained turbulent structures consistent with the grid used, and visibly finer than in our earlier work (Hedges et al. 2002). Other than that, the figure shows a strong interaction between the turbulence generated by different elements of the truck and suggests that no regular vortex shedding takes place either from the truck post or wheels.

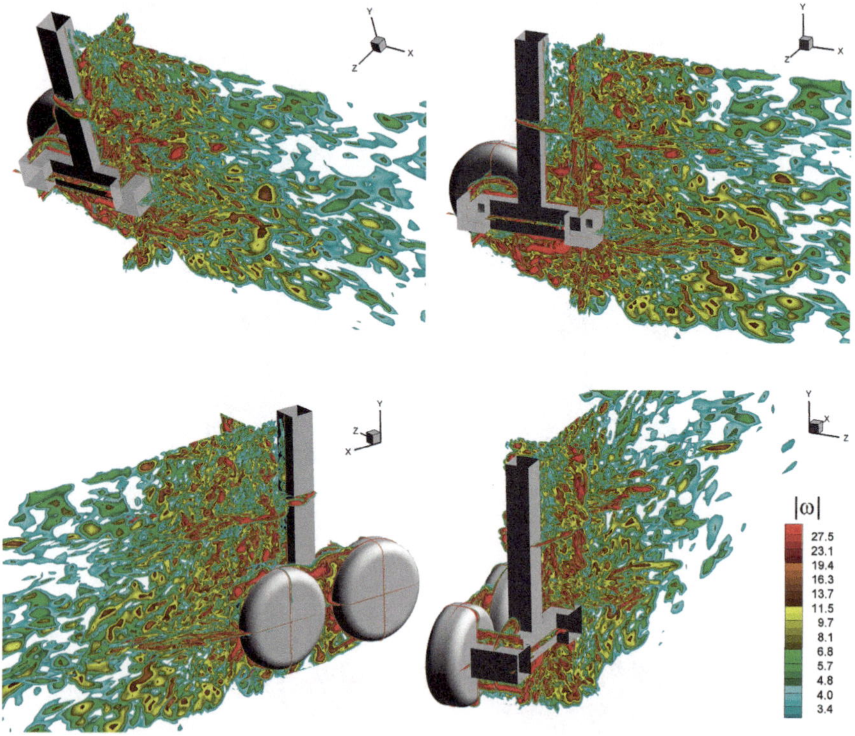

Fig. 5 Different views of instantaneous vorticity field (flow along x-axis)

Figure 6 shows different views of the instantaneous surface pressure fluctuations, $p' = (p - \langle p \rangle)/(\rho U_\infty^2)$ where the time averaged pressure is defined by averaging of the pressure field over the time sample of 70 convective time units, D/U_∞, started after a transient period of 90 convective units. The figure reveals a significant unsteadiness of the surface pressure, everywhere except for the windward side of the front wheels, post, and forward axle.

Finally, Fig.7 presents spectra of the pressure fluctuations at several points on the truck surface (location of the points on the windward and leeward surfaces is shown in the upper part of the figure). One can see that the simulation resolves a rather wide frequency range, from St=0.1 up to St=10 (the Strouhal number corresponding to the Nyquist frequency is 25). All the spectra are nearly broadband (i.e., do not have any clearly expressed shedding peaks).

Figures 8, 9 display contours of the mean surface pressure coefficient defined as $C_p = 2\cdot(\langle p\rangle - p_{ref})/(\rho U_\infty^2)$ where p_{ref} is the inflow pressure, and CFD "oil flow" patterns. Figure 10 shows a history of the drag coefficient from the simulation and compares it with a similar history from Haase et al. (2009) for the Lazos SLG. Consistently with the difference between the two geometries, the drag of the RLG is more than twice as high as that of the SLG (the corresponding values of the mean drag coefficient are 1.68 and 0.70).

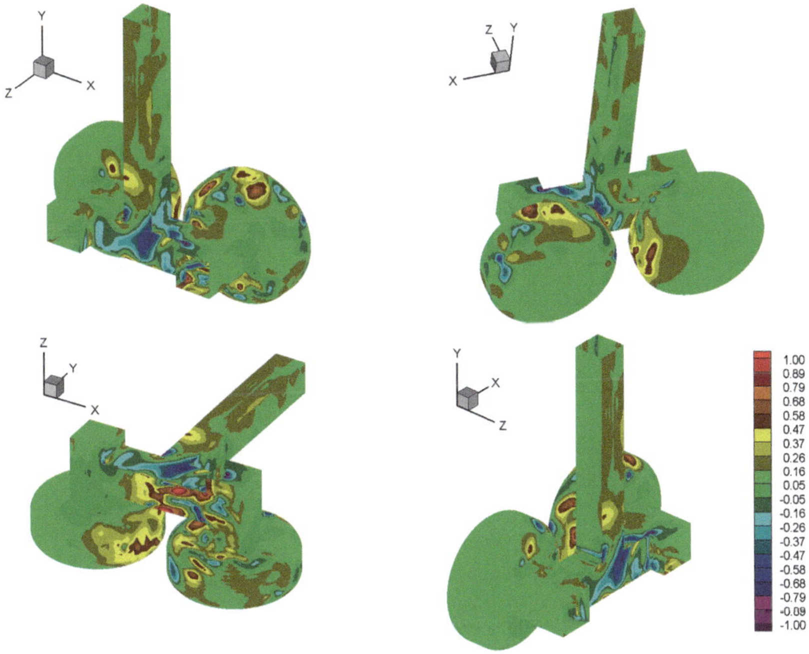

Fig. 6 Different views of instantaneous surface pressure fluctuation (flow along x-axis)

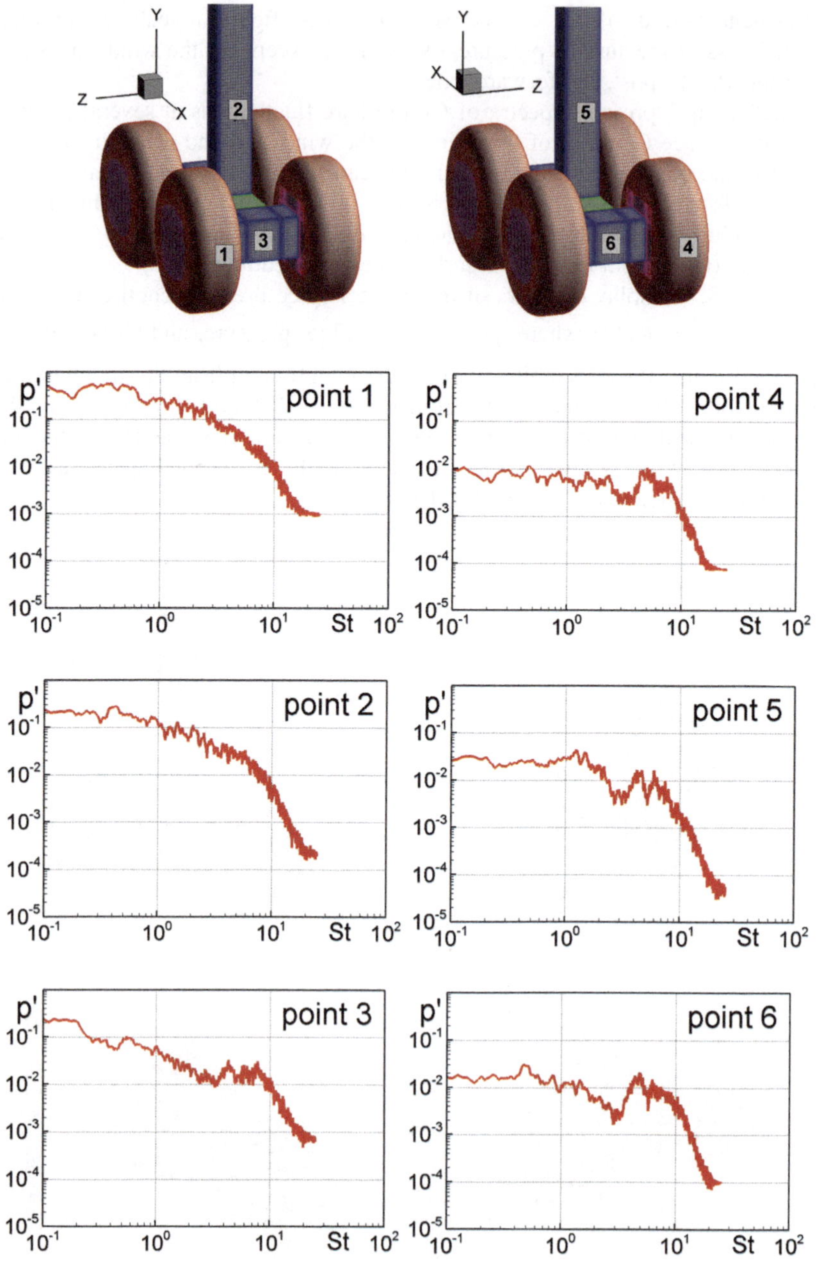

Fig. 7 Location of "probe" points and spectra of surface pressure fluctuations at the 6 different points on leeward (left column) and windward (right column) sides of track surface

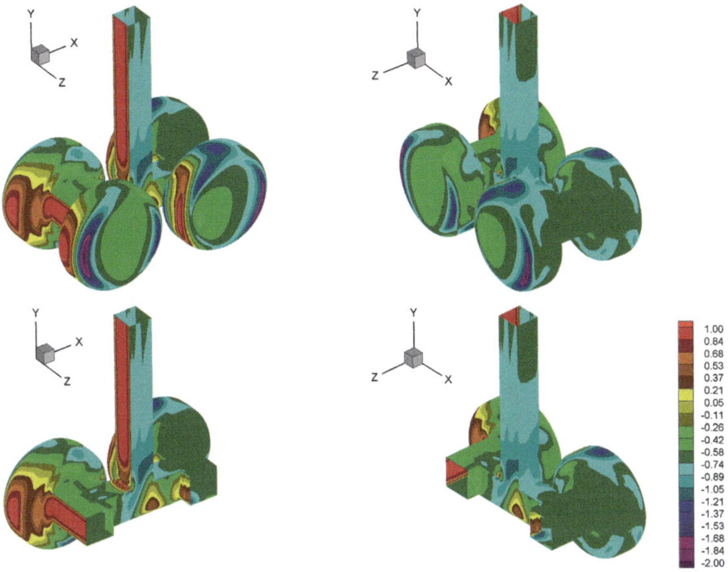

Fig. 8 Contours of surface mean pressure coefficient (flow along x-axis)

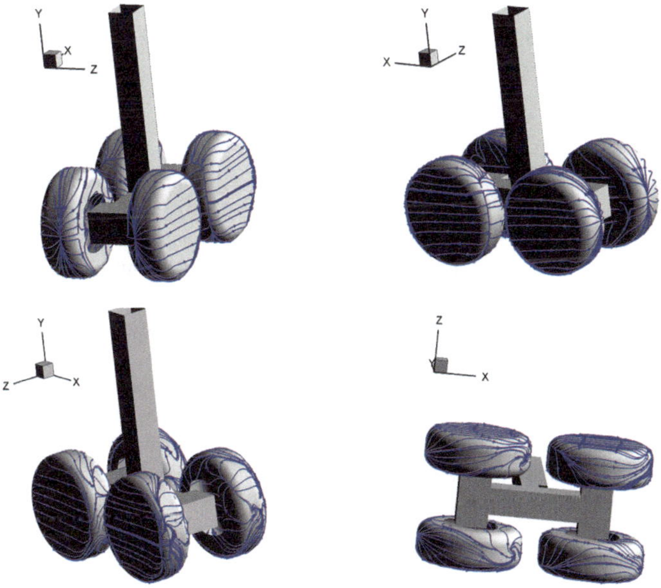

Fig. 9 Different views of wall streamlines (flow along x-axis)

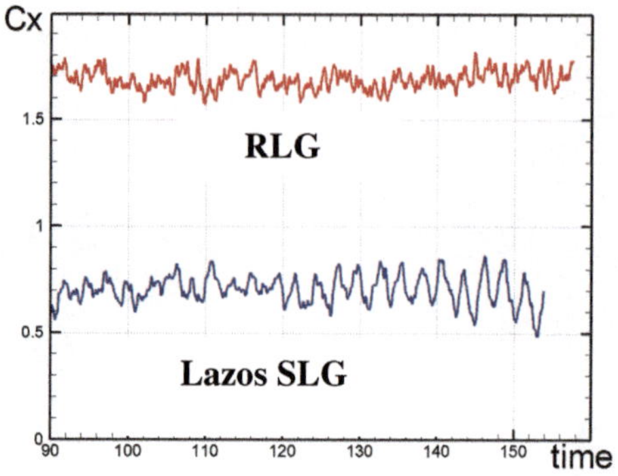

Fig. 10 Comparison of drag coefficient time histories from current DDES of the RLG and of DES of the SLG of Lazos, in Haase et al. (2009), Chapter 4.20

Summarizing we can conclude that, in general, the DDES prediction of the RLG flow looks quite plausible, although of course a final evaluation of the simulation accuracy can be made only after a comparison of the CFD database with the measurements of mean flow and unsteady surface pressures. The correct prediction of their spectra is a necessary, but not sufficient, condition for successful noise prediction.

Acknowledgements. This work has been primarily funded by Boeing Commercial Airplanes and, partially, by the Russian Basic Research Foundation (grant No. 09-08-00126) and by the EU project ATAAC, Contract No. ACP8-GA-2009-233710-ATAAC.

References

Lazos, B.S.: Reynolds Stresses Around the Wheels of a Simplified Four-Wheel Landing Gear. AIAA J. 42, 196–198 (2004)

Haase, W., Braza, M., Revell, A. (eds.): DESider - A European Effort on Hybrid RANS-LES Modelling. Notes on Numerical Fluid Mechanics and Multidisciplinary Design, vol. 103. Springer, Heidelberg (2009)

Hedges, L.S., Travin, A., Spalart, P.R.: Detached-Eddy Simulations over a Simplified Landing Gear. J. Fluids Eng. 124, 413–423 (2002)

Manoha, E., Bulte, J., Ciobaca, V., Caruelle, B.: LAGOON: further analysis of aerodynamic experiments and aeroacoustics results. AIAA Paper-2009-3277 (2009)

Spalart, P.R., Deck, S., Shur, M., Squires, K., Strelets, M., Travin, A.: A new version of detached-eddy simulation, resistant to ambiguous grid densities. Theoretical and Computational Fluid Dynamics 20, 181–195 (2006)

Detached Eddy Simulation of Turbulent Flow in a Highly Offset Intake Diffuser

M. Chevalier and S.-H. Peng

Abstract. Turbulent flow past an S-duct intake configuration was studied by means of Reynolds–averaged Navier–Stokes simulations (RANS) and detached eddy simulations (DES). One of the main purposes with the study was to make comparative studies of the two modelling approaches for internal flows with separation in comparison with available experimental measurements. The computations were conducted using an in-house unstructured flow solver. A comparison is given for the results obtained with the RANS and DES modelling approaches. It is shown that the DES method has produced a flow separation bubble larger than that computed by RANS methods. The computational discrepancies and the reason behind the inaccuracies in the modelling are discussed.

1 Introduction

Improving stealth capabilities is one of the significant aspects in modern military aircraft designs. This has usually been considered by shielding the engine face from direct observation with a more compact design, in which the air intake diffuser is offset from the engine face plane. However, an offset shape (usually, an "S" form) may induce inappropriate aerodynamic features, which can degrade the engine performance with decreasing thrust and reducing surge margin. This is one of the most significant drawbacks of an S-shaped air-intake and may be caused by the presence of a separated boundary layer located in the curve. If separation occurs, the total pressure is decreased for the gas entering the compressor. Moreover, the flow separation over the curved surface forms a recirculation bubble with counter-rotating vortices. The flow separation is responsible for the downstream non-uniform distribution of total pressure at the compressor face, which has often been referred to as compressor face distortion. This distortion may affect the engine performance and, beyond certain limits, may cause the engine to surge, leading to severe loss of thrust and potential engine damage.

Obviously, the prevention of flow separation in the air-intake duct is significant in improving the performance of the propulsive system and, consequently, the overall efficiency of the vehicle. In order to avoid flow separation, much work has been contributed to flow control of offset air intake using different control devices, see,

Mattias Chevalier · Shia-Hui Peng
Department of Aeronautics and Systems Technology,
Swedish Defence Research Agency (FOI), Sweden
mattias.chevalier@foi.se, shia-hui.peng@foi.se

S.-H. Peng et al. (Eds.): Progress in Hybrid RANS-LES Modelling, NNFM 111, pp. 111–121.
springerlink.com

e.g. Anderson and Gibb (1996), Erbslöh et al (2004) and Tournier et al (2005, 2006), based on experimental measurements and numerical simulations. Nevertheless, for improved flow control of air-intake diffusers, it is recognized that a comprehensive understanding is desirable of the turbulent flow physics in an S-duct configuration. This has thus motivated a number of experimental and CFD studies on the flow properties in baseline S-ducts, which have often been taken as references in analysis of flow-control effectiveness. One of the typical examples is the work done by Berrier and Allan (2004), who performed both experimental and computational evaluation of four S-duct inlet configurations in order to set up a database of CFD validation and to provide a baseline S-duct inlet for flow-control analysis.

For the S-duct flow, the incoming boundary layer flow is often detached shortly after the duct throat and reattached on the downstream bottom surface of the duct, forming subsequently a separation bubble. Modelling of the flow separation is essential in order to achieve accurate predictions of downstream flow properties, and to give perspectives on the location of flow control devices in order to diminish the flow separation. Lopes et al (2003) carried out even LES for the flow in an S-duct to highlight the details of resolved flow structures in relation to the flow separation and reattachment. Anderson et al (1993) made a comparative study of full and reduced Navier–Stokes analysis in RANS computations of a diffusing S-duct inlet. Using the RAE M2129 S-duct studied in AGARD Fluid Dynamics Panel Working Group 13 (1991), Mohler (2003) validated a CFD solver in RANS computations for the S-duct flow with and without vortex generators (VG) using structured and unstructured grids. The same S-duct configuration was also used by Jirasek (2006) in validation of modelling flow control devices (VG) in RANS analysis.

In the present work, the M2129 S-duct configuration is revisited in computations using both RANS and DES (detached eddy simulation) modelling approaches. One of the main purposes is to highlight the application of DES in modelling of internal separating flows in comparison with RANS methods. It should be noted that the M2129 S-duct included in this report is a test case in the GARTEUR Action Group 43 (AG43), where involved participants have computed this flow using their own flow solver with the same computational mesh as used in this work.

2 Numerical Details

2.1 RAE M2129 S-Duct Geometry

For the RAE M2129 S-duct configuration, the centerline and the diameter are defined by

$$\begin{aligned} y &= 0.0, \\ z &= -0.15L(1.0-\cos(\pi x/L)), \\ d &= d_t + (d_{ef}-d_t)(3.0(1.0-x/L)^4 - 4.0(1.0-x/L)^3 + 1.0), \end{aligned} \tag{1}$$

where $x \in [0, L]$ and L is the duct length. Moreover, in equation (1) d_t is the throat diameter and d_{ef} is the diameter at the engine face location x_{ef}. This gives an offset in the z-direction of $1.065d_t$. The duct has an upstream lip and parallel extension as well as a downstream parallel extension that includes the engine face and the outflow boundary domain. All geometrical parameters are given in Table 1, and a sketch of the configuration is depicted in Figure 1.

Table 1 Geometrical parameters defining the M2129 S-duct (SI-units)

Quantity	Value	Description
d_t	0.1288	Throat diameter
d_e	0.1440	Highlight (entry) diameter
d_{ef}	0.1524	Engine face diameter
x_{ef}	0.4839	Engine face position
L	0.4572	Duct length

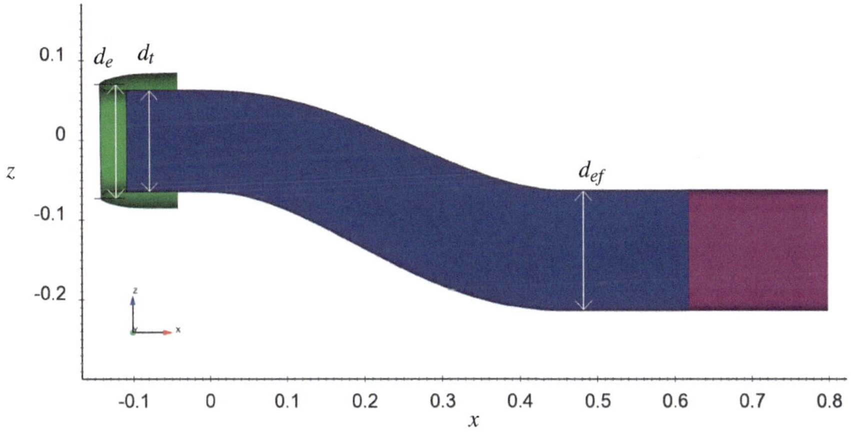

Fig. 1 Geometry of the M2129 diffuser. The three white arrows indicate three different diameters. From left to right the arrows represent the entry diameter (d_e), the throat diameter (d_t) and the engine face diameter (d_{ef})

2.2 Computational Setup

The computational grid is one of the crucial aspects in order to model (in RANS mode) and/or to resolve (in LES mode) the flow physics in the separated area downstream of the S-bend. Structured grids, generated by EADS, have been employed in the present computations.

The far-field region upstream of the duct inlet entry and in the lateral direction around the cowl has an extension of about ten duct diameters. The freestream computational domain has a round shape and an O-H-grid-structure was established in order to best fit the geometry. In Figure 2 the computational domain is shown together with the intake geometry.

To the outlet of the S-duct, a downstream extension was added keeping the outlet cross section constant over approximately one S-duct length with gradual (but rapid) coarsening of the grid up to the outflow boundary. In general, the outflow boundary should be placed far enough downstream from the flow separation bubble. For the evaluation of the results, the original compressor entry plane is kept as a block boundary and all vertexes lie in the same plane. Since the flow in the engine face plane should not be influenced by the flow immediately downstream, the cell size in the longitudinal downstream direction increases gradually.

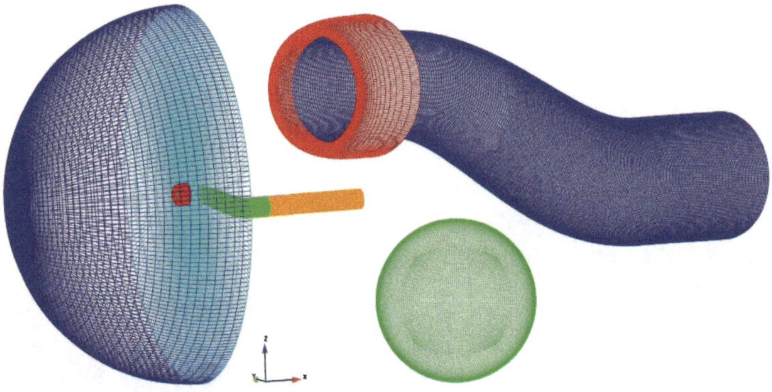

Fig. 2 Illustration of grid used for S–A RANS and DES computations. The cutplane is taken at the engine face

The computations have been carried out using the FOI unstructured CFD solver Edge, which is a node-based compressible Navier–Stokes solver system using finite volume method, see Eliasson (2001). In RANS computations, the convective flux is approximated with a second-order central scheme for the Navier–Stokes equations, and a second-order upwind scheme, of Roe's flux difference splitting type, is employed for the turbulence transport equation. In the DES modelling, the second-order central scheme has been used for the convection terms in both the momentum and turbulence transport equations. The viscous flux has been estimated using a second-order approximation. The time-dependent simulation is advanced using a dual-time stepping method, where a global physical time step is employed and a local time step (the pseudo time step) is used in the subiterations based on an explicit three-stage Runge–Kutta scheme.

The DES approach by Spalart et al (1997) has been used, in which the maximum cell size is estimated using the maximum edge of a node, see Peng (1993). For comparison, RANS computations have also been conducted using the Spalart–Allmaras (S–A) one-equation model, Spalart and Allmaras (1994), and the $k-\omega$ EARSM Wallin and Johansson (2000). A fine grid with about 6.13 million nodes has been used in the computations with the S–A RANS and DES models, while the EARSM computation has been conducted on a coarse grid with about 1.3 million nodes.

In the DES computation, two different time steps have been tested, namely, $\Delta t = 5\times10^{-6}$ and $\Delta t = 2\times10^{-5}$ with no major difference found. The result presented here have been obtained with $\Delta t = 2\times10^{-5}$ with a statistical sampling time frame of about five cycles of air going through the duct.

The static pressure, P_{out}, at the outflow boundary determines the mass flow rate in the duct. The value of P_{out} was thus estimated in such a way that the resulting mass flow rate in the computation was approximated to the measured value in the experiment. A number of RANS computations were performed initially to determine an appropriate value of P_{out}, which depends on the turbulence model used in the simulation. For S-A DES computation, the value of P_{out} was taken from the S-A RANS simulation. In Table 2, the values of P_{out} used for three computations are given. As shown in the table, the mass flow ratio (MFR) obtained from the computations is comparable to the measured value. With different models, the effect on the simulation should be insignificant due to the difference between MFR_{cfd} and MFR_{exp}. On all pipe walls no-slip adiabatic boundary conditions have been used.

Table 2 Computational information

Model	Nodes	P_{out}	MFR_{cfd}	MFR_{exp}
EARSM	1 299 456	89 850	1.9993	2.0425
S–A RANS	6 129 664	88 700	2.0155	2.0425
S–A DES	6 129 664	88 700	1.9776	2.0425

3 Results and Discussion

The experimental freestream parameters and mass flow ratio (MFR) are given in Table 3. The mass flow ratio is defined as

$$MFR = A_0/A_c \tag{2}$$

where A_c is the intake highlight area and A_0 is the cross sectional area of the captured streamtube at infinity upstream which is defined as

$$A_0 = \dot{m}/\rho_\infty U_\infty \tag{3}$$

where $\dot{m}$ denotes the mass flow in the duct. In addition, the pressure recovery is defined as

$$PRA = P_{0_{ef}} / P_0 \tag{4}$$

where $P_{0_{ef}}$ and P_0 are the total pressures at the engine face and in the freestream respectively. In order to quantify the pressure distribution over the engine face, we use the static distortion coefficient DC_{60}, defined as

$$DC_{60} = (P_{0_{ef}} - P_{0_{max60}}) / Q_{ef} \tag{5}$$

where $P_{0_{ef}}$ and $P_{0_{max60}}$ are the total pressure averaged at the engine face and the highest averaged total pressure of all 60 degree sectors at the engine face respectively. The Q_{ef} represents the average dynamic pressure at the engine face.

Table 3 Parameters used in the experimental measurements (SI-units)

Ma	MFR	$\dot{m}$	P_0	T_0	ρ_0
0.207	2.0425	2.8727	103011.2	282.6	1.2701

In the DES computation, the S–A RANS solution has been used as the initial field. Figure 3 illustrates the evolution of DC_{60} as a function of time obtained from the DES modelling. Note that the initial part has been discarded due to the transient development of DC_{60} before the flow has been fully developed. It is shown that DC_{60} fluctuates dramatically around a mean value of about 0.5. This suggests that the unsteadiness in the flow reattachment (after the separation bubble) is rather extensive, as resolved with the S–A DES model.

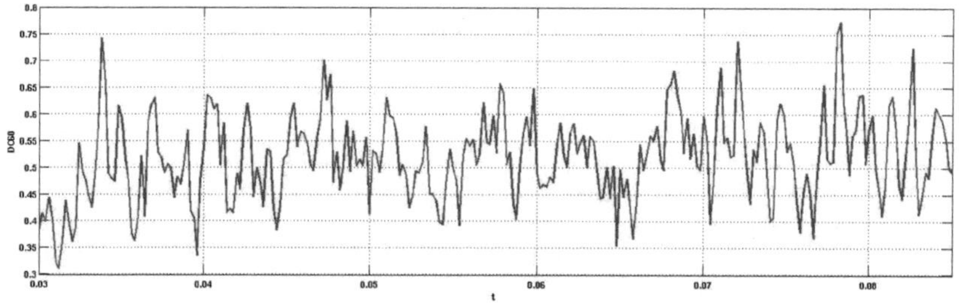

Fig. 3 Time development of the DC_{60} value based on data extraction at the experimental probe positions

In Figure 4, the DES-resolved instantaneous flow features are highlighted. As shown, the resolved shear layer remains rather stable after the flow separation and, consequently, resulting in an extended separation bubble. Also displayed in Figure 4 is the resolved instantaneous pressure recovery at the engine face, where the DES solution has pronounced a strong flow distortion.

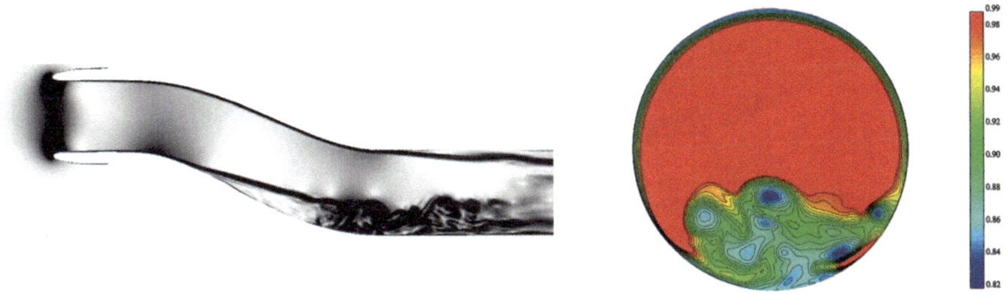

Fig. 4 DES-computed instantaneous flow features. Left: Schlieren-like visualization (magnitude of the density gradient) of the flow separation; Right: Pressure recovery at the engine face

Figure 5 presents the time-averaged streamlines plotted on the symmetric *xz*-plane, in which the flow separation due to the duct offset is illustrated. A comparison is made using the results computed by the S–A RANS and DES model, respectively. Both models have produced three-dimensional flow separation over the lower duct surface. As compared with the S–A RANS modelling, the S–A DES model has produced a somewhat earlier flow separation on the lower wall surface and a more extensive separation bubble, for which the reattachment location has been moved further downstream. As summarized in Table 4, the RANS result is actually closer to the experimental observation. The relatively large discrepancies in the DES computations may be attributed to the resolved free shear layer, which has shown undesirably delayed instabilities. Similar inaccurate predictions for the free shear layer have been observed in modelling using different hybrid RANS-LES methods by other partners in the present project work. This may partly be due to the inflow turbulent condition, which is not appropriately accommodated by the upstream modelling using the current grid resolution in the domain around the duct inlet. Moreover, the delayed instabilities of the free shear layer may also partly be associated to the function of the LES mode incorporated in the DES model. Obviously, the resolved turbulent diffusion by the LES mode of the free shear layer is insufficient. In other words, the modeled eddy viscosity in the free shear layer has been too large either due to the subgrid-scale model or due to a too coarse grid resolution, or due to both. Furthermore, a separation on the upper wall is declared in the DES computation, which is unreasonable. This is attributed to the grid used, which has undesirably made the LES-region penetrated into the wall boundary layer and causing the so-called "grid-induced separation" (GIS). One of the remedies to this problem would be to use the delayed DES approach, as proposed by Spalart et al (2006).

Figure 6 presents a comparison of the simulations for the time-averaged pressure recovery at the engine face, computed by the DES and S–A RANS model, respectively. The delayed and slightly bigger separation zone in the DES case can be clearly seen in this figure as there is a larger low pressure recovery zone. Note also that it is desired to have an even longer time period for statistic analysis to get

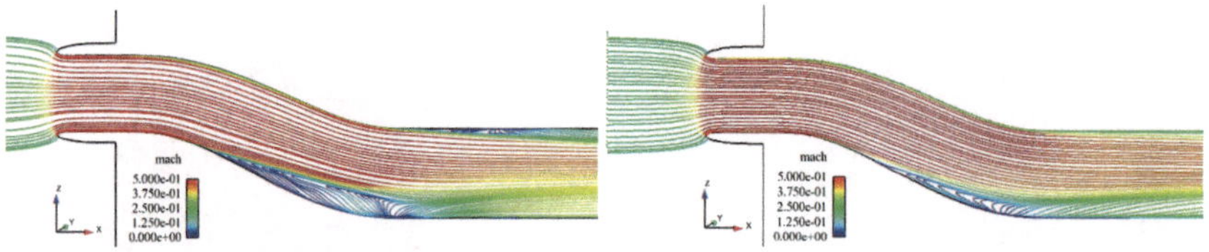

Fig. 5 Streamlines of the mean flow features showing the separation bubble. Left: S–A DES result; Right: S–A RANS computation

the averaged mean flow quantities for the DES computation to improve the expected symmetry in Figure 6.

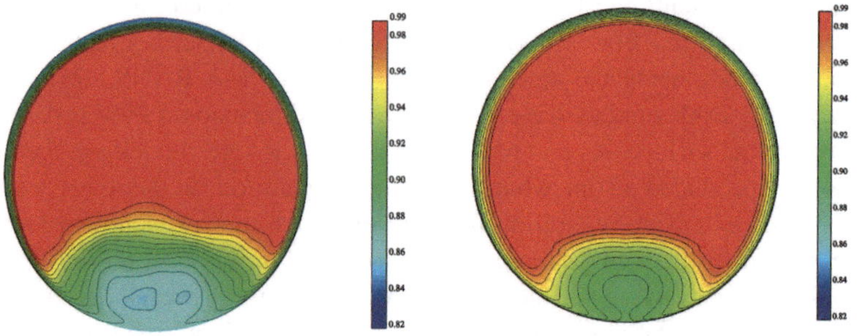

Fig. 6 Time-averaged pressure recovery at the engine face. Left: S–A DES result; Right: S–A RANS computation

The computed pressure distributions along the duct walls at four different azimuthal positions (top, bottom, left and right) have been extracted and are compared in Figure 7. As shown, the pressure distributions are nearly identical on the left and right side walls. It should be noted that the lip extension of the duct has made large pressure values at upstream positions. In spite of the difference in grid resolutions, both RANS models (the S–A model and EARSM) have produced similar pressure distributions.

As can be seen in Figure 7, the major difference occurs in the DES computation for the pressure distributions on the lower and upper walls. The over-estimated separation by DES has produced a fairly flat distribution in the separation zone on the lower wall surface. The GIS problem has caused a strong pressure drop on the upper wall, by which the flow tends to be separated.

In Table 4, the computations are summarized and compared with experimental measurements. To make the comparison between experiments and simulations as similar as possible the same data extraction software for the engine face data has been used. This means that only data from the probe locations (8 equally spaced arms with 5 pitot pressure probes on each) on the rake arms in the experiment have

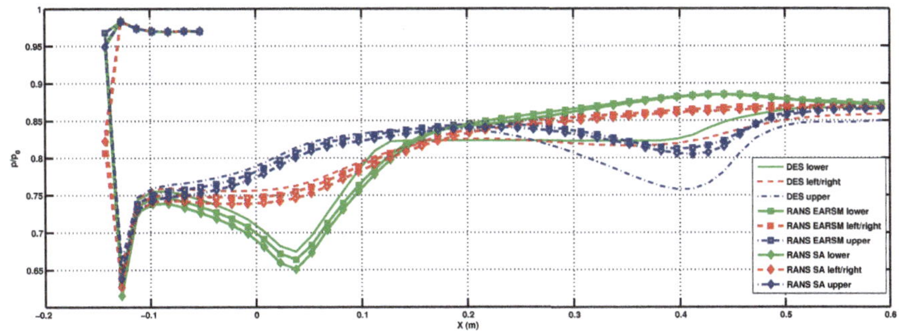

Fig. 7 Wall pressure distributions at upper, lower, left and right azimuthal location of the duct. The pressures along the left and right side coincide and are thus represented by one line

been used to compute the flow characteristics at the engine face. As shown in Table 4, the computed DC_{60} values by the RANS computations is much better than the DES computation, in comparison with the experiment. This is in line with the predictions of the flow properties due to different modelling approaches, as discussed above. It is obvious that an accurate modelling of the separation bubble over the duct offset is essential to obtain accurate predictions of the flow features at the engine face.

Table 4 Comparison between experimental and numerical results. The simulation cases are defined in Table 2

Quantity	Experiment	EARSM	S–A RANS	S–A DES
$\dot{m}$	2.8727	2.8050	2.8366	2.7794
MFR	2.0425	1.9993	2.0155	1.9776
DC_{60}	0.270	0.2275	0.3377	0.4976
HRA	0.9798	0.9824	0.9805	0.9697
PRA	0.8701	0.8687	0.8655	0.8483
Ma	0.395	0.4160	0.4157	0.4219

4 Conclusions

The flow in the generic M2129 S-duct has been studied numerically at a subsonic freestream Mach number of 0.207 with a mass flow rate of 2.87 kg/s, using both RANS and DES modelling approaches. The predicted flow features are presented and discussed. For the DES computation, two different time steps, $\Delta t = 5 \times 10^{-6}$ and $\Delta t = 2 \cdot 10^{-5}$ second, respectively, have been used. The difference is however insignificant due to these different time resolutions.

The inherent LES mode has enabled the DES computation to resolve rich large-scale vortex motions, which are triggered by the breakdown of the resolved free shear layer, and evolving downstream and inside the separation bubble. As compared with the RANS computation, the DES model has pronounced an earlier separation onset over the lower wall surface, and predicted a more extended separation bubble. In comparison with available experimental data of gross parameters measure at the engine face, large discrepancies have been observed in the DES computations. Due to the over-predicted separation bubble, for which the downstream flow has a relatively slow recovery, the distortion coefficient DC_{60} has been over-estimated by nearly 85% in the DES computation.

For both RANS and DES computations, the loss in total pressure, as well as the Mach number, are in good agreement with the experimental data at the engine face. The pressure distributions along the left and right side walls are very similar, on which the difference is insignificant due to different modelling approaches. The differences are sensible, however, in the pressure distributions along the lower and upper walls with the DES results, due to the over-predicted separation zone over the lower wall and the grid-induced separation in boundary layer near the upper wall.

In addition to the grid arrangement near the upper wall and the grid resolution in the free shear layer, the inaccuracy nestled in the DES computation has been attributed to the resolved evolution of shear layer, of which a delayed instability has been predicted by DES. Furthermore, the effect of inflow turbulent condition may also be a significant issue to address in DES computations for internal aerodynamic flows. In future work, all these issues will be investigated for appropriate use of DES.

Acknowledgment. This work was sponsored by FOI and FMV which is gratefully acknowledged. A large part of the work was completed in the GARTEUR project AG43, in which EADS, FOI, NLR, ONERA and QinetiQ had participated.

References

AGARD Fluid Dynamics Panel Working Group 13: Air Intakes for High Speed Vehicles. AGARD-AR-270 (AGARD Advisory Report 270) (1991)

Anderson, B.H., Gibb, J.: Vortex generator installation studies on steady state and dynamic inlet distortion. AIAA Paper 96-3279 (1996)

Anderson, B.H., Reddy, D.R., Kapoor, K.: Comparative Study of Full Navier–Stokes and Reduced Navier–Stokes Analyses for Separating Flows Within a Diffusing Inlet S-duct. AIAA Paper 93-2154 (1993)

Berrier, B.L., Allan, B.G.: Experimental and Computational Evaluation of Flush-Mounted, S-duct Inlets. AIAA Paper 2004-0764, Reno (2004)

Eliasson, P.: EDGE, a Navier–Stokes Solver for Unstructured Grids. Scientific Report FOI-R–0298–SE, Computational Aerodynamics Department, Aeronautics Division, FOI (2001)

Erbslöh, S.D., Growther, W.J., Frutos, J.R.: Control of Compressor Face Total Pressure Distortion on a High Bypass Turbofan Intake using Air-Jet Vortex Generators. AIAA Paper 2004-2206 (2004)

Jirasek, A.: Design of Vortex Generator Flow Control in Inlets. J. Aircraft 43, 1886–1892 (2006)

Lopes, A.S., Piomelli, U., Palma, J.M.L.M.: Large eddy simulation of the flow in an S-duct. AIAA Paper 2003-0964 (2003)

Mohler, S.R.: Wind-US flow calculations for the M2129 S-duct using structured and unstructured grids. NASA CR-2003-212736 (2003)

Peng, S.H.: Algebraic hybrid RANS-LES modelling applied to incompressible and unsteady RANS. AIAA Paper 2006-3910 (1993)

Spalart, P., Deck, S., Shur, M.L., Squires, K.D., Strelets, M., Travin, A.: A new version of DES, resistant to ambiguous grid densities. Theoretical and Computational Fluid Dynamics 20, 181–195 (2006)

Spalart, P.R., Allmaras, S.R.: A one-equation turbulence model for aerodynamic flows. La Rech Aerospatiale 1, 5–21 (1994)

Spalart, P.R., Jou, W.H., Strelets, M., Allmaras, S.R.: Comments on the Feasibility of LES for Wings, and on a Hybrid RANS/LES Approach. In: Advances in DNS/LES, 1st AFOSR Int. Conf. on DNS/LES, pp. 137–148 (1997)

Tournier, S.E., Paduano, J.D., Pagan, D.: Flow Analysis and Control in a Transonic Inlet. AIAA Paper 2005-4734 (2005)

Tournier, S.E., Paduano, J.D., Pagan, D.: Flow Control in a Transonic Inlet. AIAA Paper 2006-3883 (2006)

Wallin, S., Johansson, A.V.: An explicit algebraic Reynolds stress model for incompressible and compressible turbulent flows. J. Fluid Mech. 403, 89–132 (2000)

DES for UCAV Weapon Bay Flow

S.J. Lawson and G.N. Barakos

Abstract. This paper demonstrates the Detached–Eddy Simulation approach for the computation of flows around uninhabited combat air vehicles. This new family of aircraft may feature weapon bays to enhance stealth characteristics and improve aerodynamic performance. However, during operation with the bay open, a highly energetic flow-field develops that can dramatically change the aerodynamics of the aircraft. For this reason detailed CFD analyses are needed to provide insight in the change of loads encountered when weapon bays are exposed. In contrast to previous studies where idealised, isolated cavities are used as model problems, a realistic aircraft geometry is used in this work.

1 Introduction

Uninhabited Air Vehicles, such as the global hawk, have shown to be effective reconnaissance platforms and their use avoids exposure of pilots to hostile environments. From an aerodynamics point of view, Uninhabited Combat Air Vehicle's (UCAV) present a new set of challenges. One of these may be the presence of exposed weapon bays during parts of the flight, which results in highly unsteady flow-fields and is associated with radiation of noise, vibration and turbulent flow. The following paragraphs present a short summary of studies in the area of UCAVs and cavity flows.

Published works on UCAVs and their performance are relatively rare, especially on those designs coming into production such as Boeing's X-45. However, studies on similar configurations have been published. Wong *et al.* (2006) presented results from The Technical Cooperation Program (TTCP) studying a design designated 1303 UCAV. NASA, DSTO, AFRL, DSTL and IAR were involved in a collaboration that resulted in experimental and numerical studies. These results showed that the Reynolds–Averaged Navier-Stokes equations (coupled with various turbulence closures) gave a reasonable overall prediction for the external flow at modest angles of attack. Cummings *et al.* (2008) experimentally and numerically investigated a similar configuration designated 1301. Similar to the studies on the 1303 UCAV, agreement between the experimental and numerical data sets was only gained at angles of attack below 15 degrees.

S.J. Lawson · G.N. Barakos
Department of Engineering, University of Liverpool, Liverpool, L69 3GH, United Kingdom
G.Barakos@Liverpool.ac.uk

S.-H. Peng et al. (Eds.): Progress in Hybrid RANS-LES Modelling, NNFM 111, pp. 123–135.
springerlink.com

Internal carriage of stores has been a major issue and much research has been performed in order to fully understand the problems associated with it. Since the pioneering works in the 1950's and 1960's (Krishnamurty, 1955; Roshko, 1955, Rossiter, 1964), numerous experimental investigations have been performed on cavity flows in an attempt to better understand their complex aerodynamics and aero-acoustics. Experimental studies for cavity flows have been performed on both idealized geometries and those installed in aircraft configurations. For example, Shaw *et al.* (1988) used a scaled F-111 model with weapons bay installed to study the effects of different passive flow control devices.

Numerical studies, however, have been largely limited to idealised geometries. Early studies of cavity flows began with utilising the Unsteady RANS (URANS) equations. However, it became apparent that URANS was not capable of predicting the effect of the range of turbulent scales that exist in cavity flows. Therefore studies moved towards Large Eddy Simulation (LES) (Larchevêque et al, 2004; Nayyar, 2007) and Detached Eddy Simulation (DES) (Barakos et al, 2009; Lawson, 2009; Nayyar, 2007), which provided much better agreement with experimental data. Complex configurations, in terms of the shape of the cavity, have been studied (Shipman et al, 2007; Ukeiley et al, 2007), although the geometry was still symmetrical about the cavity centreline.

In view of the above, this paper attempts to obtain a set of numerical results for the 1303 UCAV. The authors have carried out numerous computations on idealised cavity flows for both URANS and DES models using the Multi-block solver of Liverpool (Barakos et al, 2009; Lawson et al, 2009). This experience is now used for the analysis of the much more complex geometry of the UCAV with the cavity installed. The grid generation is fully explained in paragraph 2 and includes two configurations. The first is the baseline UCAV with no cavity, which was used to demonstrate that accurate results could be gained for the external flow (paragraph 3). The cavity was then added to the UCAV model and the results are shown in paragraph 4. To the authors' knowledge, this is the first time computations have been attempted for a cavity installed in a complex UCAV geometry.

2 Grid Generation

The complex shape of the 1303 UCAV is a challenge for the grid generation task. The employed solver is compatible with structured grids, which have advantages with respect to unstructured meshes when it comes to the accuracy of the obtained results, especially near the body of the UCAV. The following paragraphs detail the complex multi-block topologies for the clean UCAV and the UCAV with cavity cases.

The geometry for the 1303 UCAV was modelled on the wind-tunnel experiments by Bruce and Mundell (2003) and is also detailed in Wong *et al.* (2006). For the CFD computations, the geometry was non-dimensionalised by the maximum wing chord length, C. Due to the symmetry of the UCAV along its centerline, a grid for only half of the geometry was created. The far-field boundaries were then located at a distance of $15C$ away from the model in all three directions. The blocking for the baseline UCAV had to incorporate several very important features. The first issue was that the sections along the wing of the UCAV varied greatly. At the apex of the model, the top and bottom surfaces came to a sharp leading edge. Further along the wing, the leading edges become blunt, although the wing had a relatively small thickness. The trailing edges all along the wing were modelled to be sharp. These features pointed towards the use of the H-grid topology for the main structure of the blocking of the wing. In total, 108 cells were used along the section of the wing, the cells at leading edge of each aerofoil section has a first cell size of $10^{-5}C$ and expanded using a hyperbolic distribution. The cells were also clustered at the trailing edge and had a cell size of $5\times10^{-5}C$. Above and below the wing, there were two layers of blocks between the wing surface and the far-field. 30 cells spanned the edges on the blocks next to the wing surface, with a first cell size of $10^{-5}C$ and expanded using an exponential distribution. The edges on the block stretching to the far-field contained 20 cells and used a hyperbolic distribution.

In order to create the sharp wing tip, the blocks making up the H-grid topology for the wing sections were collapsed. This was another reason for employing the H-grid topology, as this would not have been possible if the traditional C-grid blocking was used. 170 cells in total were used across the length of the wing, with hyperbolic distributions used throughout and a first cell size of $10^{-5}C$ at the tip.

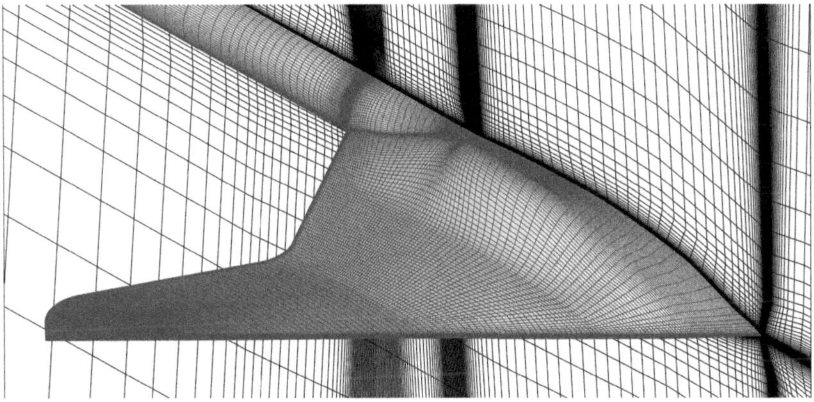

Fig. 1 Surface mesh for the baseline 1303 UCAV

In producing the grid, two iterations of testing and refinement was performed, with the final grid containing 4.3 million cells and is shown in Figure 1. The areas of refinement at the leading and trailing edge can be seen, with an extra section of clustering where the wind-tunnel sting meets the main body of the model. The grid is also comparable to other grids produced for TTCP activities (Wong et al, 2006).

The UCAV wind tunnel model had two weapons bays installed and so was still symmetrical about its centerline. Therefore, as with the baseline case, only half the UCAV was modeled. The blocking topology for the baseline configuration was used as a starting point and then extra slices were made in order to accommodate the cavity. To achieve the refinement needed inside the cavity, the number of cells along the chord of the wing section had to be increased from 108 to 282. These extra cells, plus cells due to the O-grids used for the door hinges gave a total of 270 cells along the length of the cavity.

Down the wing, the number of cells was increased from 170 to 198, with some sections of the wing coarsened slightly in order to keep the total number of cells down. This gave 80 cells along the width of the cavity, which was then matched for the depth to keep the cells inside the cavity as regular as possible. First cell sizes near all solid surfaces of the cavity were $10^{-5}C$, with exponential stretching distributions. The extra refinement gave 1.7 million cells inside the cavity and although a grid convergence study was not conducted, the density of the grid was selected based on previous experience (Barakos et al, 2009; Lawson et al, 209; Nayyar et al 2007).

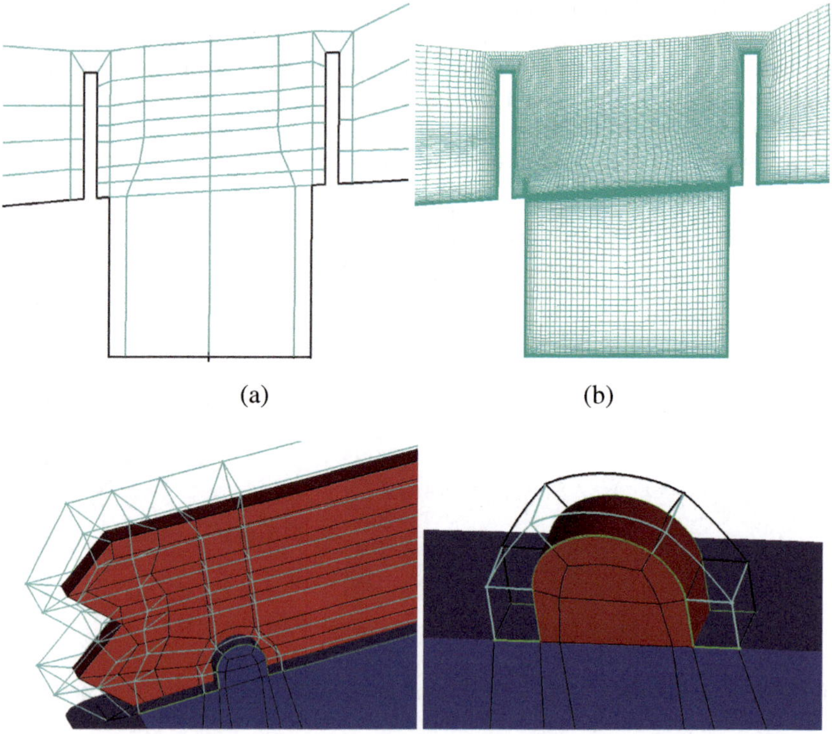

(a) (b)

(c) (d)

Fig. 2 Blocking (a) and mesh (b) for a slice through the UCAV cavity and blocking for the cavity doors (c) and door hinges (d). Meshes were generated using ICEM–CFD Hexa

A spanwise slice through the centreline of the cavity is shown in Figures 2 (a) and (b), which highlight further differences between this geometry and the idealised cavity. Although the ceiling of the cavity is straight and level, the opening follows the original surface of the UCAV and so the cross-section was not square. Extra complexity in the blocking topology was needed around the cavity in order to model the doors and their hinges (Figures 2 (c) and (d)). Due to the presence of the door hinges, the inside surface of the door was set back away from the cavity edge. This allowed a full O-grid to encompass the doors, meaning the refinement needed to describe the flow around the surface of the doors did not propagate into the far-field. Extra cells were needed for the height of the doors and to also capture structures created within the area between the doors. In total, 54 cells were used for the height of the doors and 14 cells were used for the width, again with a $10^{-5}C$ cell size next to the solid surfaces of the doors. Due to the chevrons on the door ends, O-grids were used to achieve good mesh quality between the doors. This also added further amounts of cells between the doors, as can be seen in Figure 2(b). To include the hinges, half O-grids were used for the curved top of each hinge (Figure 2 (d)). These blocks started and ended at the base of the cavity and so allowed extra cells to be added inside the cavity without the cells propagating into the far-field. Overall, the refinement for the cavity increased the mesh size from 4.3 to 12.4 million cells.

3 Baseline Configuration

Details of the flow solver can be found in Lawson and Barakos (2009). Computations for the baseline configuration at Mach and Reynolds numbers (based on the maximum chord) of 0.25 and 9.02 million respectively were used as a validation case since results from experiments and other CFD codes had already been presented (Wong et al, 2006). Steady state results were gained using the baseline k–ω turbulence model (Menter, 1993) at angles of attack from -6 to 14 degrees while an unsteady computation was performed at 25 degrees.

In order to the compare the UCAV with cavity results to the baseline configuration, two extra computations were performed at zero incidence for a Mach number of 0.85 using the URANS k–ω (Menter, 1993) and S–A (Spalart, Allmaras, 1994) models. The use of the S–A model was necessary to facilitate comparisons against UCAV with cavity DES S–A results as discussed later.

Figure 3 presents upper surface C_P distributions and streamlines for 3 different angles of attack: 4, 8, and 12 degrees. At an incidence of 4 degrees, low C_P can be seen along the leading edge of the wing and the flow was fully attached. At 8 degrees of incidence, the flow started to separate near the wing tip and with an increase to 12 degrees, separation can be seen at half span of the aircraft. The streamlines show that a vortical structure originates at approximately 50% span and travels down the remaining section of the wing. However, a second vortex also exists which originates from the apex. As the incidence was increased further, this second structure became more prominent and the first vortex moved inboard.

Figure 4 presents the lift, drag and pitching moment coefficients. The steady state computations show that, although the values of C_L were slightly under-predicted at low and negative alphas compared to the experiment, the lift slope was very well predicted. Similar trends were seen for the drag, where the values at negative alphas were under-predicted. Generally, the pitching moment was slightly under-predicted compared to the experiment.

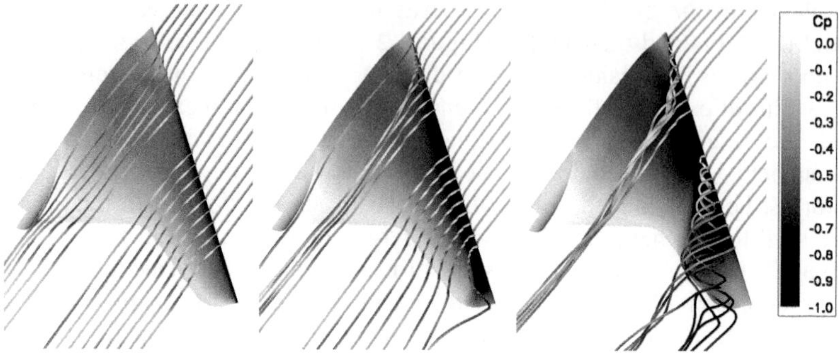

(a) (b) (c)

Fig. 3 Upper surface C_P distributions of the UCAV 1303 baseline configuration at 4 (a), 8 (b) and 12 (c) degrees angle of attack

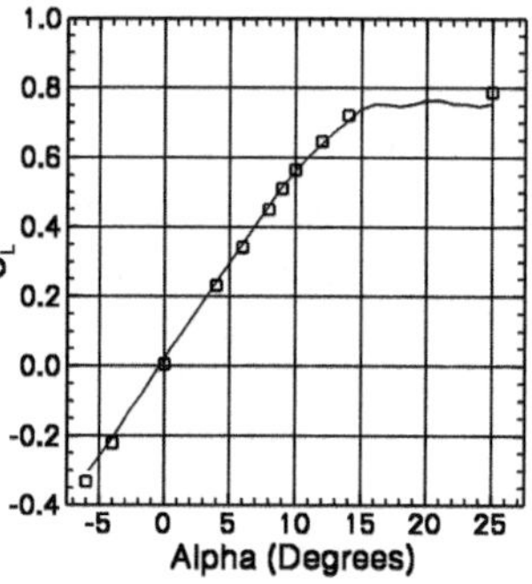

(a)

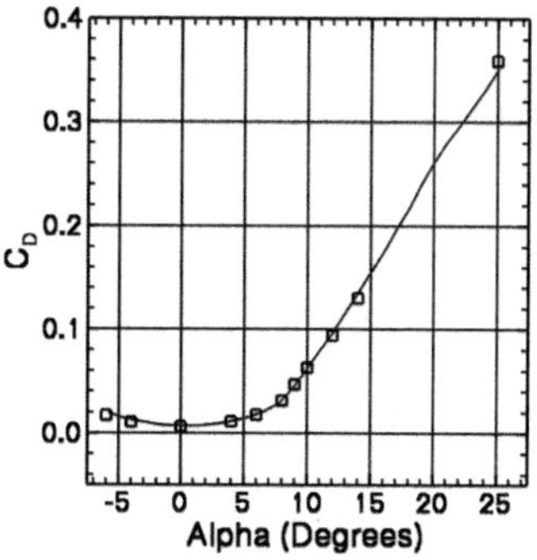

(b)

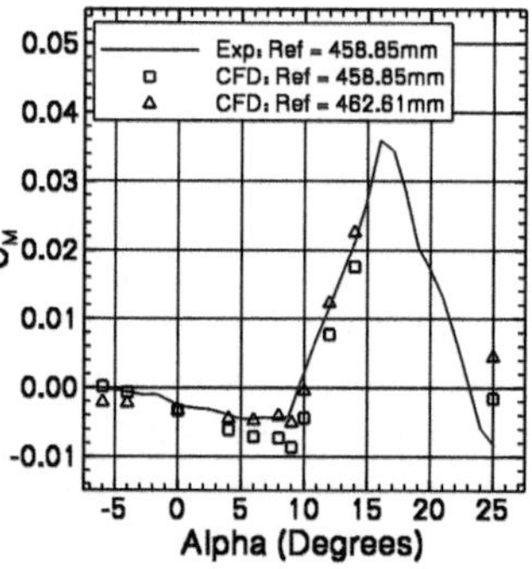

(c)

Fig. 4 The baseline 1303 UCAV results for the lift (a), drag (b) and pitching moment (c) coefficient versus alpha. Steady state computations were performed at angles -6 to 14 degrees; an unsteady computation was performed at 25 degrees. Comparisons are made against experiments by Bruce and Mundell (2003)

However, it can be seen that both the location of the first pitch break and the slope of the curve after the pitch break were well predicted. The unsteady computation at 25 degrees showed that the method was capable of predicting the complex flow-field at high angles of attack.

The sensitivity of the results on the location of the pitching moment axis was also studied. It was found that a movement of 2mm (0.37% of the mean aerodynamic chord) aft resulted in large changes to the pitching moment. This result would also suggest that the pitching moments would be very sensitive to the grid around the leading edge. Although it was necessary to use the H-topology and great care was taken to ensure the high grid quality, some of the differences between the experimental and numerical data could also be attributed to the topology employed.

4 UCAV with Cavity

The computation for the UCAV with cavity was performed at Mach and Reynolds numbers (based on the maximum chord) of 0.85 and 9.02 million, respectively and used the S–A DES (Spalart et al, 1997) turbulence model. Based on the cavity length, the Reynolds number was approximately 2.61 million. The cavity had a length to depth ratio (L/D) and a width to depth ratio (W/D) of approximately 5.45 and 1.32 respectively. The computations used a non-dimensional time-step of 0.001 (3.14×10^{-6} seconds in real time) and continued until 32,000 time-steps (0.1 seconds of real time) were obtained. This signal length was equivalent to over 32 periods of the first Rossiter mode (calculated using Rossiter's formula (Rossiter, 1964)) and was processed as reported in Lawson and Barakos (2009).

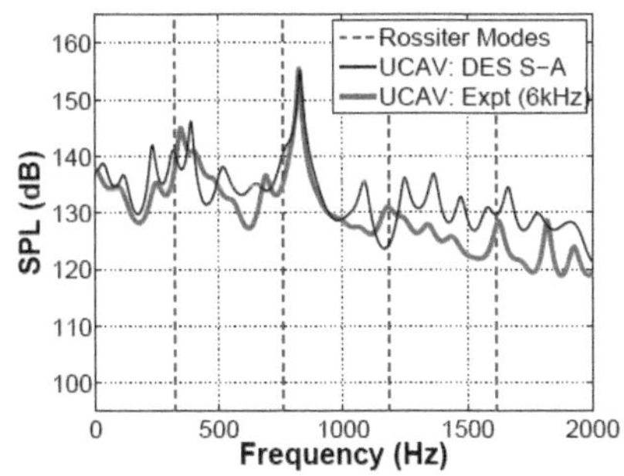

(a)

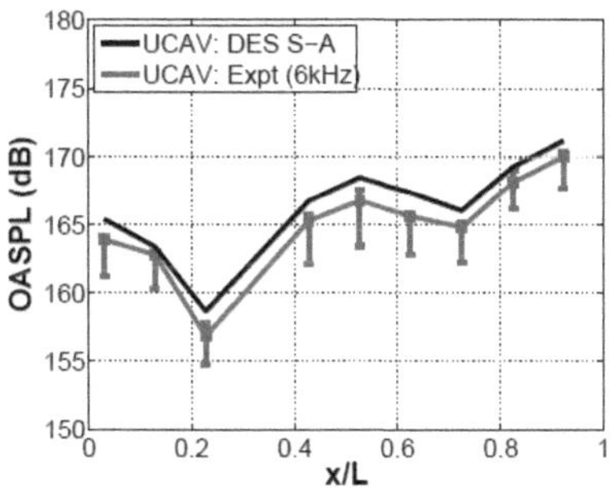

(b)

Fig. 5 Power Spectral Density (PSD) at the rear of the cavity (a) and Overall Sound Pressure Level (OASPL) along the cavity floor (b). Vertical lines on PSD represent Rossiter (1964) modes

Figure 5 (a) shows the PSD on the cavity floor near the aft wall. The largest tone can be seen at a frequency slightly higher than the prediction of the second Rossiter mode. There was a slight peak around the first mode but no tones can be seen above the broadband level for the third or fourth modes. This trend is similar to the idealised cavity case with doors attached, although in that case a tone around the third mode was also seen (Barakos et al, 2009). In comparison to the experimental data, the frequencies and magnitudes of the first two tones were predicted well. The peak level of the second mode was within 2 dB of the experimental data at all locations along the cavity floor. Also, the general level of the broadband noise was predicted well, though a slight over-prediction can be seen towards the rear of the cavity.

The Overall Sound-Pressure Level (OASPL) across the cavity floor was also computed and is shown in Figure 5 (b). As the experimental data was shortened to the numerical signal length, the minimum and maximum fluctuations in the full signal are represented by the error bars. These fluctuations were computed by splitting the experimental signal up into sections of length equal to the numerical data, with the sections also having a 90% overlap. The OASPL was then computed for each section of the maximum and minimum values found. A small over-prediction in the OASPL can be seen all along the cavity floor and has a maximum of 3 dB at around the mid-point of the cavity. This is consistent with other computations for idealised cavity flows, where over-predictions as high as 5 dB are reported in Lawson and Barakos, (2009). The error bars on Figure 5 (b) show fluctuations in the experimental data are as high as 5 dB. It is therefore more important that the shape of the OASPL curve is consistent with the experimental data, as this shows that the flow structures captured by the computations are similar to those in the experiments.

In order to see how the flow-field varies with time, Joint Time-Frequency Analysis (JTFA) is employed (Barakos et al, 2009; Kegerise et al, 2004; Larchevêque et al, 2004). This method is used to calculate the Band Integrated SPL (BISPL) and has shown that mode switching occurs in idealised cavity flows. This is where energy is exchanged between the modes over time and so negates the concept of a single dominant mode. From the BISPL (Figure 6), it can be seen that fluctuations in the modes still occur. However, in this flow the second tone is much larger in magnitude than any other. Therefore in both the numerical and experimental data there were few instances where the second mode was not the largest in magnitude.

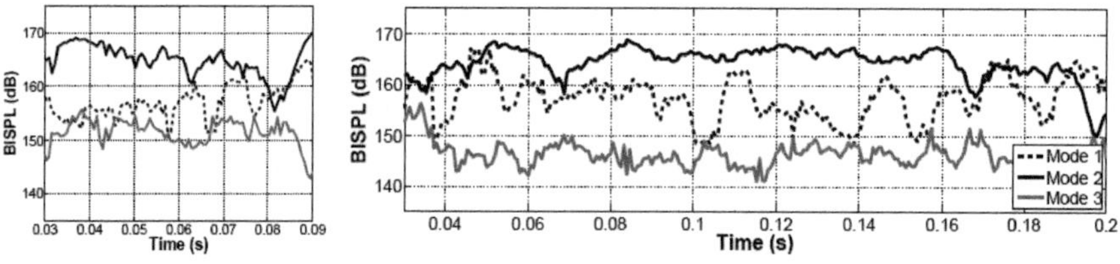

(a) (b)

Fig. 6 Joint time-frequency analysis for the numerical (a) and experimental (b) UCAV cavity at x/L=0.95 on the cavity floor. The three modes are integrated in bands of 200 Hz, centered on the Rossiter mode frequencies

As no flow-field data is available for this case, the averaged and instantaneous flow-fields were compared to the M219 idealized cavity with doors on (Barakos et al, 2009; Nayyar et al, 2007). The M219 cavity had an L/D ratio of 5 and a W/D ratio of 1 and like the current study, the free-stream Mach number was 0.85, although the Reynolds number based on the cavity length was one million. Previous studies on the M219 configuration have shown favorable comparisons to PIV data and also that Reynolds number had little effect on the cavity flow-field (Barakos et al, 2009; Nayyar et al, 2007).

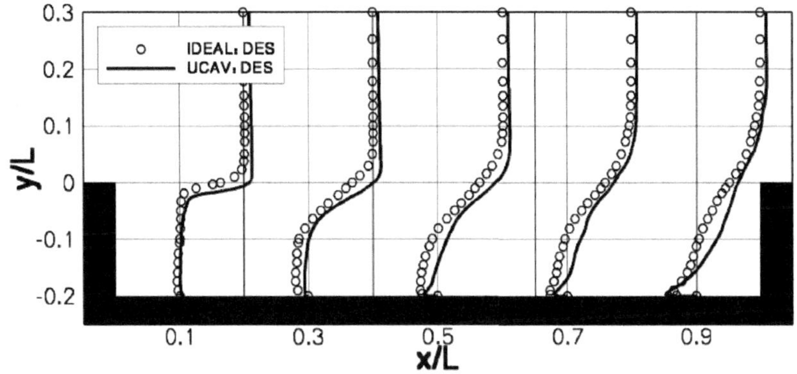

Fig. 7 Averaged streamwise velocity profiles. DES results for an isolated cavity are compared against DES for the complete case

Figure 7 compares the averaged streamwise velocity profiles for the two cases. The profiles at locations towards the front of the cavity were very similar, but differed slightly for stations near the aft wall. Also, although the free-stream Mach number for the current study is the same as for the idealised cavity, the velocity above the cavity is higher in the UCAV case. The shear layer of the averaged flow-fields can be further analysed using the momentum thickness. The incompressible definition is used and the influence of the recirculation region is neglected (Barakos et al, 2009). Previous analyses of turbulent free mixing-layers (Larchevêque et al, 2004) suggest that the growth rate of the momentum thickness falls within the range of 0.03 to 0.04, although Larchevêque *et al.* (2004) reported a value of 0.035 for the LES computation and our work (Barakos et al, 2009) gave a value of 0.037 for a DES computation. The growth rate for the averaged momentum thickness in the UCAV cavity was approximately 0.041, which was slightly higher than the values reported in literature. The differences could be due to the influence of the doors, the door hinges and the lack of an upstream flat plate.

To better identify structures within the flow and to show the level of unsteadiness within the cavity, instantaneous iso-surfaces of Q-Criteria are shown in Figures 8 and 9. Hunt *et al.* (1988) proposed the Q-Criteria to identify vortex cores and is defined as the positive second invariant of the velocity gradient tensor. The door hinges and the shape of the doors at the leading edge in the UCAV cavity generated large amounts of small structures, which make a comparison between the UCAV and idealized cases difficult. However, Figure 9 shows similar sized structures existed inside the cavity region for both cases.

Fig. 8 Iso-surfaces of instantaneous Q-criteria for the UCAV cavity configuration. Also shown is a comparison of lower surface C_P between the UCAV baseline and cavity configurations. The left half represents the baseline configuration and the right half the cavity configuration

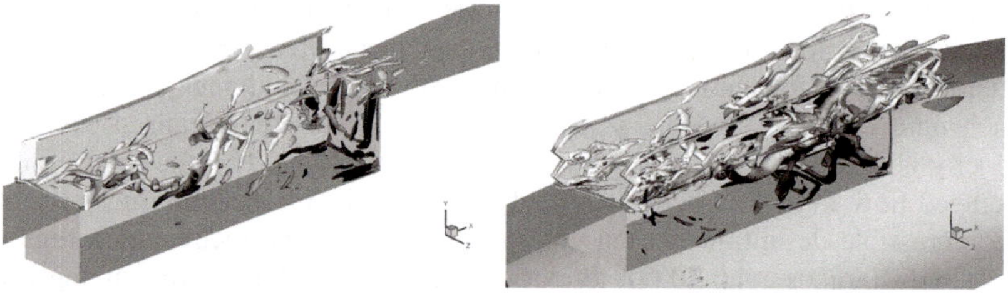

(a) (b)

Fig. 9 Iso-surfaces of instantaneous Q-criteria for the ideal cavity configuration (a) and the UCAV cavity configuration (b). Iso-surfaces are at Q=2000

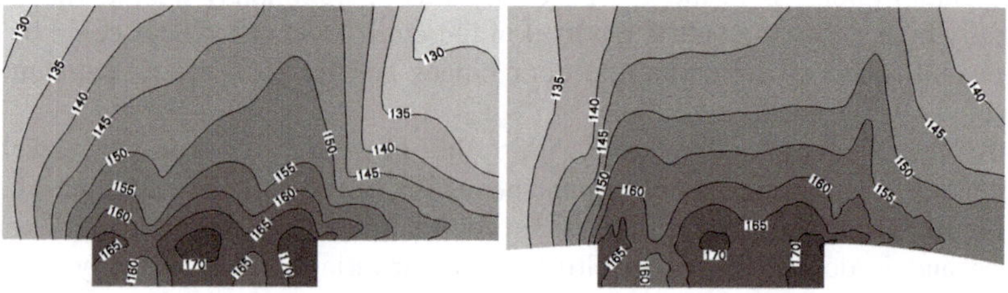

(a) (b)

Fig. 10 Contours of OASPL for a slice along the mid-plane of the cavity for the ideal cavity configuration (a) and the UCAV cavity configuration (b)

Although the grid in the far-field was not designed for studying the acoustic field, the OASPL gives an idea of how the pressure waves propagate outside the cavity (Figure 10). It can be seen that for the UCAV cavity, levels above 170 dB existed all across the aft wall and levels were in excess of 165 dB for the aft two thirds of the cavity opening. Close to the front of the cavity, levels reduced to between 155 and 165 dB; however, at the area around the front lip of the cavity the levels were above 165 dB. There are strong similarities to the idealised cavity case, especially in the front half of the cavity and around the aft wall. Outside the cavity, levels between 140 and 150 dB are seen approximately one cavity length above the cavity mouth. This is also very similar to the clean cavity case.

Averaged force and moment coefficients for the UCAV with cavity are presented in Table 1 and compared to the baseline UCAV results computed using the S–A model. It can be seen that the components mostly affected by the cavity were the lift (C_{Fy}), drag (C_{Fx}) and rolling moment (C_{Mx}) coefficients. Although the UCAV generated very little lift at zero angle of attack, the averaged value for the UCAV with the cavity was 155% higher than the value predicted without the cavity. The drag coefficient was not affected as much, with the value 86% higher than the value for the baseline configuration.

Table 1 Integrated loads for the half of the 1303 UCAV baseline and cavity configurations at M=0.85, Re=9.02 million and α=0. Baseline computation shown is with the S–A turbulence model. UCAV with cavity data was averaged and the difference between the URANS S–A is shown for the delta values

	C_{Fx} (10^{-3})	C_{Fy} (10^{-3})	C_{Fz} (10^{-3})	C_{Mx} (10^{-3})	C_{My} (10^{-3})	C_{Mz} (10^{-3})
Pressure, BG	1.0639	5.4207	16.7822	2.2282	0.4984	-1.1907
Pressure, CG	4.8638	13.8588	18.1992	5.6061	-0.3828	-0.8829
Friction, BG	2.9500	0.0050	-0.0438	-0.0005	1.4520	0.0345
Friction, CG	2.6550	-0.0245	-0.0276	-0.0070	1.3576	0.0261
Total, BG	4.0439	5.4257	16.7384	2.2277	1.9504	-1.1562
Total, CG	7.5188	13.8343	18.1716	5.5991	0.9748	-0.8568
Delta	3.4749	8.4086	1.4332	3.3714	-0.9756	0.2994
% Increase	85.9	155.0	8.6	151.3	-50.0	-25.9

BG Baseline Geometry, *CG* Cavity Geometry, C_F Force Coefficient, C_M Moment Coefficient, *x* Streamwise Axis, *y* Normal Axis, *z* Spanwise Axis.

5 Conclusions

An evaluation of DES has been conducted for a realistic UCAV geometry with an open weapon bay. Advanced multi-block topologies had to be used to properly represent the planform of the UCAV and all the details of the cavity, including doors and hinges. In addition, high mesh quality was needed to properly capture the loads on the vehicle and the flow inside the weapons bay. The employed CFD method captured well the aerodynamic loads of the baseline UCAV, and comparisons against wind tunnel data are very encouraging given the complex vortical flow present on the suction side of the aircraft. Comparisons between an idealised cavity and the weapons bay on the UCAV show many similarities in terms of the flow structure and their spectrum. Unsteady pressure measurements available for both cases show that the methods is capable of resolving most of the cavity flow physics and captured the acoustic tones present. The effect of the cavity on the integrated loads of the vehicle was also investigated. With the bay exposed, the drag was increased and a rich harmonic content was added over a baseline value. In the future, UCAVs with stores inside the bay are to be investigated to further ascertain the ability of DES to provide engineering data for this complex and interesting flow.

Acknowledgements. This work was supported by EPSRC grant EP/C533380/1 and the 2nd UK Applied Aerodynamics Consortium. The authors would like to extend their gratitude to Trevor Birch of DSTL for providing the experimental data.

References

Barakos, G.N., Lawson, S.J., Steijl, R., Nayyar, P.: Numerical Simulations of High–Speed Turbulent Cavity Flows. Flow Turbulence and Combustion 83(4), 569–585 (2009), doi:10.1007/s10494-009-9207-1

Bruce, R., Mundell, A.: Low Speed Wind Tunnel Tests on the 1303 UCAV Concept. Technical report QinetiQ/FST/TR025502/1.0. QinetiQ, Farnborough, UK (2003)

Cummings, R., Morton, S., Siegel, S.: Numerical Prediction and Wind-Tunnel Experiment for a Pitching Unmanned Combat Air Vehicle. Aerospace Science and Technology 12, 355–364 (2008), doi:10.1016/j.ast.2007.08.007

Hunt, J., Wray, A., Moin, P.: Eddies, Streams and Convergence Zones in Turbulent Flows. In: Proceedings of the Summer Program N89-24555. Center for Turbulence Research, pp. 193–207 (1988)

Kegerise, M., Spina, E., Garg, S., Cattafesta III, L.: Mode-Switching and Nonlinear Effects in Compressible Flow Over a Cavity. Physics of Fluids 16(3), 678–686 (2004)

Krishnamurty, K.: Acoustic Radiation from Two-Dimensional Rectangular Cutouts in Aerodynamic Surfaces. Technical Note 3487. National Advisor Committee for Aeronautics (1955)

Larchevêque, L., Sagaut, P., Lê, T.-H., Comte, P.: Large-Eddy Simulation of a Compressible Flow in a Three-Dimensional Open Cavity at High Reynolds Number. J. Fluid Mechanics 516, 265–301 (2004)

Lawson, S.J., Barakos, G.N.: Assessment of Passive Flow Control Devices for Transonic Cavity Flows Using Detached-Eddy Simulation. J. Aircraft 46(3), 1–2 (2009), doi:10.2514/1.39894

Menter, F.: Zonal Two Equation k–ω Turbulence Models for Aerodynamic Flows. In: 24th Fluid Dynamics Conference, Orlando, FL, July 6-9, AIAA Paper 93-2906 (1993)

Nayyar, P., Barakos, G.N., Badcock, K.J.: Numerical Study of Transonic Cavity Flows using Large-Eddy and Detached-Eddy Simulation. The Aeronautical Journal 111(1117), 153–164 (2007)

Roshko, A.: Some Measurements of Flow in a Rectangular Cut-Out. NACA Technical report 3488. California Institute of Technology (1955)

Rossiter, J.: Wind Tunnel Experiments on the Flow Over Rectangular Cavities at Subsonic and Transonic Speeds. Technical Report 64037. Royal Aircraft Establishment (1964)

Shaw, L., Clark, R., Talmadge, D.: F-111 Generic Weapons Bay Acoustic Environment. J. Aircraft 25(2), 147–153 (1988)

Shipman, J., Arunajatesan, S., Cavallo, P., Birkbeck, R., Sinha, N., Ukeiliey, L., Alvi, F.: Flow Control for Enhanced Store Separation. In: 45th AIAA Aerospace Sciences Meeting and Exhibit, Reno, NV, January 8-11 (2007)

Spalart, P., Allmaras, S.: A One-Equation Turbulence Model for Aerodynamic Flows. La Recherche Aerospatiale 1, 5–21 (1994); Also AIAA paper 92-0439

Spalart, P., Jou, W.-H., Strelets, M., Allmaras, S.: Comments on the Feasibility of LES for Wings, and on a Hybrid RANS/LES Approach. In: Advances in DNS/LES, 1st AFOSR International Conference on DNS/LES, Columbus, OH, August 4-8 (1997)

Ukeiley, L., Sheehan, M., Coiffet, F., Alvi, F., Arunajatesan, S., Jansen, B.: Control of Pressure Loads in Complex Cavity Configurations. In: 45th AIAA Aerospace Sciences Meeting and Exhibit, Reno, NV, January 8-11 (2007)

Wong, M., McKenzie, G., Ol, M., Petterson, K., Zhang, S.: Joint TTCP CFD Studies into the 1303 UCAV Performance: First Year Results. In: 24th AIAA Applied Aerodynamic Conference, San Francisco, CA, June 5-8 (2006), AIAA paper 2006-2984

Control of the Unsteady Flow Inside the SOFIA Telescope Cavity by Means of a Porous Fence

S. Schmid, Th. Lutz, and E. Krämer

Abstract. The unsteady flow field around the Stratospheric Observatory For Infrared Astronomy (SOFIA), a reflecting telescope carried inside an open port of a Boeing 747SP, has been simulated by means of Detached Eddy Simulations (DES) with the CFD-solver TAU of the German Aerospace Center (DLR) in order to predict the unsteady flow-induced loads. A porous fence has been placed upstream of the cavity and simulated with DES to investigate its impact on the unsteady pressure fluctuations on the surface of the telescope assembly. The analysis shows that the deflection of the shear layer and the alteration of its stability characteristics reduce the unsteady telescope loads significantly, due to lowered overall pressure amplitudes and due to the shift of acoustic energy to lower frequencies.

1 Introduction

NASA and the German Aerospace Center DLR are working together to create and operate SOFIA (Krabbe and Röser 1999), a reflecting telescope carried on board of a Boeing 747SP aircraft. During observation in the stratosphere, a door in the fuselage opens and exposes the telescope to the atmosphere. The flow over cavities like the SOFIA telescope port is generally characterized by unsteady phenomena associated with prominent pressure fluctuations caused by amplified acoustic resonances (Ahuja and Mendoza 1995). In case of SOFIA this phenomenon causes unwanted vibrations of the telescope structure and reduces its pointing stability. The flow inside the SOFIA telescope port is basically controlled by an aperture ramp at the rear part of the opening, a slanted wall that stabilizes the shear layer's impingement point and prevents it from violent oscillations. Several wind-tunnel studies on a 3% and a 7% model have proven the concept to be effective and robust to deviations from the nominal flow state. The objective of the present numerical investigations is to estimate the remaining potential to further increase the performance of this passive flow-control concept by modifying the shear layer's position and its stability characteristics.

S. Schmid
Deutsches SOFIA Institut, Universität Stuttgart, NASA Ames Research Center, USA
Sven.Schmid@gmx.de

Th. Lutz · Ewald Krämer
Institut für Aerodynamik und Gasdynamik, Universität Stuttgart, Germany

S.-H. Peng et al. (Eds.): Progress in Hybrid RANS-LES Modelling, NNFM 111, pp. 137–146.
springerlink.com

2 Cavity Flow Physics

The flow over cavities in general is characterized by self-sustained pressure fluctuations. The shear layer spanning the opening of the cavity amplifies flow disturbances that are convected downstream and are scattered into acoustic waves at the downstream edge. The excited waves propagate upstream inside and outside the cavity and trigger further disturbances in the shear layer to create a feedback loop. Frequencies with a phase lag of a multiple of 2π are being amplified in particular, yielding a selection of discrete modes. Rossiter (1964) found that the frequencies can be represented by the semi-empirical equation

$$f = \frac{U}{L}\frac{(m-\gamma)}{\frac{1}{K}+M}, \qquad m = 1,2,3... \tag{1}$$

where f is the frequency of the mode m, L is the cavity length, M the free stream Mach number, U is the velocity and γ and k are empirical constants. γ represents the phase delay of shear-layer disturbances that are scattered at the downstream edge, their convection speed is k times the free-stream speed. The existence and the magnitude of these Rossiter modes depend mainly on the stability characteristics of the shear layer that evolves from the boundary layer in front of the cavity, basically the momentum thickness δ_2 (Rowley 2001). Small values in general lead to higher shear-layer disturbance amplification and hence to higher fluctuation levels inside the cavity. If the Rossiter modes' frequencies coincide with acoustic cavity resonant frequencies, standing waves can be excited, leading to a so called "lock in" at the characteristic resonant frequencies (Block 1976). This effect enhances the feedback of disturbances and thereby increases the pressure amplitudes.

3 Modeling Approach

DES-simulations have been carried out with the Finite Volume CFD-solver TAU, that was developed by the German Aerospace Center, DLR (Gerhold 2005). The cell-vertex based code solves the three-dimensional Reynolds-Averaged Navier-Stokes Equations on unstructured and hybrid grids. The central-scheme according to Jameson was used for spatial discretization in combination with the implicit LUSGS-scheme. Turbulence was modelled by the Spalart Almaras Subgrid Scale model, the transition from the RANS mode to the LES mode is based on the wall distance and the local mesh size (natural DES). The DDES-approach was applied to guarantee that the LES-region is far enough from viscous walls to avoid problems resulting from grid induced separation. The computational model represents the left hand side of the entire SOFIA-aircraft including the wings, the fuselage, the whole cavity and the telescope structure. The computational grids of the different configurations were built with the commercial software GRIDGEN. The DES-grid of the baseline configuration consists of 46E+06 cells in total. Boundary layers evolving on the no-slip walls of the fuselage, the wings, the cavity and the telescope were resolved by 6E+06 prismatic cells (38 layers in wall normal direction) extruded from the surfaces. The region around the shear layer was discretized by 15E+06 hexahedral cells (see Fig. 1 left) that are highly isotropic in order to allow for the resolution of the shear-layer vortices (see Fig. 1 right). The remaining computational domain was discretized by 25E+06 tetrahedra that allow for automated mesh generation of complex geometries.

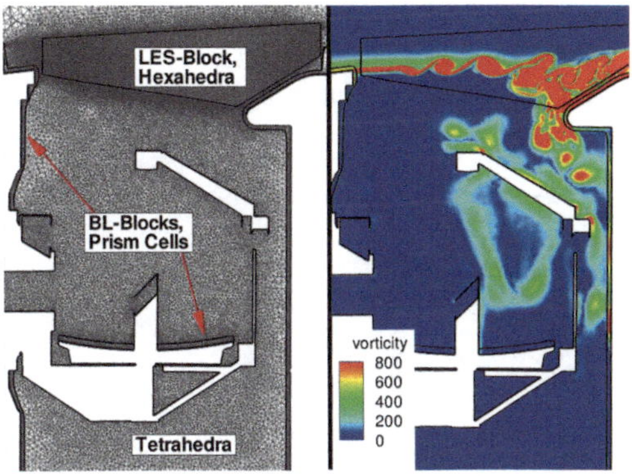

Fig. 1 Cut through the SOFIA telescope port: DES mesh setup (left), shear-layer visualization (right)

A mesh with increased grid density in the shear layer regime was built for this configuration as well in order to study the impact of the spatial discretization error. This finer version contains 58E+06 cells in total (6E+06 prisms, 22E+06 hexahedra and 30E+06 tetrahedra). The comparison of the predicted pressure fluctuations with the coarse mesh solution shows a minor difference as can be seen in Fig. 5. In order to limit the computational effort, the coarse mesh was used for the consecutive simulations. The time-dependant solution was gained by employing the Dual-Time-Stepping scheme with 40 inner iterations per physical time step. About 8500 physical time steps were run to initialize the computation until transient pressure fluctuations disappeared. After this initial period approximately 7000 time steps were computed with a physical time step size of 1E-04 s. As the convection velocity of the shear layer is about 140 m/s, it takes about 200 physical time steps for a particle to travel across the length of the cavity. In order to investigate the discretization error of the time step, a separate simulation with a step size of 6E-05 s was carried out as well. The comparison of both simulations shows that the decrease of the time step size had a nearly negligible impact (see Fig. 5). In order to save computational resources, the time step size of 1E-04 s was chosen for all consecutive simulations.

After the assessment of the spatial and temporal discretization error, the CFD mesh was extended by a porous fence for shear-layer control. This fence was designed by NASA as a contingency concept and will be installed in case of adverse resonance conditions that might have a negative impact on the telescope performance. The fence is made up of a perforated plate, tilted by 30° against the fuselage surface. The height of the fence is about 152mm, corresponding to approximately 30% of the local boundary layer thickness. The size of the holes and the distances in between are identical to the original design. In the current DES simulations the fence was modelled as a baffle surface with zero thickness. Viscous no-slip boundary conditions were specified on its surface. The cell size in the vicinity of the surface holes (hole diameter: 20mm) was chosen to 6mm in order to resolve each hole with about 10 cells. The support elements that connect the fence to the fuselage surface were neglected in the CFD model. Figure 2 shows the surface mesh in the vicinity of the aperture. The telescope is marked as green and the fence is marked as orange. The simulations were run on 1020 processor cores in parallel on the supercomputer Altix-4700 at the Leibniz Rechenzentrum in Garching, Germany.

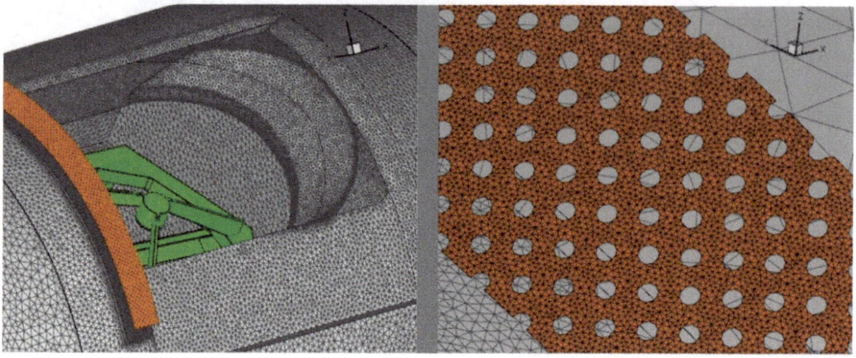

Fig. 2 CFD surface mesh in the vicinity of the SOFIA telescope port showing the aperture ramp, the telescope (green) and the shear-layer fence (orange)

4 Results

In previous wind-tunnel investigations performed by NASA in the 14 by 14ft transonic wind tunnel at the NASA Ames Research Center (McIntyre 1999), the unsteady pressure fluctuations were measured at 56 locations (see Fig. 3) on the telescope assembly surface of a 7% wind-tunnel model. The PSD plots and the SPL values that are predicted by the simulation at the same locations compare well with experimental data. The average SPL measured in wind tunnel and scaled to flight conditions is 127dB, whereas the DES simulation yields 124.5 dB. The correct representation of the boundary layer in the computational model is crucial for a physically meaningful simulation as the state of the boundary layer determines the stability characteristics of the shear layer. Boundary layer profiles predicted by DES at different locations upstream of the cavity show an excellent agreement with measurement.

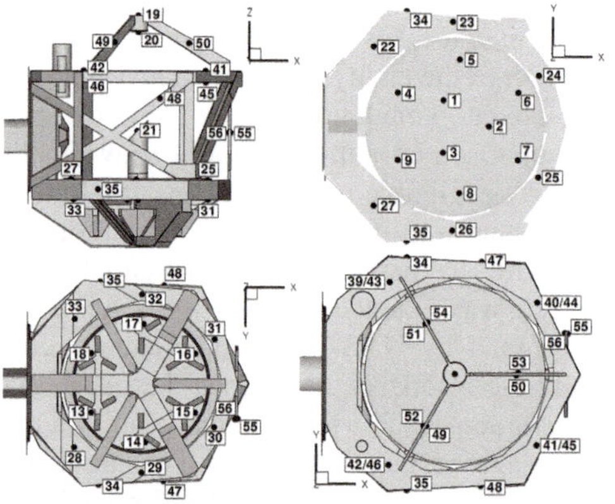

Fig. 3 Positions of pressure sensors on the SOFIA telescope assembly

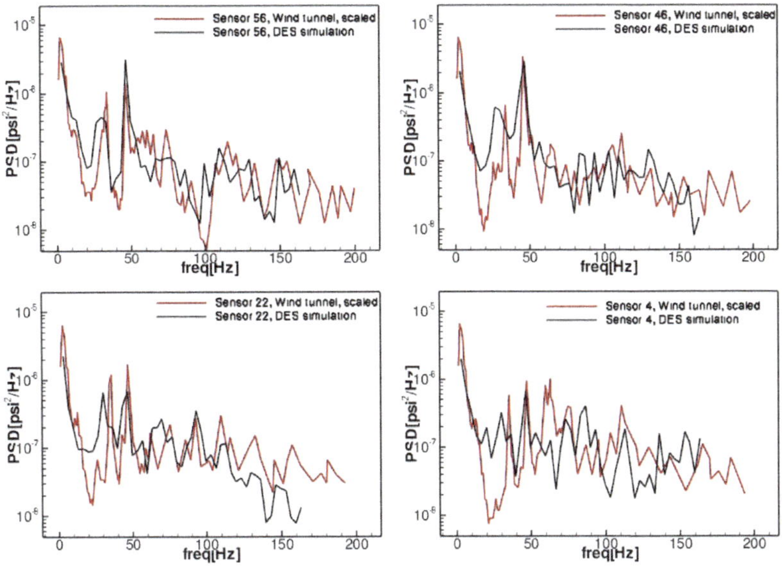

Fig. 4 Comparison between wind-tunnel data and DES data

Figure 4 compares DES-based pressure PSD plots with experimental data at four different positions on the telescope assembly under reference conditions (M=0.85, 40° telescope elevation, α=2.5°, β=0°). The Reynolds number based on the cavity length was $Re_{L,cav}$=2.8E+06 in the wind tunnel, the Reynolds number of the aircraft under flight condition is $Re_{L,cav}$=17.3E+06. In the DES-simulation the flight Reynolds number was specified. Characteristic peaks in the PSD plot at ~3.5Hz, ~30Hz and ~47Hz are well predicted by the computational model, even though the Reynolds number of the actual SOFIA configuration is by a factor of 6 larger than in the wind tunnel and the geometry of the cavity walls and the telescope assembly is slightly different. The low frequency peak at 3.5Hz is caused by a Helmholtz mode, the peaks at ~30Hz and ~47Hz are the footprints of acoustic resonance modes excited by Rossiter modes. Compared to URANS simulations carried out prior to the DES simulations by Schmid et al. (2008, 2009), the amplitudes match significantly better at frequencies above 80Hz. The URANS simulations show a significant drop of the amplitudes beyond 80Hz due to shortcomings of the turbulence modelling.

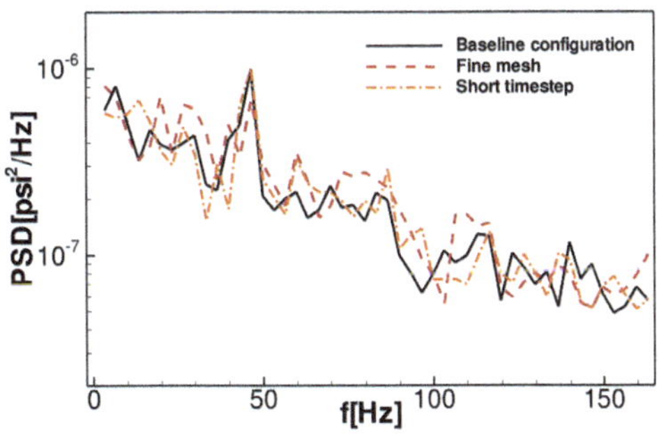

Fig. 5 Impact of temporal and spatial discretization, DES simulaton

Figure 5 shows the results of the grid sensitivity study and the impact of the finite computational time step. The three lines in Fig. 5 represent the average of the PSDs of all 56 pressure transducers. The comparison shows that neither a finer mesh nor a smaller time step have a significant impact on the frequency and the amplitude of the characteristic peak at around 47 Hz. Slight differences can be observed between 10Hz and 40Hz.

5 Impact of the Fence on the Unsteady Cavity Flow

The free shear layer spanning the telescope port is characterized by coherent Kelvin-Helmholtz structures that evolve at the upstream part and decay after a certain distance. Using Linear Stability Theory (LST) and assuming a hyperbolic-tangent velocity profile, Michalke (1965) predicted the most amplified frequency of a free shear layer to be $f_0=0.132 \cdot u_{mean}/\delta_\Omega$ with u_{mean} being the arithmetic mean of both freestream velocities and δ_Ω the vorticity thickness of the shear layer. According to this formula the most amplified frequency of the SOFIA shear layer is f_0=320Hz, what is in good agreement with the shedding frequency of f_{sh}=300Hz predicted by the DES. The shear-layer fence in front of the cavity has a significant impact on the steady and unsteady flow features in and around the telescope port. Sound-pressure levels inside the cavity are reduced by three main effects. First of all the fence directs the flow away from the aperture. The interaction of the shear-layer disturbances with the aperture ramp is weakened, less acoustic energy is scattered back into the cavity that perpetuates the unsteady pressure fluctuations. Second, the fence thickens the shear layer and increases its defect. Thicker shear layers are generally characterized by more benign stability characteristics and lead to lower sound-pressure levels inside the cavity as the feedback mechanism is weakened. Third, the fence decreases the spatial coherence of the shear layer and the inherent turbulent structures. Figure 6 shows the Mach number (upper figures) and the Λ_2 value (lower figures) in the vicinity of the shear layer. The shear layer on the left side (SOFIA configuration with fence) is significantly shifted away from the cavity compared to the right side (SOFIA configuration without fence). The Mach number in the region between the aperture ramp and the secondary mirror spider-arm of the telescope assembly (marked by a yellow arrow

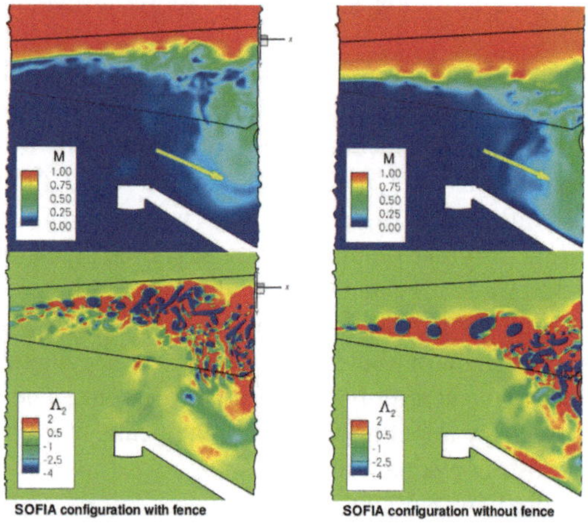

Fig. 6 Visualization of coherent structures in shear layer; left: with fence – right: without fence

in the upper figures) is remarkably decreased by the fence, indicating a lower interchange of energy between the outer flow and the cavity interior. The distribution of the Λ_2 value inside the shear layer as depicted in the lower figures also shows a lower interaction of the shear-layer vortices with the cavity interior. Also, it can be seen that the vortex structures in the shear layer are smaller in size and less organized. Figure 7 shows the evolution and the decay of the coherent structures in the cavity shear-layer. Ashurst and Meiburg (1988) showed by means of a Lagrangian vortex method that both the development and the breakup of these Kelvin Helmholtz vortices are due to inviscid vortex dynamics. The decay of the spanwise structures is triggered by a secondary instability that manifests itself in streamwise vortex braids that connect two consecutive spanwise vortices. As Fig. 7 reveals, these streamwise vortex braids are more numerous and occur at an earlier stage further upstream due to the fence.

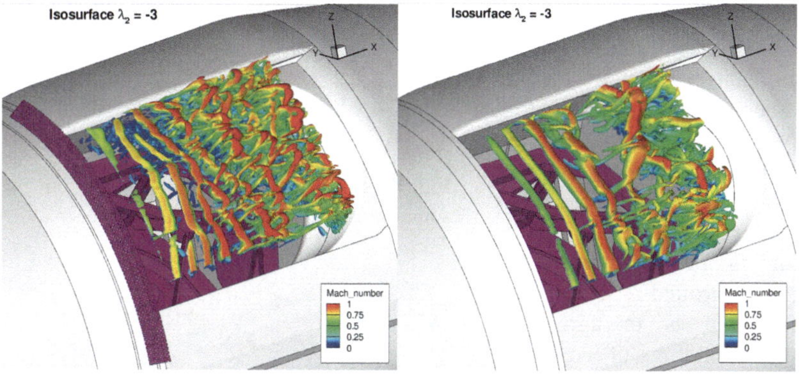

Fig. 7 Visualization of coherent shear-layer structures; left: with fence, right: without fence

The alteration of the shear-layer dynamics by the fence influences the unsteady flow inside the telescope port signifycantly. The Detached Eddy Simulation of the configuration with fence shows an increased oscillation at a very low frequency close to the Helmholtz mode frequency of the cavity. This phenomenon can be explained with the formula by

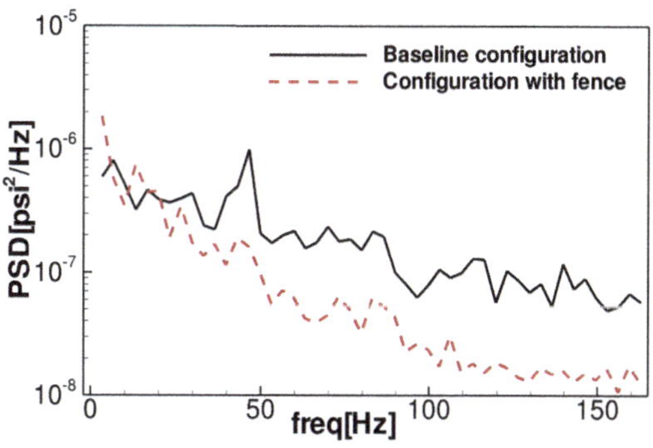

Fig. 8 PSD – average over all 56 sensors on the telescope assembly

Michalke stated earlier in this document. The formula yields a decrease of the most amplified frequency with increasing vorticity thickness. Figure 8 reveals the impact of the fence on the frequency content of the pressure fluctuations inside the cavity. The black line indicates the average PSD on the telescope assembly of the baseline configuration without fence, the red line represents the average PSD on the telescope assembly of the configuration with fence. The plot reveals a significant drop of acoustic energy above 20Hz. Most prominent is the decrease of the 47Hz peak that is caused by the coupling of the second Rossiter mode with a natural acoustic cavity-resonance mode (Schmid et al. 2007). As the telescope structure shows a couple of significant elastic vibration modes at frequencies around 50Hz, the reduction of this peak is expected to yield a significant reduction of the telescope vibrations and an improvement of the pointing stability. The PSD plot also reveals that the fence increases the energy of the Helmholtz mode in the low frequency regime around 3Hz. This observation is in agreement with the findings of Tramel et al. (2005), who investigated the impact of a porous fence on the unsteady flow inside the open port of the Widebody Airborne Sensor Platform (WASP).

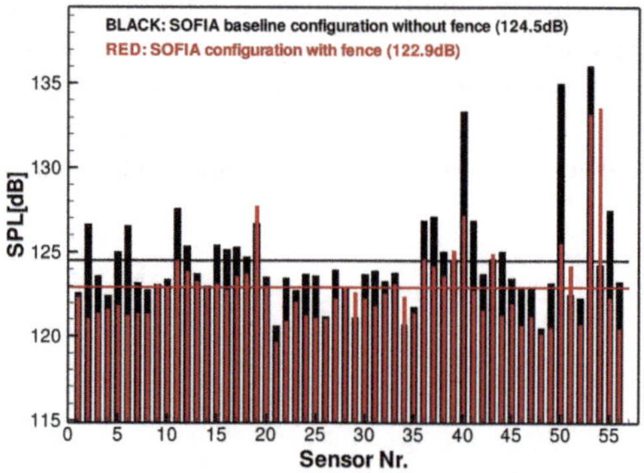

Fig. 9 Impact of shear-layer fence on overall sound-pressure levels

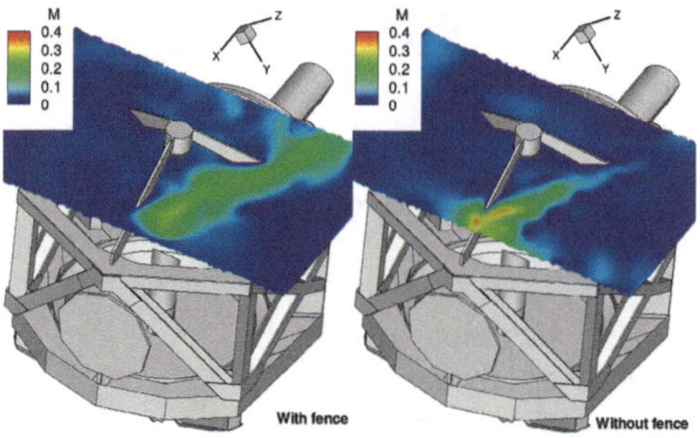

Fig. 10 Mach contour plot above telescope assembly's headring

The impact of the fence on the overall sound pressure levels on the SOFIA telescope is depicted in Fig. 9. Almost all 56 pressure sensors show SPL-values that are distinctively lower than in the baseline configuration. The average SPL of all sensors drops by approx. 2dB (indicated by the horizontal black and red solid lines). A few sensors show slightly higher SPL values due to the presence of the fence, like sensor No. 19, 29, 34 and 51. A major increase of the SPL value due to the fence can be seen at sensor 54 that is located on the secondary mirror spider-arm. This increase is caused by the altered mean flow inside the cavity. The velocity of the down-flow into the cavity that is caused by the flow separation on the front lip of the aperture ramp is reduced and the stream is deflected further to the front. As can be seen in the contour plot in Fig. 10, the maximum Mach number inside the core of the stream is lower in the configuration with fence compared to the baseline configuration. But due to the deflection of the stream, the spider arm on the right hand side experiences a stronger impact, yielding a higher SPL value for sensor No. 54. Figure 11 reveals the implications of the mean-flow change as well. The rear spider arm and the telescope-assembly headring show increased pressure fluctuations due to the down-flow, caused by flow separation on the aperture ramp of the SOFIA baseline configuration (right figure). The wall-streamlines and the distribution of the SPL values indicate a prominent stagnation point in the right aft part of the primary mirror. The presence of the fence shifts this stagnation point further to the front of the mirror and decreases the magnitude of the pressure fluctuations on the surface (left figure). The global flow-induced loads on the telescope assembly have been calculated by the integration of the predicted surface pressure. The fence reduces the RMS values of the unsteady moments around the line of sight, the elevation- and the cross elevation axes of the telescope by 53%, 43% and 30% respectively. Besides the general reduction of the pressure fluctuation amplitudes, this decrease can be attributed to the shift of acoustic energy to lower frequencies. The acoustic mode at 47Hz that is decreased by the fence correlates well with the telescope structure as its wavelength is in the same order as the telescope dimensions. However, as the wavelength of the Helmholtz mode is far beyond the telescope dimensions, its slight growth caused by the fence does not affect the moments significantly.

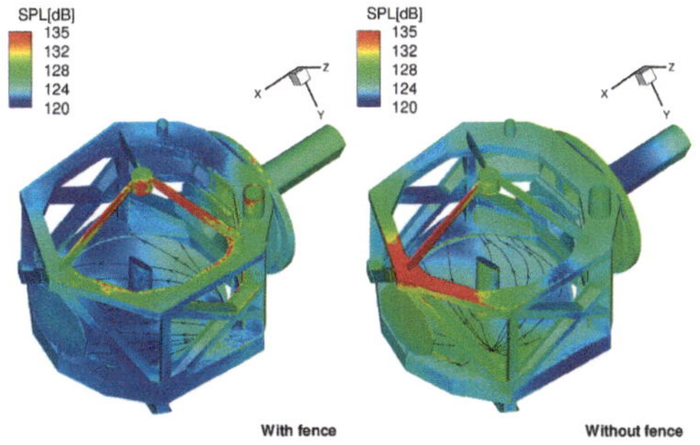

Fig. 11 SPL values and streamlines on telescope-assembly surface

6 Conclusions and Outlook

The DES simulations carried out within the present study reveal a significant improvement of the steady and unsteady flow inside the SOFIA telescope cavity by the shear-layer fence. The drop of the overall SPL values and the shift of acoustic energy from higher to lower frequencies yield a significant reduction of unsteady loads on the telescope assembly. The flow-induced excitation at around 47Hz is reduced significantly. Furthermore the simulations show that the fence also reduces the magnitude of the separation-induced down-flow on the telescope assembly, yielding a desirable decrease of the interaction between the flow and the telescope assembly.

Acknowledgement. SOFIA, the "Stratospheric Observatory for Infrared Astronomy" is a joint project of the Deutsches Zentrum für Luft- und Raumfahrt e.V. (DLR; German Aerospace Centre, grant: 50OK0901) and the National Aeronautics and Space

Administration (NASA). It is funded on behalf of DLR by the Federal Ministry of Economics and Technology based on legislation by the German Parliament the state of Baden-Württemberg and the Universität Stuttgart. Scientific operation for Germany is coordinated by the German SOFIA-Institute (DSI) of the Universität Stuttgart, in the USA by the Universities Space Research Association (USRA).

References

Ahuja, K.K., Mendoza, J.: Effects of cavity dimensions, boundary layer, and temperature on cavity noise with emphasis on benchmark data to validate computational aeroacoustic codes. NASA Contractor Report 4653 (1995)

Ashurst, W.T., Meiburg, E.: Three-dimensional shear layers via vortex dynamics. Journal of Fluid Mechanics 189, 87–116 (1988) ISSN 0022-1120

Block, P.J.W.: Noise Response of Cavities of Varying Dimensions at Subsonic Speeds, NASA TN D-8351 (1976)

Gerhold, T.: Overview of the Hybrid RANS Code TAU. In: Kroll, N., Fassbender, J.K. (eds.) Notes on Numerical Fluid Mechanics and Multidisciplinary Design, vol. 89, pp. 81–92. Springer, Heidelberg (2005)

Krabbe, A., Röser, H.-P.: SOFIA-Astronomy and Technology in the 21st Century. In: Schielicke, R.E. (ed.) Reviews of Modern Astronomy, Astronomical Instruments and Methods at the Turn of the 21st Century, Hamburg, Astronomische Gesellschaft, vol. 12 (1999)

McIntyre, M.: SOFIA V Pressure Sensor Power Spectral Density Plots, NASA Ames Research Center, SOFIA TN MJM-007 Rev. 0 (1999)

Michalke, A.: On spatially growing disturbances in an inviscid shear layer. J. Fluid Mech. 23, 521–544 (1965)

Rossiter, J.E.: Wind-Tunnel Experiments on the Flow over Rectangular Cavities at Subsonic and Transonic Speed, R&M No. 3438, Aeronautical Research Council, London, UK (1964)

Rowley, C.W.: Modelling, Simulation, and Control of Cavity Flow Oscillations. PhD-Thesis, California Institute of Technology Pasadena, California (2001)

Schmid, S., Lutz, T., Krämer, E.: Numerical Simulation of the Flow Field Around the Stratospheric Observatory For Infrared Astronomy. In: Tropea, C., Jakirlic, S., Heinemann, H.J., Henke, R., Hönlinger, H. (eds.) Notes on Numerical Fluid Mechanics and Multidisciplinary Design, vol. 96. Springer, Berlin (2007)

Schmid, S., Lutz, T., Krämer, E.: Simulation of the Flow Around the Stratospheric Observatory For Infrared Astronomy SOFIA Using URANS and DES. In: Wagner, S., Steinmetz, M., Bode, A., Brehm, M. (eds.) Transactions of the Third Joint HLRB and KONWIHR Status and Result Workshop, December 3-4, 2007. Leibniz Supercomputing Centre, Garching/Munich, Germany (2009)

Schmid, S., Lutz, T., Krämer, E.: Simulation of the Unsteady Cavity Flow of the Stratospheric Observatory For Infrared Astronomy. In: Braza, M., Hourigan, K. (eds.) Proceedings of the IUTAM Symposium on Unsteady Separated Flows and their Control, Corfu, Greece, June 18-22, 2007 (2009)

Schmid, S., Lutz, T., Krämer, E., Kühn, T.: Passive Control of the Flow Around the Stratospheric Observatory For Infrared Astronomy. AIAA-Paper 2008-6717 (2008)

Tramel, R., Rock, S., Ellis, J., Sharpes, D.: Comparison of Large Cavity Aeroacoustic Computations with Flight Test Results, AIAA Paper 2005-2800 (2005)

Zonal-Detached Eddy Simulation of a Civil Aircraft Engine Jet Configuration

V. Brunet and S. Deck

Abstract. This paper presents an advanced simulation on an aircraft powerplant configuration at a realistic Reynolds number. The modelling approach retained is the Zonal-DES (ZDES) approach which allows attached boundary layers to be protected and a rapid switching into LES mode away from attached boundary layers. The simulation was performed on a well adapted grid for such method and comparisons with experiments are carried out. It is shown that the aerodynamic field is accurately predicted with ZDES. The results also provide an insight into the unsteady nature of the flow. Future upgrades for this kind of complex simulation necessary to reach an even higher level of accuracy are presented too.

1 Introduction

The diameter of the turbofan engines for civil transport aircraft has increased in the last decades to improve the propulsion efficiency and reduce the fuel consumption. Even for long range aircraft, for which two big engines or four smaller engines are proposed, the two engine versions are often selected for global economic reasons. This tendency, coupled with the different integration constraints at aircraft level (ground clearance, short pylon preferable for heavy engines for structural reasons ...), leads to relative engine positions closer to the wing. For the next generation of aircraft projects, the intensity of the interactions between the engine and the airframe could be important and generate critical phenomena. In addition, when the engine is installed closer and closer to the airframe, the evolutions of the interactions are no more linear and it is difficult to extrapolate the actual conventional solutions in term of engine installation.

Furthermore, because of necessary noise level reduction of aircrafts, it becomes now mandatory to use advanced CFD simulations that could allow an accurate noise prediction of realistic aircraft configurations.

Consequently, the prediction of such phenomenon thanks to CFD tools around a whole civil aircraft is a problem of outstanding importance. This study focuses on the capability of advanced modelling to simulate accurately a civil aircraft engine jet configuration, based on an accurate wind tunnel database acquired during the JEDI project between ONERA and Airbus. Thanks to ONERA's

V. Brunet · S. Deck
ONERA, Applied Aerodynamics Department, Meudon, France
vincent.brunet@onera.fr

S.-H. Peng et al. (Eds.): Progress in Hybrid RANS-LES Modelling, NNFM 111, pp. 147–156.
springerlink.com

previous experience on jet simulations, the selected numerical approach is the so-called ZDES method allowing an accurate modelling of both the near flow around the aircraft thanks to a RANS approach and a precise, detailed simulation of the jet thanks to a LES-like approach.

2 Turbulence Modelling – ZDES Approach

Hybrid RANS/LES methods have during the last years received increasing attention among turbulence modelling researchers and CFD engineers. As an example, the Detached Eddy Simulation (DES) method, first introduced in 1997 by Spalart *et al* has gained widespread acceptance.

However, standard DES introduces a significant dependency into the RANS part of the simulation which requires grid spacing for the wall in tangential direction that is larger than the boundary layer thickness at that location. This grid resolution may be easily violated here due to the thick boundary layer. More precisely, if the switching in LES mode occurs inside the RANS boundary layer, this will result in an underestimation of the skin friction coefficient (Caruelle, 2000). The grey-area needs careful monitoring but novel approaches are emerging (Menter *et al.*, 2003 and Spalart *et al.*, 2006) to overcome this limitation. To avoid this problem in the incoming attached boundary layer, we used a "Zonal-DES" approach (Deck, 2005) where attached boundary layer regions are explicitly treated in RANS mode regardless to the grid resolution. In effect, the interest of this ZDES approach is that the user can focus his grid refinement on regions of interest without corrupting the boundary layer properties farther upstream or downstream. Furthermore, it allows to handle as safely as possible complex geometrical configurations.

Besides, some modifications concerning the formulation of DES have been used because the near-wall functions in LES mode have been removed and we have used the cube root $\Delta=(\Delta_x\Delta_y\Delta_z)^{1/3}$ as length scale for LES. This choice also avoids that the damping functions of the RANS model interpret the low eddy viscosity levels typical of resolved LES regions as closeness to the wall with corresponding fast non-linear drop of sub-grid viscosity. The modification of the subgrid length scale prevents from a delayed development of instabilities in the shear layer and a subsequent late transition to fully turbulent flow. In other words, ZDES switches very quickly to the LES mode, thus limiting the grey area, responsible for the delay in the formation of instabilities. This is an essential feature to get the right development of the mixing layers. As an example, standard DES failed to reproduce the LES development in the mixing layer in a jet (Shur *et al*, 2006) due to an excess of eddy viscosity. Conversely, ZDES has been successfully used to study the mixing enhancement process in a supersonic mixing layer of a sonic underexpanded jet (see Chauvet *et al*, 2007).

3 Test Case

The investigated configuration is a wall-to-wall swept wing model equipped with a pylon and an engine, as presented in Fig. 1. The wing model is representative of the external part of a civil aircraft type wing and was tested in the S3Ch wind tunnel of ONERA.

The engine installation is slightly different from the one of a present aircraft: the nacelle has a much bigger size relative to the wing and is much closer to the wing. Such configuration represents what could be the engine installation of future aircraft. The nacelle was studied and manufactured to be able to simulate both the engine fan and core jets, but also to simulate the higher temperature of the core jet. An air supply system, was installed upstream of the nacelle and delivered the pressurized air for the jet generations (See Fig. 1). This system allowed variations of the total pressure ratios and of the core jet temperature. A boundary layer develops at the surface of this system and contaminates the nacelle. In order to diminish the thickness of this incoming boundary layer, a suction system, has been installed just upstream of the nacelle.

The wing span is 0.8 m and the reference aerodynamic chord is 0.44 m. The diameter of the nacelle D corresponds to 0.1 m at the trailing edge. Reynolds number based on the aerodynamic mean chord is equal to about 6 million.

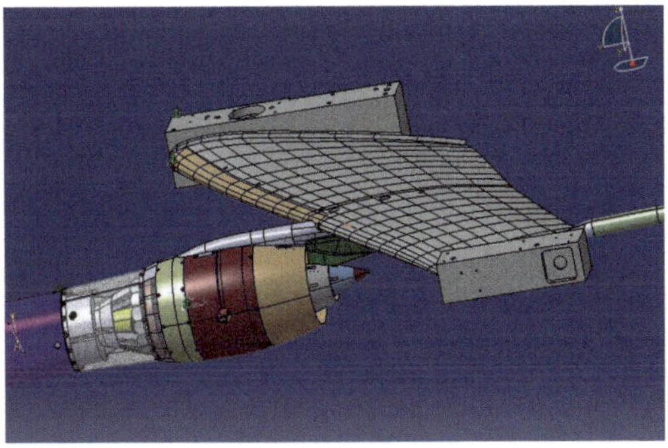

Fig. 1 Visualisation of the model

The simulations are carried out in the wind-tunnel test conditions at a freestream Mach number of about 0.84 and at atmospheric flow conditions. Nozzle pressure ratios for the fan and the core jets are respectively equal to 2.7 and 1.8.

4 Numerical Methods

Based on the ***elsA*** code (Cambier, 2002), this simulation uses "classical" numerical methods for transonic RANS computations. This software relies on a finite volume cell centred formulation with multi-block structured grids. Time integration is performed with a second order Gear scheme with 5 Newton sub-iterations. Implicit LU-SSOR is also used on each sub-iteration. The advective fluxes are discretized with the second order accurate scheme of Jameson with artificial dissipation ($\kappa_2 = 0.25$ and $\kappa_4 = 0.008$ as parameters for first and third order of artificial dissipation, which corresponds to very levels).

In this simulation, the physical time step is equal to 10^{-6} s which corresponds to a maximum CFD number based on maximum acoustic velocity $u+a$ equal to 50. This latter is confined in the trailing edge region of the fan.

5 Grid

The grid generation constitutes an important issue in DES because the cell extension controls which wavelengths can be resolved.

Hence, the grid used for the present ZDES computations is presented in Fig. 2. It is note worthy that the wing and the engine installation are explicitly treated in a RANS mode. The engine jet and wing wake are treated with a ZDES approach. The interest of this approach is that the user can focus his grid refinement on regions of interest (*e.g.* LES regions) without corrupting the boundary layer properties farther upstream or downstream. To avoid a false modelling on the pylon and the plug, the full RANS behaviour is maintained in the ZDES region as long as $d_w < \delta^o_{RANS}$ where δ^0_{RANS} is the boundary layer thickness obtained with the initial RANS calculation.

Patched-grid method is applied in order to perform accurate simulation in ZDES region while limiting the grid density in others RANS regions. Finally two grids of respectively 20 and 40 million nodes have been built, most points being clustered in the jet region thanks to an iterative process, based on a RANS – SA results, which objective was to maximize the grid density into shear layers. The finest grid was built based on the first results obtained with the ZDES simulation with the coarse grid, and it allows most of the correct turbulent development problems to be solved just downstream of trailing edges of the fan and core jets.

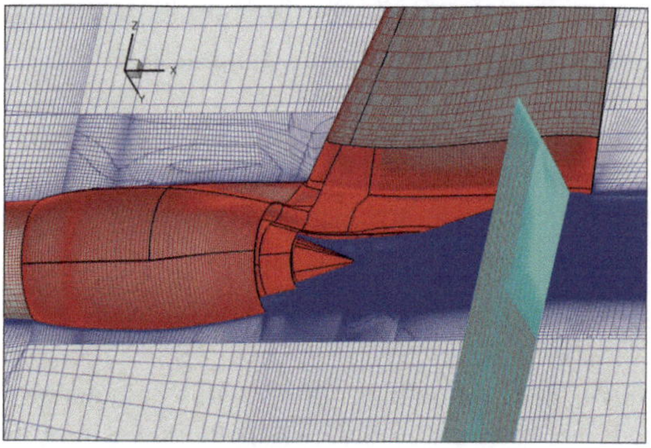

Fig. 2 Zonal-DES grid (Nxyz=40.10^6)

In order to initialise the ZDES simulation, a converged RANS computation was firstly performed with the Spalart-Allmaras model. After the transient phase, the real unsteady calculation begins. The average procedure is performed during the calculation.

6 Results and Discussion

To get further insight into the instantaneous flowfield, Fig. 3 (top) shows a positive value of the Q-criterion. One can notice the roll-up of quasi-toroïdal eddies in the early stage of the fan jet mixing layer. Further downstream, eddies become mainly streamwise due to entrainment effect of the jet. The mixing layers between the fan and core jets and between the fan jet and the outer flow as well as the shock in the fan jet are clearly underlined in the Schlieren type flow visualisation. Besides, this figure highlights the turbulence activity linked to the interaction between the pylon wake and the jet which is further displayed by the vorticity visualization. In this last figure, the complex turbulent behaviour of the flow into shear layers and interaction regions is easily visible.

Besides, from a turbulence modelling point of view, it is worthwhile to notice that there is no delay in the formation of instabilities in the mixing layer thus showing the ability of ZDES to simulate this kind of flows properly.

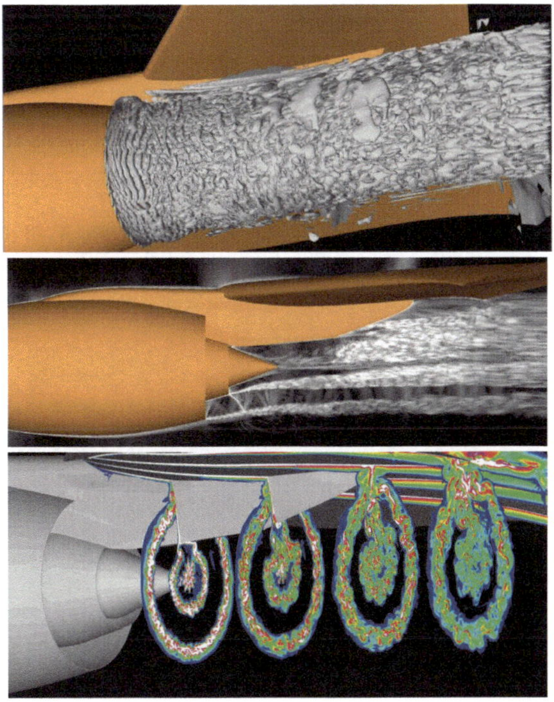

Fig. 3 Iso-surface of a positive value of the Q criterion (upper part) and numerical Schlieren (middle part) and vorticity visualization at different slices into the jet (lower part)

A first comparison between methods and experiments is presented in Fig. 4. In this figure, the streamwise velocity in the axis engine plan allows a good behaviour of the flow to be understood. Minor discrepancies are observed in the near field between all simulations. Let us be reminded that SA-RANS model was designed to treat thin shear layer flows which is the case in the present configuration. However, all calculations underestimate the diffusion into the jet far from the engine compared with the measured PIV plan. As for ZDES results, the grid resolution is clearly not sufficient in the far wake which could partially explain these differences. Another important point concerns the injected turbulence which is rather high into experiment and neglected into simulations. Nonetheless, in the near wake of the engine jet, comparisons agreed quite well.

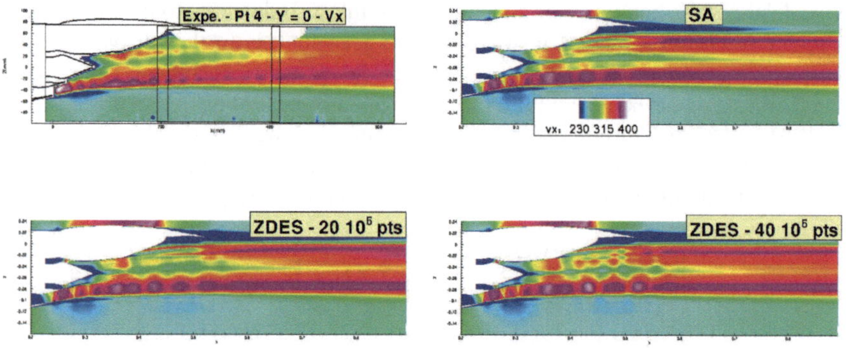

Fig. 4 Comparisons of mean streamwise velocity thanks to PIV plans acquired during the wind tunnel tests

Within the experimental campaign, LDV measurements were acquired on the nacelle and near its trailing edge. The location of the different rakes is given in figure 5.

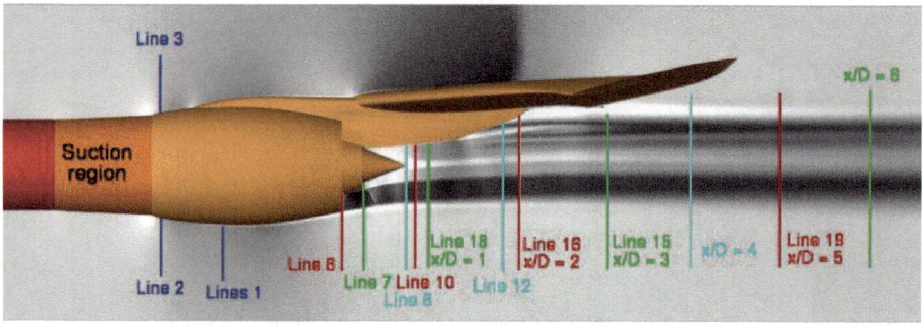

Fig. 5 Rake locations

The mean computed velocity profiles are compared with experiment in figure 6. One can notice that only minor differences are observed between RANS-SA and ZDES. Especially, the mixing layer between the fan jet and the outer flow is very well predicted (z = -50 mm). Due to quite high pressure ratios, comparisons between section 8 and 10 are quite different because of shock cells that are not properly predicted and cause strong spatial variations which can explains discrepancies between simulations and LDV measurements. In the end, the wake of the upper side of the plug (z = 0 – 20 mm) is slightly underestimated but this may be caused by wrong boundary conditions compared with "real" experimental ones. Indeed, the fan and core jet profiles at the near exit differ from those measured experimentally. These discrepancies come from the injection system into the nacelle which could not be taken into account in the simulation and was simulated with uniform injection.

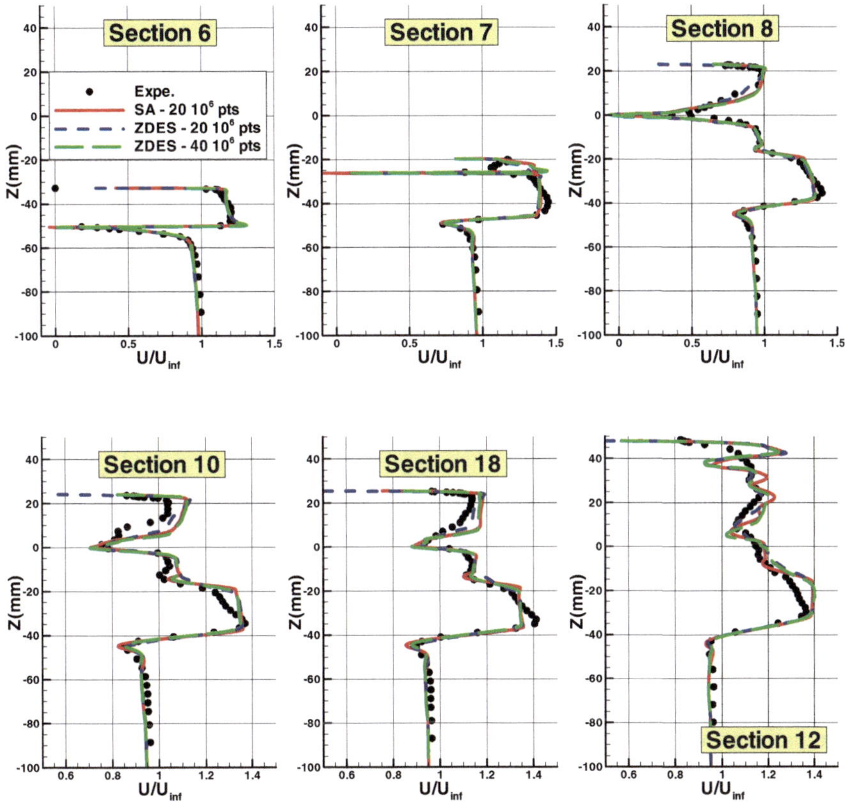

Fig. 6 Mean streamwise velocity profiles in the engine jet

At same locations into the jet, the Reynolds stresses u'w' (w corresponds to the vertical velocity component) are presented in Fig. 7. The peak in shear stresses (modelled + resolved) is slightly underestimated by all simulations into mixing layers. Besides, the shear layer levels between the fan and the core jet (Z = -20 mm) is very well predicted by the finest grid with ZDES. This particular region was modified between the two grid refinements to better simulate the fine scale turbulent structure in that region.

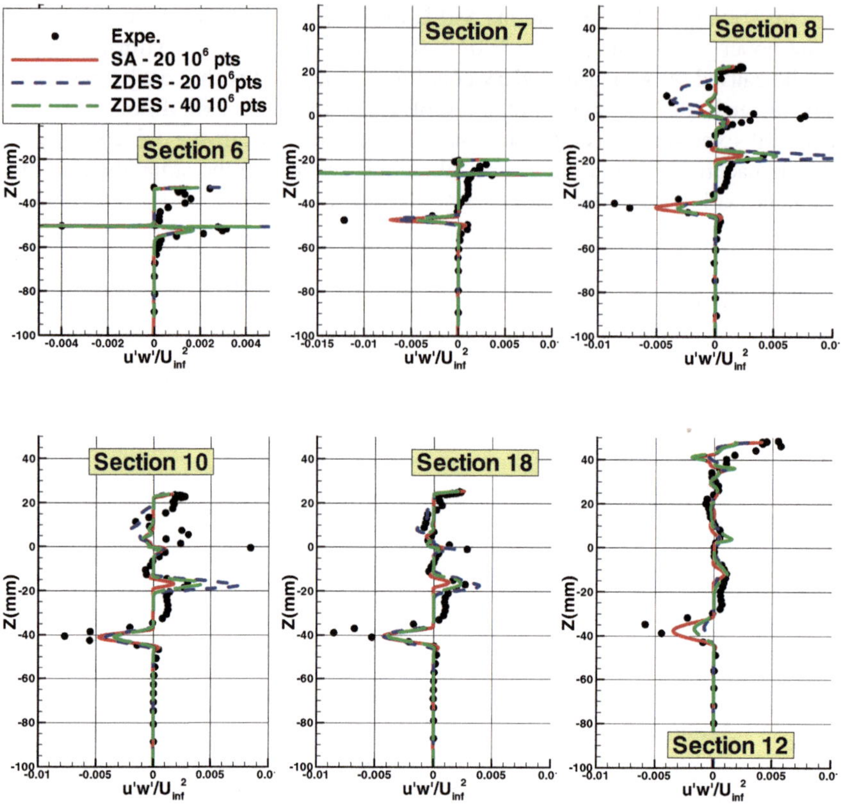

Fig. 7 u'w' / U^2_∞ profiles in the engine jet

The velocity fluctuation profiles are presented in Fig. 8. One can notice that ZDES compares fairly well with experimental data in the shear layers, especially the one between the fan and the core jets and with the finest grid. On all these comparisons, one can remark the high levels of turbulence coming from the jets inside the engine, caused by the injection system. These experimental turbulence levels inside jets are in the same order of magnitude as the one of the shear layers and so their influences on the diffusion of the jet are probably not negligible. This was not taken into account in simulations, especially the ones with the ZDES approach for which an isotropic turbulence generation has to be integrated in order to solve this problem. In the end, the region of interaction between the pylon and the jets is really complex as seen earlier and a slight underestimation of turbulence levels is observed. Let us remind that the grid refinement in that region becomes too coarse to allow enough accurate turbulence resolution and moreover, the turbulence coming from the attached boundary layers on the pylon are not resolved.

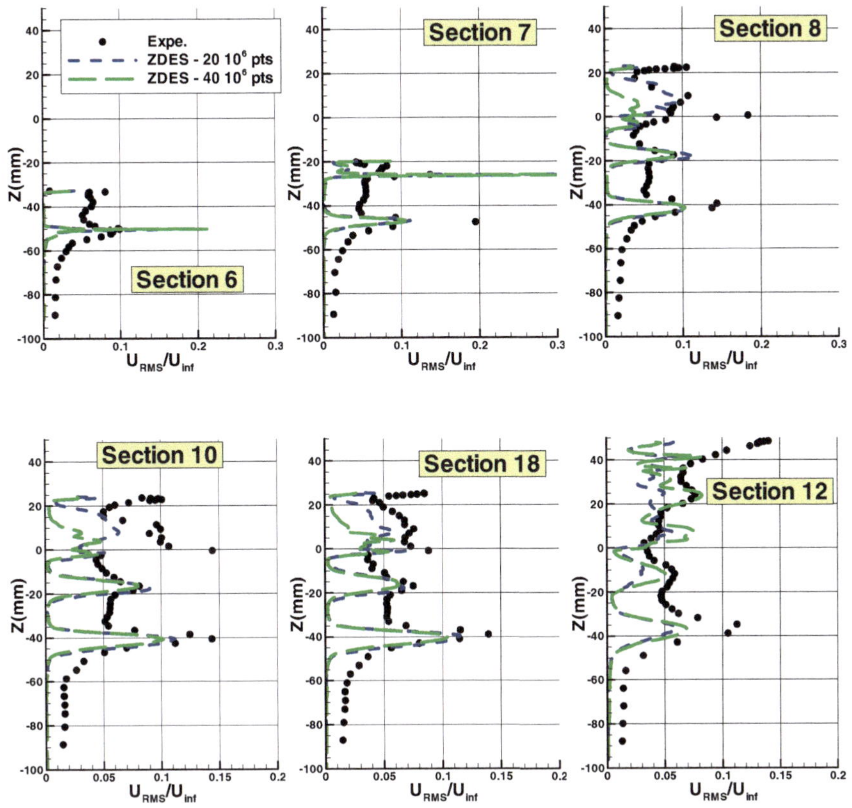

Fig. 8 u_{rms} / U_∞ profiles in the engine jet

7 Conclusion

The numerical simulation of the unsteady flow around a complete civil aircraft type powerplant configuration is a problem of outstanding importance but is a very difficult and challenging case for DES. Such a complex simulation has been possible thanks to the use of a ZDES method, allowing the reduction of the cost of the simulation compared to a LES one by limiting the extend of the DES zones while maintaining the desired level of accuracy in (U)RANS and focused regions.

In this paper, it is clearly demonstrated that a realistic industrial engine installation jet can be simulated with a hybrid RANS-LES method. Accurate comparisons are presented into the jet with measurements and with a RANS-SA solution. Concerning the mean field, RANS-SA and ZDES simulations allow both an accurate prediction of the jet characteristics and its interaction with the pylon. The interest of ZDES for such configuration is that it allows a very deep analysis of the turbulence unsteady behaviour of the jet which is mandatory for more complex analyses like acoustics, unsteady or thermal effects. Furthermore, for more complex geometries (like chevrons) featuring large scale mixing processes, the solution quality of RANS simulations are potentially poorer.

Some limitations for such studies were nonetheless pointed out. Indeed, it is necessary to know exactly the "cartography" of the fan of the core jets, including both mean velocity and pressure, as well as the turbulence level. As a matter of fact, it was highlighted that most of the discrepancies are issued from a wrong knowledge and application of the exact boundary conditions corresponding to the wind tunnel test. Otherwise, the precise description of experimental boundary conditions is of primary importance for advanced modelling methods and militates for a close collaboration between experimentalist and numericists.

Acknowledgement. The authors gratefully acknowledge Airbus-France who financed this work into the framework of the JEDI project.

References

Brunet, V., Molton, P., Bézard, H., Deck, S.: Advanced Experimental and Numerical Investigations of an Aircraft Powerplant Configuration. In: 27th AIAA Applied Aerodynamics Conference, San Antonio, 2009-4117 (2009)

Cambier, L., Gazaix, M.: An Efficient Object-Oriented Solution to CFD Complexity. In: 40th AIAA Aerospace Sciences Meeting and Exhibit, Reno, Nevada, AIAA Paper 02-01108 (2002)

Caruelle, B.: Simulations d'Ecoulements Instationnaires Turbulents en Aérodynamique: Application à la prédiction du phénomène de tremblement. PhD thesis, Institut National Polytechnique de Grenoble (2000)

Chauvet, N., Deck, S., Jacquin, L.: Zonal Detached Eddy Simulation of a Controlled Propulsive jet. AIAA J. 45(10), 2458–2473 (2007)

Deck, S.: Numerical Simulation of Transonic Buffet over a Supercritical Airfoil. AIAA J. 43(7), 1556–1566 (2005)

Deck, S.: Zonal-Detached Eddy Simulation of the Flow around a High-Lift Configuration. AIAA Journal 43(11), 2372–2384 (2005)

Menter, F.R., Kuntz, M., Bender, R.: A Scale-Adaptive Simulation Model for Turbulent Flow Predictions. In: 41th AIAA Aerospace Sciences Meeting and Exhibit, Reno, Nevada, AIAA paper 03-0767 (2003)

Nikitin, N.V., Nicoud, F., Wasistho, B., Squires, K.D., Spalart, P.R.: An Approach to Wall Modeling in Large Eddy Simulation. Physics of Fluids 12, 1629–1631 (2000)

Spalart, P., Jou, W.H., Strelets, M., Allmaras, S.R.: Comments on the Feasibility of LES for Wings and on a Hybrid RANS/LES approach. In: Proceedings of 1st AFSOR Int. Conf. on DNS/LES, Ruston, pp. 137–147 (1998)

Spalart, P.R., Deck, S., Shur, M.L., Squires, K.D., Strelets, M., Travin, A.: A New Version of Detached-Eddy Simulation, Resistant to Ambiguous Grid Densities. Theoritical and Computational Fluid Dynamics 20, 181–195 (2006)

Menter, F.R., Kuntz, M., Bender, R.: A Scale-Adaptive Simulation Model for Turbulent Flow Predictions. In: 41st AIAA Aerospace Sciences Meeting and Exhibit, AIAA 03-0767 (2003)

Shur, M., Spalart, P., Strelets, M., Garbaruk, A.: Further Stepps in LES-based Noise prediction for Complex jets. AIAA paper 2006-485 (2006)

Assessment of DES on the Flow after a Snow-Flake Orifice

H.W. Zheng, F.C.G.A. Nicolleau, and N. Qin

Abstract. To understand the mechanism of the multi-scale nature of turbulent flows, the interaction between the turbulence flow and the two special multi-scale fractal shaped geometries (SF2 and SF3) with high Reynolds numbers in a pipe is investigated by detached-eddy simulation (DES). As compared to the circular orifice, more mixing structures are generated after the fractal shape orifice. The numerical mean velocity profiles along the pipe show a good agreement with the experiment values obtained by hotwire anemometry. These show the feasibility of DES to be applied to study the problem.

1 Introduction

Turbulence flows are very interesting topics that has been intensively studied by many researchers. However, the physics of turbulent flows is still not well understood. Recently a different approach has been adopted (Queiros-Conde and Vassilicos, 2001; Seoud and Vassilicos, 2007; Hurst and Vassilicos; 2007). Rather than trying to understand turbulence from the Navier Stokes equations, these new approaches are relying on interacting with turbulence in a fractal manner; either through a fractal object or a numerical fractal forcing. In this paper, the first way, that is, the interaction between the turbulence flow and the special fractal shaped orifice is investigated.

Because of the limited experimental studies of fractal-shaped objects, there are only several numerical simulations which have been done to study this interaction. Most of them (see e.g. Laizet, S. et al., 2008) are performed by direct numerical simulation (DNS). However, the Reynolds numbers achieved in DNS cannot be very high in contrast to laboratory experiments. Hence, the numerical results can only be compared qualitatively with experimental results. Besides, DNS or LES are still too time and memory consuming for quite a long time to come.

Hence, this work is to assess the feasibility of studying the turbulence flows after a fractal orifice using detached eddy simulation (DES). It is a further investigation of the previous work (Zheng et al., 2009) where only one fractal shape orifice is studied. As compared to the fractal shape grids in other works (e.g., Seoud and Vassilicos, 2007), our orifice shape is relatively simple. The

H.W. Zheng · F.C.G.A. Nicolleau · N. Qin
The University of Sheffield, Department of Mechanical Engineering, Mappin Street, Sheffield S1 3JD, United Kingdom
H.Zheng@sheffield.ac.uk

S.-H. Peng et al. (Eds.): Progress in Hybrid RANS-LES Modelling, NNFM 111, pp. 157–165.
springerlink.com

simulation is based on an existing density-based finite volume solver (DG-DES) which has been developed by our group (Xia, 2005; Xia and Qin, 2007) for solving the unsteady compressible Farve-averaged Navier-Stokes equations and the Spalart and Allamas (S-A) equation (Spalart and Allamas, 1992).

2 The Configuration

Experiments (Chong, 2008; Abou et al., 2008) were conducted in our fluids lab using a high length to diameter ratio (about 28:1) circular Perspex tube with the outer and inner diameters of , 150 mm and 140 mm, respectively(Fig.1). The tube is connected to the outside quiescent air thorough a bell-mouth. To control the flow rate, a rotating valve is connected to a fan and placed at the other end of the pipe.

Fig. 1 Flow set-up

The orifice is put after the center position 0.22m toward the valve. Here, two snowflake fractal shapes (SF2 and SF3) are used to generate the orifice as illustrated in Fig. 2. Their constructions begin with a triangle of unit length 1 (108 mm), then the middle third of each side is removed and replaced with two lines each having the same length (1/3). For example, by repeating this procedure twice, the snowflake fractal shape (SF2) will appear. Similarly, the SF3 shape orifice could be produced.

The measurements of the velocity time series across the radius were made on several positions downstream of the orifice by using hotwire anemometry. These positions were at an interval space of 35mm (one quarter of the diameter). The first position is at 70mm from the fractal orifice. After the velocity time series were obtained, the experimental mean velocity profiles were produced by calculating the mean velocity at different points over a long time.

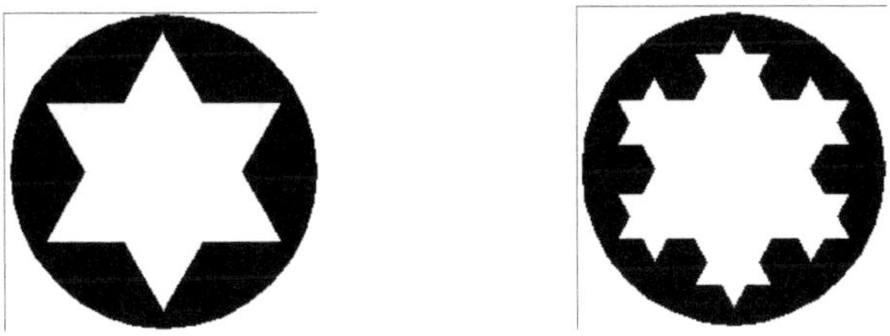

(a) SF2 (b) SF3

Fig. 2 Different shapes of orifice

3 Numerical Method

The density based finite volume method are used to solve the unsteady compressible Farve-averaged Navier–Stokes equations

$$\partial_t \int_{\Omega} U dV + \int_{S} \mathrm{F} \cdot \mathrm{n} dS = 0 \tag{1}$$

with an one-equation Spalart and Allmaras (S-A) model (Spalart and Allmaras, 1992),

$$\begin{aligned} &\frac{\partial}{\partial t}\int_{\Omega} \tilde{\upsilon} d\Omega + \int_{\partial\Omega}\left(\tilde{\upsilon}\mathrm{u} - \frac{1}{\sigma_{\tilde{\upsilon}}}(\upsilon+\tilde{\upsilon})\nabla\tilde{\upsilon} - \frac{1}{\sigma_{\tilde{\upsilon}}} C_{b2}\left(\nabla\tilde{\upsilon}\right)^2\right)\cdot\vec{n} dS \\ &= \int_{\partial\Omega}\left[C_{b1}\left(1-f_{t2}\right)\tilde{S}\right]\tilde{\upsilon} - \left(C_{w1} f_w - \frac{C_{b1}}{\varphi^2} f_{t2}\right)\left(\frac{\tilde{\upsilon}}{d}\right)^2 dS \end{aligned} \tag{2}$$

As in the standard DES97 (Spalart et al., 1997), the distance in Eq. (2) are modified as

$$\tilde{d} = \min\left(C_{DES}\Delta, d\right) \tag{3}$$

Here, $\Delta = \max\left(\Delta x, \Delta y, \Delta z\right)$ is the largest distance between the cell centroid under consideration and the cell centroid of the neighbors.

Eqs. (1-2) are discretized by using implicit dual-time marching with multistage Runge–Kutta sub-iterations (Weiss and Smith, 1995),

$$\partial_t \int_{\Omega} U dV + \frac{\partial U}{\partial W}\partial_\tau \int_{\Omega} W dV + \int_{S} \mathrm{F}\cdot\mathrm{n} dS = 0 \tag{4}$$

with the Jacobian matrix

$$\frac{\partial U}{\partial W}=\begin{bmatrix} \rho_p & 0 & 0 & 0 & \rho_T \\ u_1\rho_p & \rho & 0 & 0 & \rho_T u_1 \\ u_2\rho_p & 0 & \rho & 0 & \rho_T u_2 \\ u_3\rho_p & 0 & 0 & \rho & \rho_T u_3 \\ H\rho_p-\left(1-\rho H_p\right) & \rho u_1 & \rho u_2 & \rho u_3 & \rho_T H+\rho c_p \end{bmatrix} \tag{4}$$

In each physical time step, the equivalent equations are solved until it reaches a steady state in pseudo time.

4 Results

The calculations are performed on the UK's national High-End Computing Terascale Resource (HECToR) by using a parallelized, density-based, dynamic mesh and detached eddy simulation code which is called DG-DES (Xia, 2005; Xia and Qin, 2007). To keep a constant mass flow rate, the inlet velocity is set to 9.87 m/s, according to the experiment.

4.1 The Grids

Note that in our simulation the fractal geometries is quite complex and the thickness of the fractal-shaped orifice is very small (0.003m) when compared to the whole length of the pipe (4m). Hence, fully unstructured meshes (Fig. 3) were generated, as shown in Fig. 3 with a total cell number of about 2 million.

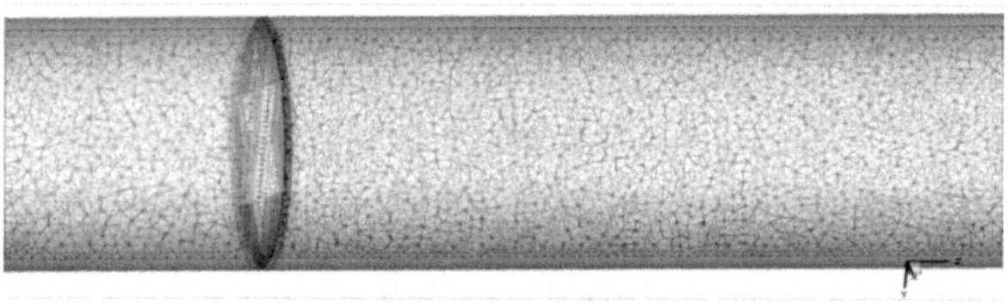

(a) Slice of the Mesh for SF2

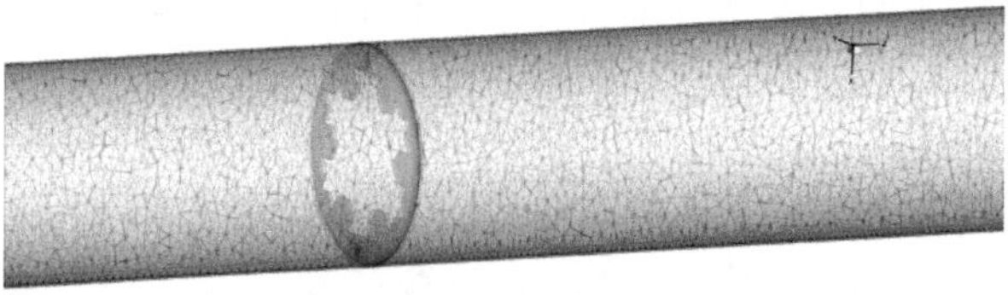

(b) Slice of the Mesh for SF3

Fig. 3 The grids

For DES, the region of RANS and LES are determined by the mesh itself as shown in Fig. 4. RANS are only performed in the near wall area. While, the other areas that is far away from the wall are calculated by LES.

4.2 Flow Features

Fig. 4 gives a snapshot of the instantaneous vortices iso-surfaces of the flow after SF2 orifice. It could be easily observed the upcoming thin boundary layer at the orifice is detached after the orifice. This shows that the detached-eddy simulation is a proper method to study this problem. Besides, the mixing flow patterns and vertical structures coming out of the orifice are clearly observed.

Fig. 4 Instantaneous iso-surface of the vortices after SF2 orifice

As a comparison, the flow pass by the normal circular orifice with the same area of SF2 are also simulated. The instantaneous vorticity iso-surfaces are plotted in Fig. 5. As expected, the mixing process is much less pronounced after the circular orifice. This demonstrates the capability of the fractal orifice in promoting turbulence mixing.

Fig. 5 Instantaneous iso-surface of the vortices after circular orifice

The instantaneous vortices iso-surfaces of the flow after SF3 orifice are plotted in Fig. 6. More thorough mixings near the orifice is observed as compared to Fig. 4, showing the importance of the fractal length scale in turbulence mixing. It is also noted that for SF3 the mixing progress very close to the wall while for SF2 the mixing is more limited to the area away from the wall.

Fig. 6 Instantaneous iso-surface of the vortices after SF3 orifice

4.3 Time Averaged Velocity Profiles

The mean velocity profile is an indicator of the time mean property of the turbulent flow, which can be relatively easily measured. Note that we compare the same quantity which is the average absolute velocity normal to the hotwire. To compare with the experimental results, we probe the numerical velocity data at the positions corresponding to the experimental results.

The time averaged velocity after the SF2 fractal orifice at sections 4, 6, and 8 are plotted as a function of the distance (y) to the centre line of the pipe in Figs. 7-8. The numerical results are compared to the hot wire (HW) measurements. The peak value and the shape are very close to the experiment results although the peak of the profile does not match the experimental result well.

It could also be easily observed that as expected the profile is influenced by the fractal shape at the positions which are close to the fractal orifice. From Fig. 7, the numerical velocity increases after the position at about 0.06m from the centre line. In fact, this position is one of the peaks of the SF2 fractal.

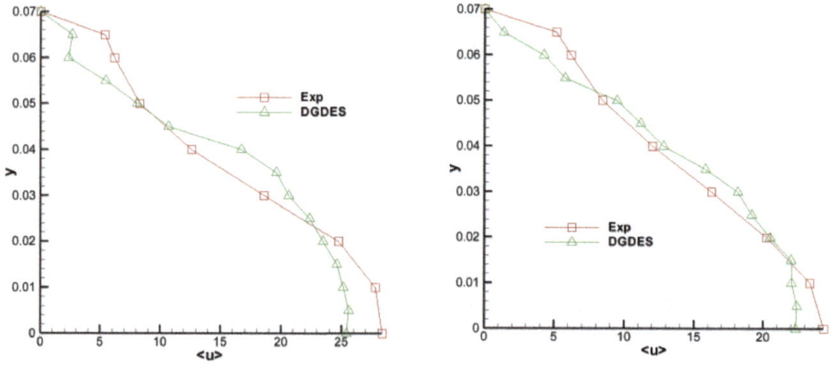

(a) section 4 (b) section 6

Fig. 7 Mean velocity profiles in the pipe with SF2 orifice

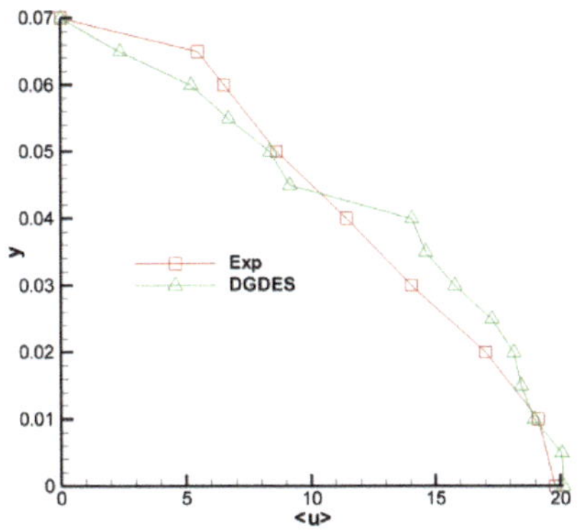

Fig. 8 Mean velocity profile in the pipe with SF2 orifice at section 8

The results of SF3 are shown in Figs. 9-10. The agreements are also observed.

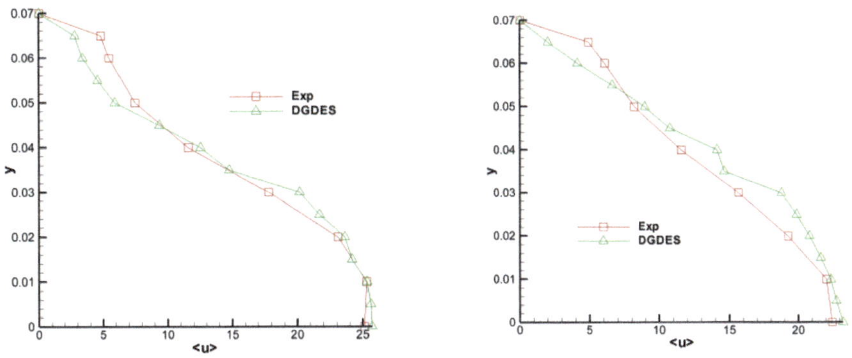

(a) section 4 (b) section 6

Fig. 9 Mean velocity profiles in the pipe with SF3 orifice

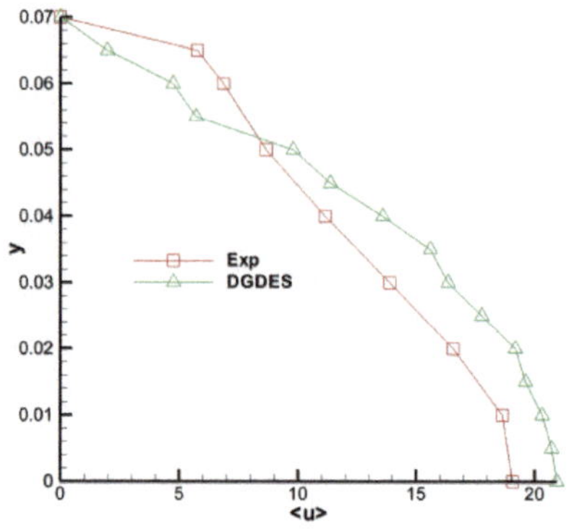

Fig. 10 Mean velocity profile in the pipe with SF3 orifice at section 8

4.4 Power Spectra Analysis

Fig. 11 presents the power spectra density which is obtained from the Fourier transformations of the velocity (in the pipe direction) signals. It illustrates the relation of the power spectra density against frequency. The energy cascade is clearly shown, in which noticeable inertial range (-5/3) and the dissipation range due to viscosity can be found.

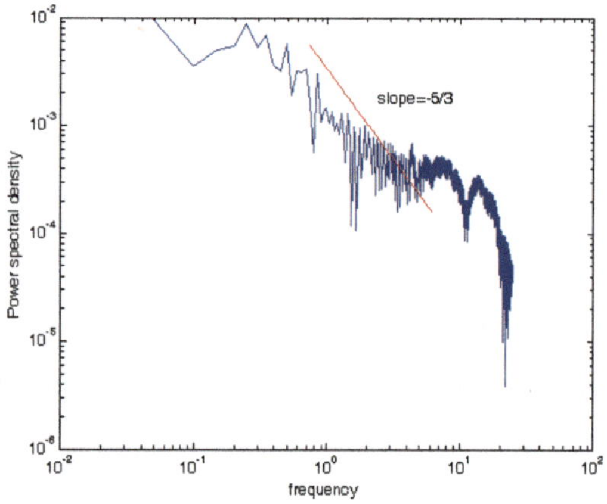

Fig. 11 Power spectra density vs. frequency

5 Conclusions

In this paper, the results of turbulent flow through snow flake fractal shape orifices (SF2 and SF3) by the detached eddy simulation method (DES) are presented. The results for the turbulent flows after the orifice are compared with those of the corresponding experiment. They show reasonable agreements in the mean velocity profiles at different positions downstream the orifice. As compared to the circular orifice, more mixing structures are generated after the fractal shape orifice. The power spectra density also shows the expected energy cascade. These show that DES could be applied to study the interaction of turbulent flow and the fractal shaped orifice. More analyses are needed to be performed in the future.

Acknowledgement. This project is funded by the Leverhulme Trust Research Grant (F/00118/AZ). The computation was conducted using the facilities of HECToR, the UK's national high-performance computing service, which is provided by UoE HPCx Ltd at the University of Edinburgh, Cray Inc and NAG Ltd, and funded by the Office of Science and Technology through EPSRC's High End Computing Programme.

References

El-Azm Aly, A., Chong, A., Nicolleau, F., Beck, S.: Experimental Study of the Pressure Drop after Fractal-Shaped Orifices in a Turbulent Flow Pipe. Proc. of World Acad. of Sc., Eng. and Tech. (2008) ISSN 1307-6884

Chong, A.C.H.: Turbulence flow generated by a fractal grid. PhD Thesis, The University of Sheffield (2008)

Hurst, D., Vassilicos, J.C.: Scalings and decay of fractal-generated turbulence. Phys. Fluids 19(3), 035103 (2007)

Laizet, S., Lamballais, E., Vassilicos, J.C.: A numerical strategy to combine high-order schemes, complex geometry and massively parallel computing for the DNS of fractal generated turbulence, Submitted to Computers and Fluids (Preprint, 2008)

Queiros-Conde, D., Vassilicos, J.C.: Intermittency in turbulence and other dynamical systems, Turbulent wakes of 3-D fractal grids. Cambridge University Press, Cambridge (2001)

Seoud, R.E., Vassilicos, J.C.: Dissipation and decay of fractal-generated turbulence. Phys. Fluids 19(10), 105108 (2007)

Seoud, R.E., Vassilicos, J.C.: Dissipation and decay of fractal-generated turbulence. Phys. Fluids 19(10), 105108 (2007)

Spalart, P.R., Allmaras, S.R.: A One-Equation Turbulence Model for Aerodynamic Flows, AIAA Paper 92-0439 (1992)

Spalart, P.R., Jou, W.-H., Strelets, M., Allmaras, S.R.: Comments on the feasibility of LES for wings, and on a hybrid RANS/LES. In: Proceedings of the 1st AFOSR International Conference on DNS and LES, Louisiana Tech., pp. 137–147. Greyden Press (1997)

Weiss, M., Smith, W.A.: Preconditioning applied to variable and constant density flow. AIAA J. 33, 2050–2057 (1995)

Xia, H.: Dynamic Grid Detached-Eddy Simulations for Synthetic Jet Flows. PhD thesis, The University of Sheffield (2005)

Xia, H., Qin, N.: Dynamic Grid and Unsteady Boundary Conditions for Synthetic Jet Flow, AIAA 2005-106 (2005)

Zheng, H.W., Nicolleau, F.C.G.A., Qin, N.: Detached Eddy Simulation of Turbulence Flows in a Pipe with a Snow Flake Fractal Shape Orifice. In: Synthetic models of turbulence. ERCFOTAC Series. Springer, Heidelberg (2009)

Detached-Eddy Simulations Applied to Unsteady Crosswind Aerodynamics of Ground Vehicles

T. Favre, B. Diedrichs, and G. Efraimsson

Abstract. Crosswind stability is an important safety issue for manufacturers of cars, buses and rail vehicles. Since side wind conditions are unsteady phenomena they require time-dependent techniques to simulate the flow. In this study, a hybrid RANS-LES methods, Detached-Eddy Simulation, is applied to evaluate headwind and unsteady crosswind situations for a simple model of car. A grid refinement study is carried out to evaluate the accuracy of the calculations. Convergence in the force coefficients while refining the grid suggests that a certain level of grid convergence is reached. A similar conclusion is drawn from the unsteady simulations.

1 Introduction

Sensitivity to crosswind increases as the vehicles tend to be lighter and designed for a minimum aerodynamic resistance. Several experimental and numerical attempts have been performed in order to characterize the vehicle behaviour under crosswind conditions. However, unsteady crosswind conditions, like gusty weather or overtaking manoeuvres, remain difficult to investigate since they require sophisticated experimental setups, see e.g. Cairns (1994) and Chadwick (2001), or large numerical resources, e.g. Tsubokura et al. (2009), Favre et al. (2008) and Demuth et al. (2006). This current research explores the use of hybrid RANS-LES techniques (D-DES) to describe unsteady crosswind aerodynamics of a simple road vehicle model.

The complexity of aerodynamics motivates the use of accurate turbulent methods. Reynolds-Averaged Navier-Stokes (RANS) seems insufficient to investigate time-dependent flow conditions. However scale resolving methods, like Large Eddy Simulations (LES) or Detached-Eddy Simulations (DES), are more suitable, Spalart (1997). DES is a relative new technique still under development and perfectibility, see Travin et al. (2002) and Spalart et al. (2006).

As mentioned above, most of the experimental work in wind tunnels has focused on steady crosswind conditions, where vehicles are usually tested for fixed yaw angles, see Hucho (1998) and references therein. However, unsteady wind conditions involve dynamic variations of the wind yaw angle of the wind

T. Favre · B. Diedrichs · G. Efraimsson
Centre for ECO^2 Vehicle Design, KTH Aeronautical and Vehicle Engineering, Stockholm, Sweden
favre@kth.se

S.-H. Peng et al. (Eds.): Progress in Hybrid RANS-LES Modelling, NNFM 111, pp. 167–177.
springerlink.com

that are challenging in experiments. A noticeable attempt, performed at the Cranfield University in UK, is reported in Cairns (1994) and Chadwick (2001), where a vehicle model is propelled through a wind tunnel exhaust which simulates a gust. Comparison with the experimental data is reported in Favre et al. (2008). A new vehicle model is tested in the present paper.

Numerical simulations are more flexible to set up and design compared to experiments. Some previous work that is worth mentioning were carried out by Docton et al. (1996) who used time dependent boundary conditions to propagate a gust in a 2D computational domain to match a similar experiment; Demuth et al. (2006) used periodic boundary condition and a Lattice-Boltzman solver to investigate wind gusts with a cosine-shape for a production car. Another production car has been used by Tsubokura et al. (2009), to investigate yaw moment dynamics variation in crosswind. Finally, the authors used time dependant boundary conditions to perform an unsteady crosswind analysis at lower Reynolds numbers compared to the Reynolds number in the present study, see Favre et al. (2008) and Diedrichs (2009).

The accuracy of the DES in unsteady crosswind is considered to have a significant impact on the calculated loads history. This issue is investigated in this study via grid refinements and using the most suited numerical scheme provided by the code. The mesh practically involves successive refinement zones, similar to those of Maddox (2004) or Tsubokura et al. (2009) to reduce the cell size to the finest area, i.e. the wake, where LES applies. The numerical schemes used for the spatial discretization of the convective fluxes are rather sensitive to the interface between two refinement zones.

The paper is organized as follows: the first part describes the computational domain, numerical setup and models used in this study. The second part contains the results of both headwind and unsteady crosswind simulations.

2 Test Case and Methods

A simple vehicle-like geometry, the Windsor model, is used in this study. The Windsor model is selected since it presents a realistic front for crosswind simulations, preventing also front separation, and sharp trailing edges that fix the separation. The vehicle is not attached to the ground and the streamline overhead strut attaching the car is not included in the CFD model. In addition, the value of the drag coefficient from the experiments of Howell (2005) are corrected by taking into account the blockage of the wind tunnel and are used to compare with the CFD simulations presented in this paper.

Figure 1 shows the computational domain and its dimensions. The streamwise length is 11L, *L* being the vehicle length, where in front and behind the vehicle, the distances are 3.5L and 6.5L, respectively. The height and width are 10h and 20h, with *h* being the height of the vehicle.

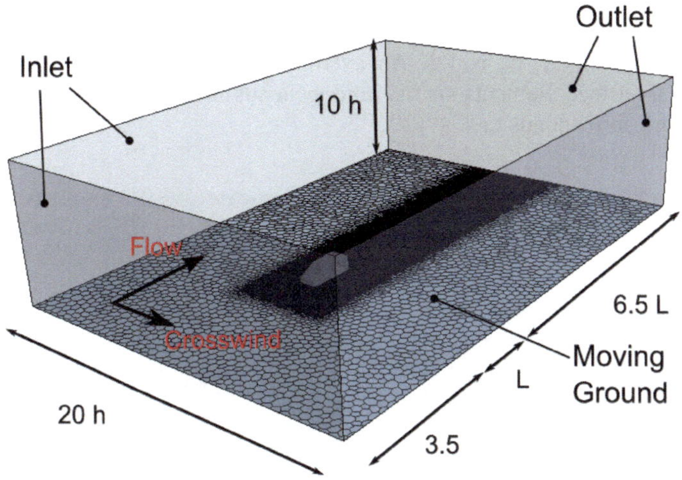

Fig. 1 The computational domain and boundary conditions considered

The computational mesh is composed of unstructured polyhedral cells. Characteristics and benefits of polyhedral cells compared to tetrahedral cells are described in Perić (2004). Polyhedral cells contain more faces than tetrahedral or hexahedral cells; hence, the total number of faces is added to the total number of elements in Table 1 to facilitate comparisons.

The mesh topology has been designed in order to perform DES simulations according to the guidelines in Spalart (2001). A homogeneous core of fine cells is placed around the vehicle. This area extends down to 1L behind the vehicle and 0.3L on the side. In this area, the flow structures are believed to be simulated using LES and demands a fine spatial discretization. Several refinement zones are created to reduce the far field cell size down to the finest core of the mesh. Three cell sizes have been tested for this study detailed in Table 1. The ratio between the finest cells of the fine grid and the coarse grid is $\sqrt{2}$ in the all three space directions.

Table 1 Mesh characteristics

Cases	Total no. cells	Cells faces	Core cell size
Coarse	6 772 861	36 838 917	0.0096 m (3.3% *h*)
Medium	9 286 689	53 716 481	0.0080 m (2.8% *h*)
Fine	11 301 528	67 705 881	0.0068 m (2.3% *h*)

Adjacent to the walls of the vehicle and the ground, 17 and 6 prismatic cell layers are installed, respectively, to properly resolve the boundary layers. The flow solver uses a hybrid method that distinguishes between high- and low-Reynolds wall modelling. This corresponds to a y^+ of 2.5 (low-Reynolds wall modelling) in average around the vehicle and 30 on the ground (high-Reynolds wall modelling). The three meshes have the same wall treatments. The height ratio between two prismatic layers, y_{n+1}/y_n, is 1.2. An average of 9 cells are below y^+=11.5. The polyhedral surface elements size is homogeneous in the directions tangent to the surface and corresponds to $x^+ \approx 120$.

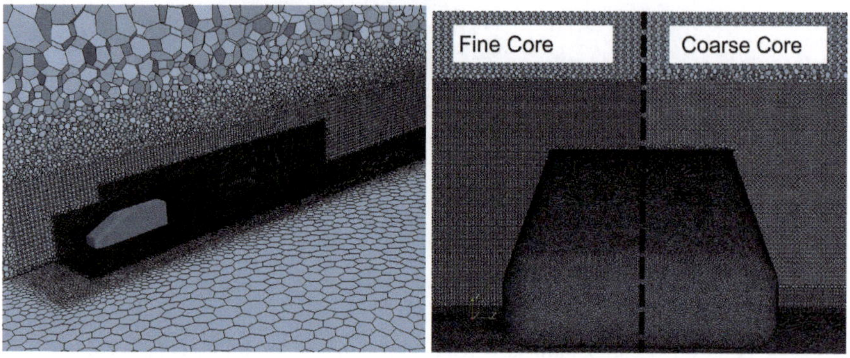

Fig. 2 Polyhedral mesh of the Windsor model. Left: the successive refinement zones. Right: finest core of the fine and coarse meshes

The boundary conditions are illustrated in Fig. 1. The domain is composed of two inlet boundaries, one located in front of the vehicle and one on the side. The latter is used in order to introduce the unsteady wind gust in the computational domain. The inflow velocity for the head wind simulations is set to 27 m/s according to the experiments of Howell (2005). The corresponding Reynolds number, Re_L, is $1.87 \cdot 10^6$ with the viscosity of air and the L as reference length. Using the square-root of the frontal area, it yields a Re_A of $6 \cdot 10^5$. The turbulence intensity is chosen to be 0.3%, corresponding to a low turbulent external aerodynamic simulation. A pressure outlet boundary condition is given at the back of the vehicle and on the remaining side. A moving ground of 27 m/*s* is used. Finally, at the top surface, a symmetrical boundary condition is given.

The wind gust, see Fig. 3, is described by a continuous function representing a jet flow or a wind tunnel exit like in the Cranfield experiment, Cairns (1994) and Chadwick (2001). The cosine functions model the mixing zone at the entrance and the exit of the wind gust and have been chosen according to the measurement presented in Schlichting (1960). The period of the cosine function is 1.5L. The maximum wind gust speed is 9.8 m/s and corresponds approximately at 20° yaw angle. The gust duration corresponds to 5L. The gust is introduced in the computational domain by the frontal inlet and propagates along the side at which an inlet boundary condition is given. Note that the velocity at both front and side inlets, respectively, changes when the gust applies.

The Delayed-DES, (D-DES), presented in Spalart (2006) is used in all DES simulations within this study where C_{DES} was set to 0.65.

First, 2000 iterations of steady state RANS are performed in order to obtain initial solution for the steady inlet DES. In addition, a brief study of the numerical schemes for the spatial discretization in RANS is performed. All simulations are fully converged. The RANS turbulence model k-ω SST from Menter (1992), is used for these purposes.

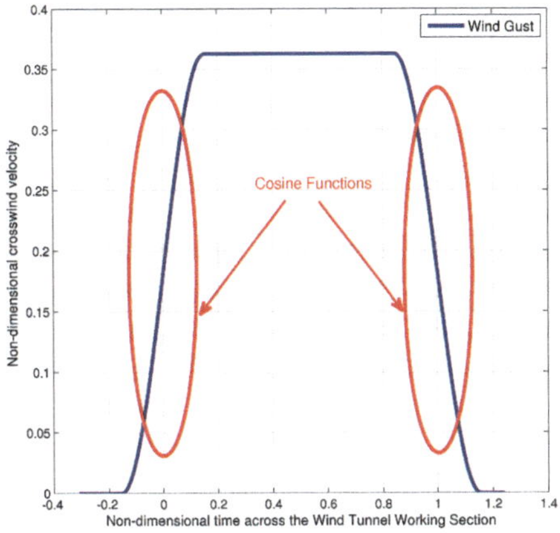

Fig. 3 The wind gust model considered. 0 and 1 in the non-dimensional time refer to the limit of the wind tunnel exhaust or wind gust

The solver used is STAR-CD v4.06 developed by CD-Adapco (www.cd-adapco.com), which solves the Navier-Stokes equations with the Finite Volume (FV) method. The numerical schemes for discretization in space used in this study are first order accurate upwind (UD), second order accurate upwind (LUD) and a blending between a second order accurate central difference and with a first order accurate upwind scheme (B_f·CD, where B_f is the blending factor equal to 0.8 or 0.9 in this study). Also, the Monotone Advection and Reconstruction Scheme (MARS) is employed. The MARS scheme is a two step bounded scheme developed by CD-Adapco where the first step is based on a Total Variation Diminishing (TVD) scheme. All DES simulations were performed using the LUD scheme.

The time step used in the present study is set to comply with the solver requirements for stability and is $0.8 \cdot 10^{-4}$ s. This corresponds to an average CFL number of 0.7.

The DES with steady inlet conditions are simulated for about 45 000 time-steps before a time dependent gust is introduced in the computational domain. This corresponds to 9 times the whole computational domain and 290 time units t* with $t^* = tU_\infty / \sqrt{A}$. The averaging process is started at t* = 30, where the flow is estimated to be sufficiently developed.

The DES were performed on two HPC clusters at KTH: cluster type 1 is equipped of 3.4 GHz 'Nocona' Xeon processors; cluster type 2 is equipped of 1.3

GHz Itanium 2 'McKinley' processors. CPU times for 45000 time steps of each DES are reported in Table 2.

Table 2 Computational time for the DES of headwind

Case	Cluster type	Time in CPU h	No. of CPU cores
Coarse	2	16 000	64
Medium	1	9 000	64
Fine	1	14 500	84

3 Results

Travin et al. (2002) suggested a solution dependant method that allows space discretization of the LES region with an "almost" Central Difference (CD) scheme and the RANS region in an "almost" Upwind scheme. This maintains high accuracy in the wake while keeping enough numerical dissipation in other regions. High order upwind schemes used in the whole domain can also show generous LES contents, see Forsythe et al. (2004) and Maddox et al. (2004) when the grid is sufficiently fine; however for a given mesh, the hybrid scheme method provide better accuracy and finer flow structures, according to Strelets (2001).

A modified version of the hybrid scheme of Travin et al. (2002) is implemented in STAR-CD and was initially tested. However, unphysical flow fluctuations rising at the interface between two successive refinements are polluting the solution, see Fig. 4, and lead to an overestimate of the loads on the vehicle. The results in Fig. 4 suggest that CD is already used upstream the vehicle and gives numerical oscillations as illustrated.

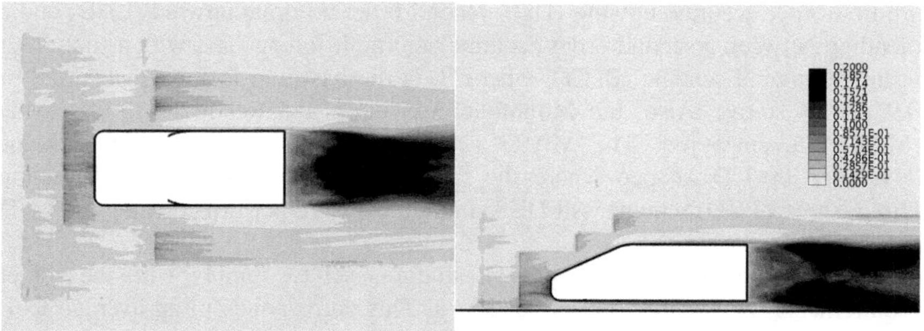

Fig. 4 Unphysical flow fluctuations at the refinement zone interfaces illustrated by the plot of streamwise fluctuations, u/U_∞

Figure 5 shows the calculated drag coefficients obtained with the RANS and DES using the numerical schemes available for the convective fluxes in STAR-CD. As expected, the 2nd order schemes (LUD and MARS) provide a better accuracy of the calculated solution. Refining the grid for a given scheme improve the accuracy and is shown for almost all the schemes available. Note that the 2nd order scheme LUD gives the most consistent results while refining the grid. This may indicate that LUD is the most consistent scheme among the stable schemes tested and is hence used in the subsequent DES. An attempt to estimate the numerical dissipation with a grid refinement is proposed here.

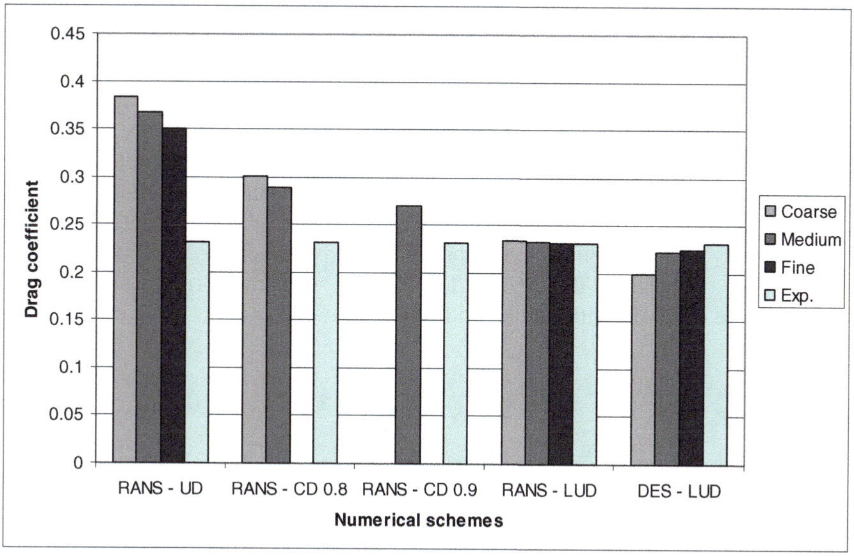

Fig. 5 Drag coefficient for different headwind simulations as a function of the different turbulence models, numerical schemes and meshes

Figure 5 also shows the average drag coefficient calculated with DES for the three grids where the 2nd order LUD scheme has been used for the whole domain. The results suggest that a grid converged drag coefficient is reached on the medium mesh.

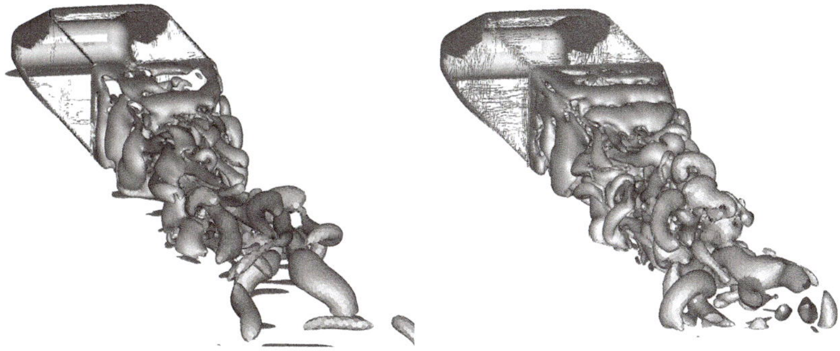

Fig. 6 Comparison of the flow structures using iso-surfaces of the 2nd invariant of the velocity gradient, Q = 10000. Left: Coarse grid. Right: Fine grid

Flow visualization is shown in Fig. 6, using the so-called second invariant of the velocity gradient, Q. The iso-surfaces Q = 10000 is plotted for two mesh resolutions, medium and very fine. It is noticeable that slightly more structures are resolved further downstream with the fine mesh.

In Fig. 8 the average streamwise velocity profile u/U_∞, the resolved Reynolds stress $<uv>/U_\infty^2$ and the resolved turbulent kinetic energy (TKE), defined as $<uu+vv+ww>/(2\cdot U_\infty^2)$, are plotted at different locations along the centerline of the wake. The locations are [0.2, 0.5, 1, 1.5]L downstream the car, respectively indicated by A, B, C and D in Fig. 7. The agreement of the results from the two finest meshes is good. An offset between the values of the coarse and the two other fine grids is present for the average streamwise velocity profile. A significant discrepancy is found at 0.5L downstream of the vehicle in the turbulent kinetic energy. A larger amount of kinetic energy is resolved with the finest grid. At 0.5L downstream the vehicle, the finest grid resolves 25% more TKE than the coarse grid. This location corresponds approximately to the end of the recirculation bubble in the wake. The overall statistics are consistent: finer structures are resolved with a finer grid.

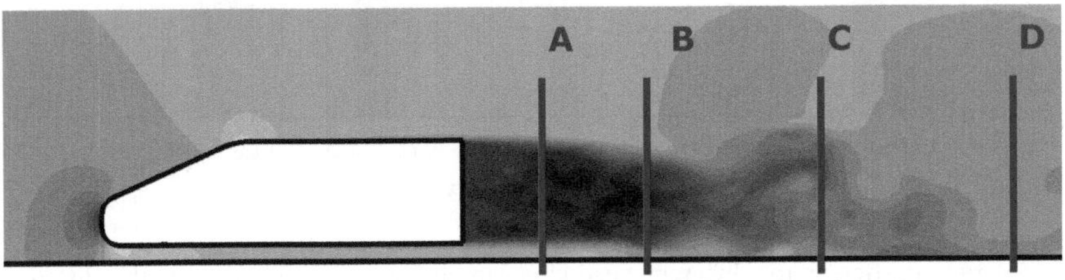

Fig. 7 Section plot of the instantaneous velocity field. The sections A, B, C and D are used in Fig. 8

When the solution is well established, a wind gust is introduced in the domain as shown in Fig. 9. The gust models the situation when a car is propelled through a jet flow of a length 5L that is perpendicular to the vehicle side. The calculated force coefficients when passing the gust are shown in Fig. 9. The forces reveal that the effect of the small scale turbulence is rather limited (fluctuation of 0.05 C_{side}), whereas the large scale effect of the gust is clearly dominating.

The side force history demonstrates a clear mesh dependency where the coarse mesh seems insufficient. Indeed, the side force calculated with the coarse mesh decreases in the gust whereas the finer meshes predict to the load a slight increase. However, Fig. 9 shows that the medium and fine meshes yield consistent time histories. These curves suggest once again that the refinement is good enough to calculate the side force coefficient.

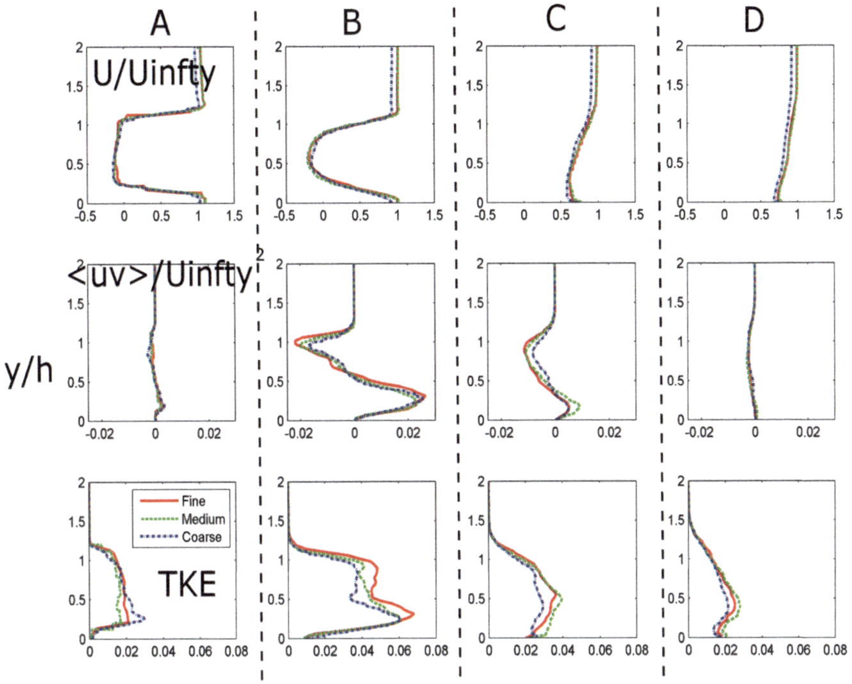

Fig. 8 Resolved flow statistics in the wake for the Fine (red), Medium (green), Coarse (blue) meshes

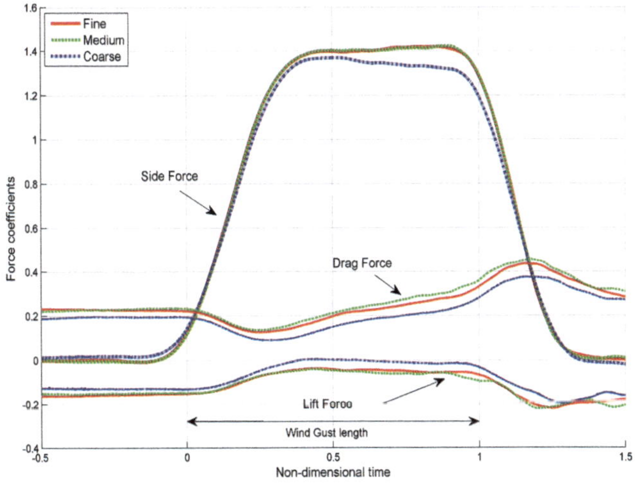

Fig. 9 Time histories of the drag, side and lift forces when passing the gust

4 Conclusions

Results of the hybrid schemes based on a central difference scheme revealed unresolved numerical issues. Therefore a 2nd linear upwind scheme was employed and the accuracy of the solution in the LES region was enhanced by a grid refinement. Further efforts to minimize numerical dissipation could be to change the grid design or to ensure that the upstream region to be discretized with LUD. The conclusions drawn from the headwind simulations regarding the grid refinement study are consistent with those of the unsteady crosswind simulations. The results demonstrate what level of grid refinement is necessary to achieve accurate results of the force coefficients regarding the Windsor model passing a gust.

Acknowledgement. This study is carried out within the crosswind project of the Centre for ECO^2 Vehicle Design. The study has benefited from computing resources at the Center for Parallel Computers, (PDC) at KTH, which are granted by the Swedish National Infrastructure for Computing (SNIC).

References

Cairns, R.S.: Lateral Aerodynamics of Motor Vehicles in Transient Crosswinds. PhD these, Cranfield University, College of Aeronautics (1994)

Chadwick, A., Garry, K., Howell, J.: Transient aerodynamics characteristics of simple vehicle shapes by the measurements or surface pressures. SAE Technical Paper, 2001-01-0876 (2001)

Demuth, R., Buck, P.: Numerical investigations on the unsteady aerodyanmics of road vehicle under gusty weather conditions. In: Sixth World MIRA International Vehicle Aerodynamics Conference, England (2006)

Diedrichs, B.: Unsteady aerodynamic crosswind stability of high-speed train subjected to gusts of various rates. In: Proceeding of Euromech Colloquium 509, Germany (2009)

Docton, M., Dominy, R.: The simulation of Transient Cross Winds on Passenger Vehicles. In: MIRA International Vehicle Aerodynamics Conference, England (1996)

Favre, T., Efraimsson, G., Diedrichs, B.: Numerical investigation of unsteady crosswind vehicle aerodynamics using time-dependent inflow conditions. In: 7th MIRA International Vehicle Aerodynamic Conference, Coventry (2008)

Forsythe, J.R., Squires, K.D., Wurtzler, K.E., Spalart, P.R.: Detached-Eddy Simulation of the F-15E at high alpha. Journal of Aircraft 41(2), 193–200 (2004)

Howell, J.: Vortex drag for a simple bluff body at incidence in ground proximity. SAE Technical Paper, 2005-01-0869 (2005)

Hucho, W.-H.: Aerodynamics of Road Vehicles, 4th edn. SAE International (1998)

Maddox, S., et al.: Detached-eddy simulation of the ground transportation system. In: Mc Callen, R., Browand, F., Ross, J. (eds.) The Aerodynamics of Heavy Vehicles: trucks, busses and trains. Springer, Heidelberg (2004)

Menter, F.R.: Improved two-equation k-ω turbulence models for aerodynamic flows. NASA Technical Memorandum 103975, October 1992 (1992)

Peric, M.: Flow simulation using control volumes of arbitrary polyhedral shapes. Ercoftac Bulletin 62, 25–29 (2004)

Schlichting, H.: Boundary Layer Theory. McGraw-Hill series in mechanical engineering (1960)

Spalart, P.R.: Comments on the feasability of LES for wings, and on appoach on a hybrid RANS/LES approach. In: Proceeding on 1st AFOSR International Conference on DNS and LES (1997)

Spalart, P.R.: Young-person's guide to detached-eddy simulation grids. NASA Technical Report, NASA/ CR-2001-211032 (2001)

Spalart, P.R., Deck, S., Shur, M.L., Squires, K.D., Strelets, M.K., Travin, A.: A new version of detached-eddy simulation, resistant to ambiguous grid densities. Theoritical Computational Fluid Dynamic 20, 181–195 (2006)

Strelets, M.: Detached eddy simulations of massively separated flows. In: AIAA 39th Aerospace Sciences Meeting, Reno, NV, AIAA Paper 2001-0879 (2001)

Travin, A., Shur, M., Strelets, M., Spalart, P.: Physical and numerical upgrades in the detached-eddy simulation of complex flows. In: Friedrich, R., Rodi, W. (eds.) Advances in LES of complex flows, pp. 239–254 (2002)

Tsubokura, M., Kobayashi, T., Nakashima, T., Nouzawa, T., Nakamura, T., Zhang, H., Onishi, K., Oshima, N.: Computational visualization of unsteady flow around vehicles using high performance computing. Computers and Fluids 38(5), 981–990 (2009)

Destabilizing Free Shear Layers in X-LES Using a Stochastic Subgrid-Scale Model

J.C. Kok and H. van der Ven

Abstract. In this paper, a stochastic term is introduced in the subgrid-scale model of the X-LES method in order to improve the computation of turbulent free shear layers. These shear layers may develop too slowly in X-LES and similar DES-type methods, even when fine grids and high-order numerical methods are used. For a plane free shear layer, the stochastic subgrid-scale model induces rapid development of fully 3D turbulence in contrast to the standard method for which the flow remains purely 2D. A comparison is made with experimental results and with results of a zonal RANS–LES method with LES content added at the RANS–LES interface.

1 Introduction

An important issue in DES-type methods such as X-LES is the development of free shear layers starting from turbulent boundary layers. Even though free shear layers are intrinsically unstable, computational results may show stable shear layers of lengths that seem unphysical (see, e.g., Spalart (2009)). In our experience, this behaviour is not necessarily caused by excessive levels of dissipation as it is also found when the shear layer is fully captured in LES mode using fine grids and high-order numerical schemes (and even if no subgrid-scale model is used). Instead, it appears that the unstable modes of the shear layer are simply not triggered because the computations are void of any disturbances, either physical or numerical. In this paper, the subgrid-scale model employed in X-LES is extended with a stochastic term, destabilizing the shear layer and leading to rapid development of fully 3D turbulence.

In DES-type methods, the turbulent boundary layers upstream of a free shear layer are typically in RANS mode and therefore do not contain any resolved turbulence. As a consequence, the development of turbulence in the shear layer will start from the intrinsic instability of the shear layer and will not be forced by turbulence coming from the boundary layers. This resembles the situation when the upstream boundary layers are laminar. In that case, however, the distance over which the initial, essentially 2D, Kelvin–Helmholtz instabilities develop into fully 3D turbulence may be very long (of the order of hundreds of initial displacement thicknesses or more, Huang & Ho (1990)).

One possible approach to resolve this issue is by adding LES content at the RANS–LES interface, forcing the development of 3D turbulence in the shear layer.

Johan C. Kok · Harmen van der Ven
National Aerospace Laboratory NLR, Anthony Fokkerweg 2, 1059 CM Amsterdam, The Netherlands
jkok@nlr.nl

S.-H. Peng et al. (Eds.): Progress in Hybrid RANS-LES Modelling, NNFM 111, pp. 179–189.
springerlink.com

This is feasible in a zonal RANS–LES approach. In X-LES and similar DES-type methods, however, the RANS–LES interface is dynamic, which complicates adding LES content. Therefore, an alternative method to speed-up the shear-layer development is considered. This method consists of adding a stochastic term to the subgrid-scale (SGS) model. Stochastic SGS models have been used in the literature to model backscatter, seen as a random forcing of the resolved scales through non-linear interactions with the subgrid scales (e.g., Leith (1990), Schumann (1995)). Here, the main goal is not to model backscatter, but to see if a stochastic SGS model can act as a disturbance to trigger 3D instabilities in the shear layer.

2 Modelling

2.1 The X-LES Method

The X-LES formulation (Kok et al., 2004) is a particular DES method (Spalart et al., 1997) that consists of a composition of a RANS k–ω turbulence model and a k-equation SGS model. Both models use the Boussinesq hypothesis to model the Reynolds or subgrid-scale stress tensor, which depends on the eddy-viscosity coefficient ν_t. Furthermore, both models are based on the equation for the modelled turbulent kinetic energy k, which depends on its dissipation rate ε. Both the eddy viscosity and the dissipation rate are modelled using the turbulent kinetic energy as velocity scale together with a length scale l_t,

$$\nu_t = l_t \sqrt{k} \qquad \text{and} \qquad \varepsilon = \beta_k \frac{k^{3/2}}{l_t}, \tag{1}$$

where l_t is defined as a combination of the RANS length scale $l = \sqrt{k}/\omega$ and the SGS filter width Δ,

$$l_t = \min\{ l, C_1\Delta \}, \tag{2}$$

with C_1 = 0.05. The RANS k–ω model is completed by an equation for the specific dissipation rate ω. The X-LES method will be in LES mode if $C_1\Delta < l$. Note that in that case the SGS model is completely independent of ω.

2.2 Stochastic Subgrid-Scale Model

In the standard X-LES method (and in most other DES-type methods), the SGS stress tensor is modelled using the Boussinesq hypothesis, which essentially means that the effect of the subgrid-scale turbulent fluctuations on the resolved scales is modelled as a diffusion process. In the representation of this diffusion process, however, the 'randomness' of the subgrid-scale fluctuations is lost. An alternative representation of diffusion, based on an analogy with the random-walk process, is given by the following stochastic PDE:

$$\varphi(x, t + \delta t) = \varphi(x, t) + \delta t \, \nu \, \xi_i \xi_j \frac{\partial^2 \varphi}{\partial x_i \partial x_j}, \tag{3}$$

where the vector components ξ_i (the random-walk direction) are independent stochastic variables with standard normal distribution N(0,1), so that $E(\xi_i) = 0$ and $E(\xi_i\xi_j) = \delta_{ij}$. Note that the expectation of this equation is the standard diffusion equation.

To incorporate this stochastic diffusion model in the SGS model, the cross terms are dropped and a single stochastic variable $\xi = N(0,1)$ is used for all three vector components. The stochastic diffusion model then reduces to the standard diffusion equation with a stochastic diffusion coefficient $\nu^* = \xi^2\nu$. In the same way, the stochastic variable can be included in the expression for the SGS eddy viscosity:

$$\nu_t = \xi^2 C_1 \Delta \sqrt{k} \,. \tag{4}$$

At each time step, a new value of ξ is drawn for every grid cell.

The stochastic SGS eddy viscosity should only be used when the X-LES method is in LES mode, not when it is in RANS mode. One option is to multiply $C_1\Delta$ with ξ^2 in the expression for l_t (equation (2)). This, however, would make the switching between RANS and LES modes as well as the dissipation rate ε dependent on the stochastic variable. Instead, the eddy viscosity is set explicitly as

$$\nu_t = \begin{cases} k/\omega & \text{, if } l \le C_1\Delta \,, \\ \xi^2 C_1 \Delta \sqrt{k} & \text{, if } l > C_1\Delta \,, \end{cases} \tag{5}$$

while the expressions for l_t and ε are left unaltered.

2.3 Zonal RANS–LES Method

As the stochastic SGS model is effective throughout the complete LES zone, it may have other effects besides triggering the shear-layer instabilities. Therefore, as a reference, also a different approach to destabilizing the shear layer is considered. A zonal RANS–LES method is used, setting the boundary layers explicitly to RANS and the shear layers explicitly to LES. The same turbulence models as in the X-LES method are used, i.e., the RANS k–ω turbulence model and the k-equation SGS model. At the RANS–LES interface, LES content is added to the solution in the LES zone. This added LES content may destabilize the shear layer, while the original SGS model is maintained in the complete LES zone. Developing methods for introducing LES content is currently an active research area, for example, using precursor data (Schlüter et al. (2004)) or synthetic turbulence (Batten et al. (2004), Jarrin et al. (2006)). Here, a simple model for the LES content is chosen, based on white noise and introducing isotropic resolved stresses. Although such an approach requires a significant distance to develop realistic turbulence, it is considered sufficient for our present purpose (i.e., reference method for destabilizing the shear layer).

To introduce the LES content, the following flux is added to the LES momentum equation at the interface:

$$\rho(\overline{u}_i u'_j + u'_i \overline{u}_j + u'_i u'_j + \tau^r_{ij}) n_j \,, \tag{6}$$

with the ρ density, $\overline{u}_i$ the time-averaged velocity vector, u'_i the added resolved turbulent velocity fluctuations, τ^r_{ij} the added resolved turbulence, and n_i the unit vector normal to the interface. This flux was derived by subtracting the RANS equations from the LES equations. For the time-averaged velocity, the instantaneous RANS velocity is used. The turbulent velocity fluctuations are computed as

$$u_i' = \sqrt{\tfrac{2}{3}k^r}\,\xi_i\,, \tag{7}$$

where k^r is the desired resolved turbulent kinetic energy. Again, the vector components ξ_i are independent stochastic variables with standard normal distribution N(0,1). This approach introduces isotropic resolved stresses, given by

$$\tau_{ij}^r = -\overline{u_i'u_j'} = -\tfrac{2}{3}k^r\delta_{ij}\,. \tag{8}$$

Alternatively, a non-isotropic stress tensor can be introduced by including the Choleski decomposition of the resolved stress tensor in the model for u_i' (see Batten et al. (2004)). Note that the expectation or time average of the additional momentum flux equals zero, so that no net momentum source is created.

The desired resolved turbulent kinetic energy (k^r) is equal to the modelled turbulent kinetic energy in the RANS zone (k^R) minus the time-averaged SGS turbulent kinetic energy (k^s),

$$k^r = k^R - k^s\,, \tag{9}$$

where for k^R the solution of the k-equation at the RANS side of the interface is used. The value of k^s is determined by comparing the RANS and SGS models of the dissipation rate,

$$\varepsilon^R = \frac{\beta_k\left(k^R\right)^{3/2}}{l} \quad \text{and} \quad \varepsilon^s = \frac{\beta_k\left(k^s\right)^{3/2}}{C_1\Delta}\,. \tag{10}$$

Because the dissipation of resolved stresses can be neglected (viscous dissipation taking place at the Kolmogorov scales), the two modelled dissipation rates should be equal. This gives the following relation between k^s and k^R:

$$k^s = \alpha k^R \quad \text{with} \quad \alpha = \left(\frac{C_1\Delta}{l}\right)^{2/3}, \tag{11}$$

so that finally

$$k^r = (1-\alpha)k^R\,. \tag{12}$$

In regions where $\alpha > 1$, the value of k^r is set to zero.

3 Results

3.1 Plane Free Shear Layer

As a basic test case for investigating the stability issue, the experiment of Delville (1995) for a plane free shear layer is considered. The free shear layer starts from the trailing edge of a flat plate with free-stream velocities u_1 = 41.54 m/s and u_2 = 22.40 m/s at the different sides of the flat plate. At the trailing edge, the turbulent boundary layers are fully developed with the momentum and displacement thicknesses equal to θ_1 = 1.0 mm and δ_1* = 1.4 mm at the high-speed side and θ_2 = 0.73 mm and δ_2* = 1.0 mm at the low-speed side. The θ-based Reynolds number at the high-speed side is $Re_{\theta 1}$ = 2900 at the trailing edge. The shear layer develops in a 0.3 m × 0.3 m square test section of length 1.2 m. A self-similar flow with fully developed turbulence is reached well within the test section.

A computational domain is used with a width of 0.15 m (z-direction) and a height of 0.3 m (y-direction). To capture the correct boundary-layer profiles at the trailing edge, the flat plate has a length of 0.5 m on the high-speed side and 0.3 m on the low-speed side and fine-tuning has been done by varying the transition locations. The grid is stretched in the y-direction to capture the boundary layers with approximately y^+ = 1 for the first grid cell. A computational 'test section' is defined with a length of 1 m after the trailing edge and with a uniform grid in the x- and z-directions, followed by a buffer zone of 1 m length with a stretched grid in x-direction.

Two grid levels are used, G1 and G2, with 1.29 and 10.3 million grid cells. Grid G2 has 96 cells in z-direction and 640 cells in x-direction in the computational test section, giving a mesh size h = 1.5625 mm in both directions (similar to Tenaud (2005)). The filter width Δ is set equal to this mesh size. For grid G2, a dimensionless time step of $\delta t^+ = u_1 \delta t / L = 0.0016$ is used (with the length of the test section, L = 1m, as reference length), giving a CFL number $u_1 \delta t / h = 1$. For grid G1 the mesh sizes, the time step, and the filter width are all doubled compared to G2.

Computations are performed with a fourth-order finite-volume method that has low numerical dissipation and dispersion and that preserves the skew-symmetry of convection (Kok (2009)). This method was designed for the LES regions of hybrid RANS–LES computations and was shown to significantly improve the grid convergence with a fixed filter width (Kok et al. (2008)).

First, standard X-LES computations are performed on grid G1. The computations are started from the solution of a steady X-LES computation, which was not converged and therefore contains 3D perturbations of the shear layer. Initially, instabilities in the shear layer are visible shortly after the trailing edge (Fig. 1a), but as the computation continues, the shear layer stabilizes with the first visible instabilities shifting down to a location of approximately 0.4 m after the trailing edge (Fig. 1b). Moreover, even though the solution originally contains 3D disturbances, it becomes quickly two-dimensional (Fig. 2a). The solution essentially displays the behaviour of an initially laminar shear layer: growth of a 2D Kelvin–Helmholtz instability followed by vortex pairing. For such a shear layer, fully developed 3D turbulence is only reached after a distance of at least 10 λ_0 / R with λ_0 the initial instability wavelength and $R = (u_1 - u_2) / (u_1 + u_2)$ (Huang & Ho (1990)). In this case, $\lambda_0 \approx 60$ mm and R = 0.3, giving a distance of at least 2 m.

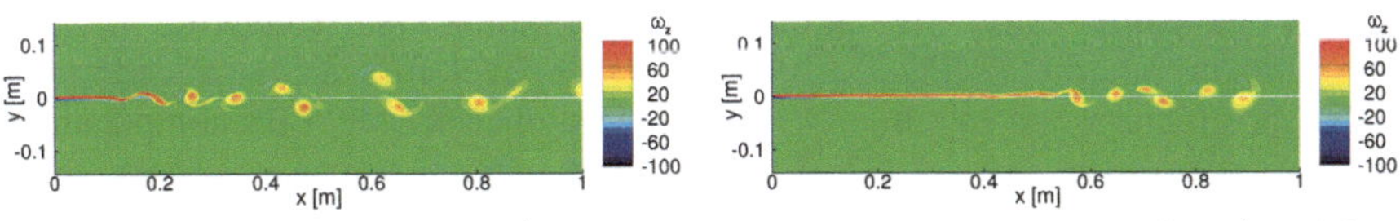

a) After 500 time steps (time t^+ = 1.6)   b) After 3500 time steps (time t^+ = 11.2)

Fig. 1 Instantaneous contours of vorticity (y-component) for standard X-LES computation of plane shear layer on grid G1 (1.29 million cells)

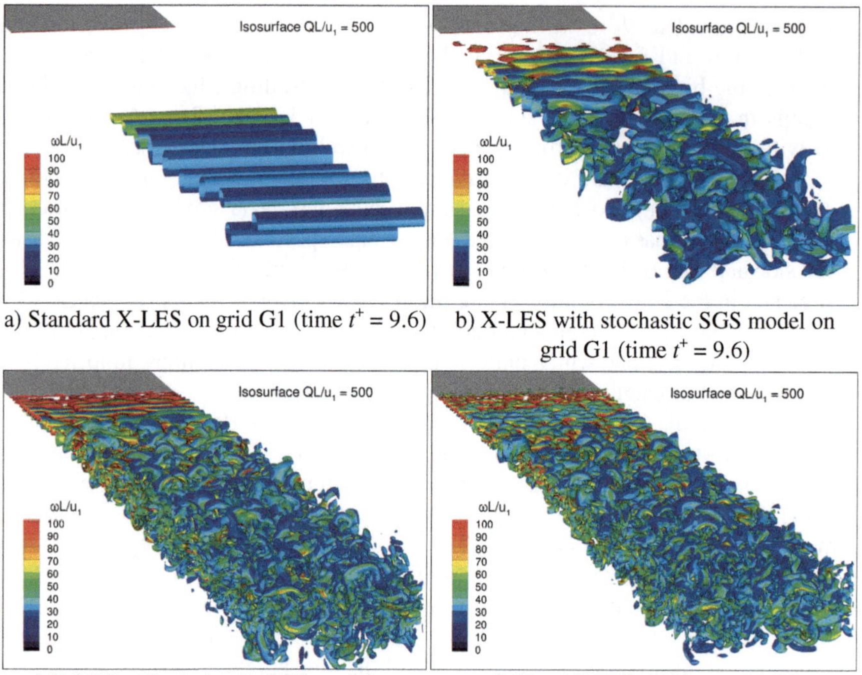

a) Standard X-LES on grid G1 (time $t^+ = 9.6$)

b) X-LES with stochastic SGS model on grid G1 (time $t^+ = 9.6$)

c) X-LES with stochastic SGS model on grid G2 (time $t^+ = 4.8$)

d) Zonal RANS–LES on grid G2 (time $t^+ = 1.6$)

Fig. 2 Instantaneous isosurfaces of $Q = \omega^2 - S^2$, coloured with vorticity magnitude ω, for X-LES and zonal RANS–LES computations of plane shear layer

A possible cause for slow shear-layer development is an excessive level of dissipation. The presented computation, however, uses a fourth-order finite-volume method with low numerical dissipation on a grid with high resolution in all directions. It has been verified that the shear layer is fully captured in LES mode. Switching off the SGS model completely in the shear layer or continuing the computation on the finer grid G2 makes no essential difference. Therefore, it is concluded that, in this case, the slow development is not caused by high dissipation. Instead, the lack of any kind of disturbance in the computation, either physical or numerical, is perceived as a possible cause.

Using the stochastic SGS model in X-LES, results in a dramatic improvement for the plane free shear layer. The initial, spanwise vortices appear much closer to the trailing edge and already show 3D disturbances (Fig. 2b). The flow then rapidly develops fully 3D turbulence. Continuing the stochastic X-LES computations on grid G2 (Fig. 2c), the shear layer develops even quicker and finer turbulent structures are captured. As a reference, zonal RANS–LES computations are also performed on grid G2 (Fig. 2d). The initial, spanwise vortices show stronger 3D disturbance than stochastic X-LES, but the subsequent development of the shear layer appears similar, showing similar fine-scale structures.

The mean velocity and resolved stresses are computed by averaging in time (over at least 2000 times steps) and in the spanwise direction. Self-similar solutions are obtained for stochastic X-LES and for zonal RANS–LES within the test section. Fig. 3 compares these self-similar solutions at three stations towards the end of the test section with the experimental results. The similarity coordinate is given by $\eta = (y - y_{1/2}) / \theta$, with $y_{1/2}$ the location where $u^+ = (u - u_2) / (u_1 - u_2) = ½$ and θ the shear-layer momentum thickness. The profiles of the mean velocity and the resolved shear stresses of the computations are close to each other and to the experiment. For the resolved normal stresses, the stochastic X-LES and zonal RANS–LES results on grid G2 are close to each other and show only a small difference with the experiment, whereas the stochastic X-LES results on grid G1 deviate more strongly.

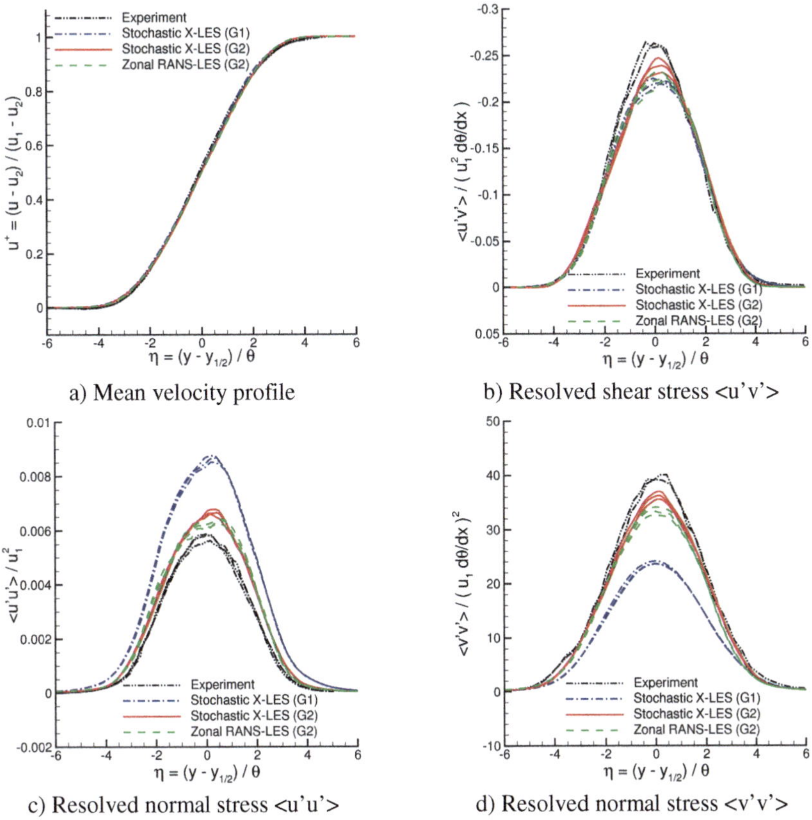

Fig. 3 Self-similar solution for X-LES and zonal RANS–LES computations of plane shear layer compared to experiment of Delville (1995)

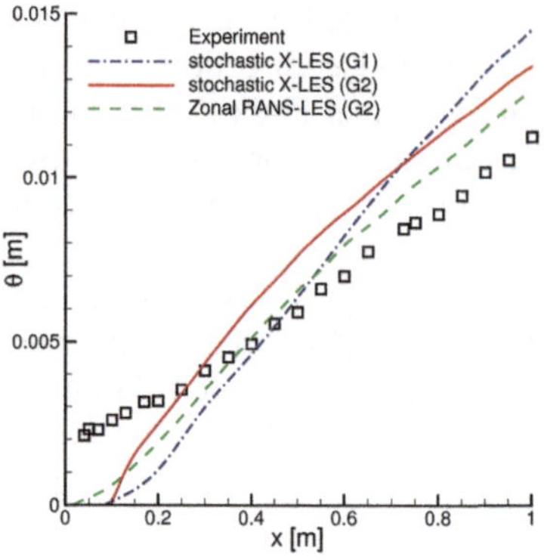

Fig. 4 Momentum thickness for X-LES and zonal RANS–LES computations of plane shear layer compared to experiment of Delville (1995)

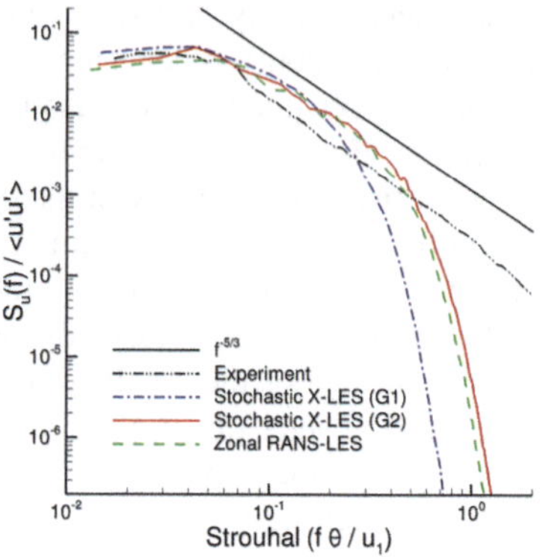

Fig. 5 Energy spectrum for X-LES and zonal RANS–LES computations of plane shear layer compared to experiment of Delville (1995)

The main difference between the computations and the experiment is seen in the growth of the shear layer in terms of the momentum thickness (Fig. 4). The experiment shows a linear growth over a large part of the test section, which is consistent with a self-similar solution. The computations on grid G2 initially show a larger growth rate ($d\theta/dx$) but tend towards the same rate as the experiment near the end of the test section. Again, the strongest difference with the experiment is found for the stochastic X-LES results on grid G1 with a significantly larger spreading rate over the entire test section.

Note that the present methods do not aim to resolve a significant part of the turbulence coming from the boundary layers. Therefore, the initial development of the shear layer differs from the experiment, which results in different initial momentum thicknesses. The best one can expect is a rapid development towards the same similarity solution with a similar spreading rate.

Finally, the temporal energy spectrum of the x-component of velocity is shown in Fig. 5 (at location $x = 0.8$ m, $y = 0$ m). To compute the spectrum, Hanning windows are used of 500 time steps (0.02 s) with an overlap of 250 time steps and the spectrum is averaged in the spanwise direction. On grid G2, the spectrum follows the −5/3 law up to a frequency that is about a factor two higher than on grid G1, as expected. More importantly, the spectra from the stochastic X-LES and the zonal RANS–LES computations are close to each other, which indicates that the stochastic term has no detrimental effect on the energy cascade.

3.2 Supersonic Base Flow

Finally, as a less academic application, the supersonic base flow of a cylindrical afterbody is considered (experiment of Herrin & Dutton (1994), free-stream Mach number $M_\infty = 2.46$, Reynolds number based on diameter $Re_D = 3.3{\cdot}10^6$). A grid of 4.4 million cells, with a typical mesh size $h = 0.015\ D$, and a time step of

$\delta t^+ = u_\infty \delta t / D = 0.0033$ are used. In this case, the shear-layer stability issue is less severe, because unstable modes may be triggered by the resolved turbulence that is convected in the wake of the base towards the shear layer onset.

In a standard X-LES computation, the shear layer develops too slowly, leading to an overprediction of the size of the wake. In fact, as the computation continues, the level of resolved turbulent kinetic energy in the shear layer keeps dropping, its peak keeps shifting down stream, and the size of the wake keeps growing. Even the final solution (after 19,500 time steps, i.e.,. a time interval $t^+ = 64.4$) seems not to have settled yet (and therefore has been averaged over a relatively small time interval of 6,000 time steps). With the stochastic X-LES method, the computation does settle well within 17,000 time steps (and has been averaged over the last 10,000 time steps). As a result, the stochastic method predicts higher levels of resolved turbulent kinetic energy (Fig. 6a) that compare better to the experiment (Fig. 6b) and a smaller size of the wake, also closer to the experiment (Fig. 7).

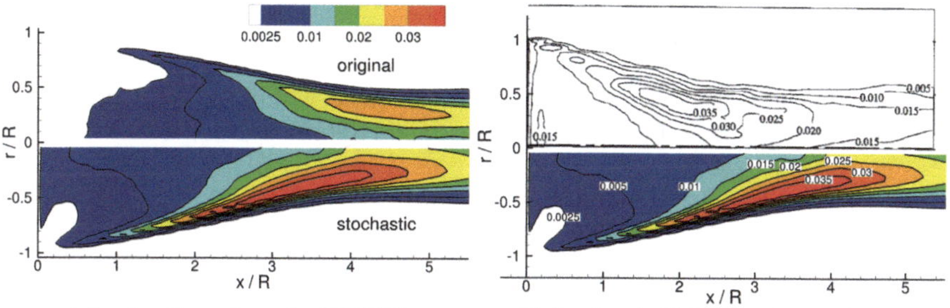

a) Stochastic and standard X-LES b) Stochastic X-LES and experiment

Fig. 6 Resolved turbulent kinetic energy for X-LES computation of supersonic base flow of a cylindrical afterbody compared to experiment of Herrin & Dutton (1994)

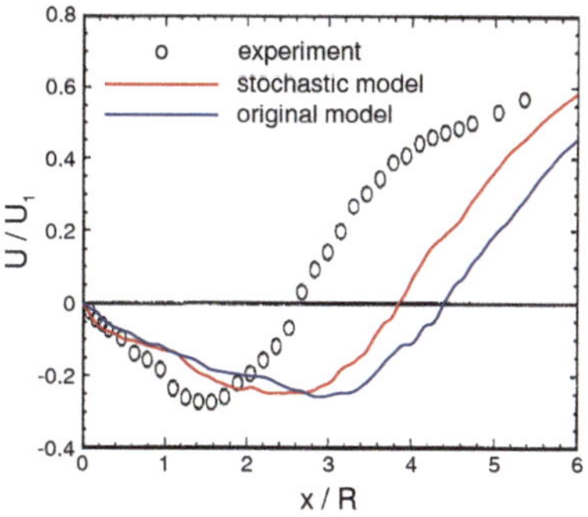

Fig. 7 Mean axial velocity component along wake axis for X-LES computation of supersonic base flow of a cylindrical afterbody compared to experiment of Herrin & Dutton (1994)

Despite these improvements, a clear overprediction of the size of the wake remains, compared to the experiment. The initial development of the shear layer still appears to be too slow. Due to the chosen grid topology (O-type), the grid in the initial shear layer is relatively coarse compared to other DES computations (Garbaruk et al. (2009)) that give results closer to the experiment. Locally refining the grid may possibly improve the results.

4 Conclusion

Turbulent free shear layers as computed in DES-type methods such as X-LES may be stable over a length that is inconsistent with experiments. Here, it has been shown that this is not necessarily due to excessive levels of dissipation that stabilize the shear layer, but that it can also occur for computations on fine grids using fourth-order numerical methods. Apparently, the computations are void of any disturbances that can induce the rapid growth of shear-layer instabilities. Instead, the shear layer develops only slowly, similar to an initially laminar shear layer. With the introduction of a stochastic term in the subgrid-scale model, the shear layers in X-LES computations are destabilized and realistic, full 3D turbulence develops more rapidly.

Acknowledgement. This work has been performed within the NLR programme 'Kennis als Vermogen, Platformtechnology and Flight Physics'.

References

Batten, P., Goldberg, U., Chakravarthy, S.: Interfacing statistical turbulence closures with large-eddy simulation. AIAA Journal 42(3), 485–492 (2004)

Delville, J.: La décomposition orthogonal aux valeurs propres et l'analyse de l'organisation tridimensionnelle de écoulements turbulents cisaillés libres. PhD thesis, Université de Poitiers (1995)

Garbaruk, A., Shur, M., Strelets, M., Travin, A.: Supersonic base flow. In: Haase, W., Braza, M., Revell, A. (eds.) DESider – A European Effort on Hybrid RANS-LES Modelling, ch. IV-10, Notes on Numerical Fluid Mechanics and Multidisciplinary Design, vol. 103, pp. 197–206. Springer, Heidelberg (2009)

Herrin, J.L., Dutton, J.C.: Supersonic base flow experiments in the near wake of a cylindrical afterbody. AIAA Journal 32(1), 77–83 (1994)

Huang, L.-S., Ho, C.-M.: Small-scale transition in a plane mixing layer. Journal of Fluid Mechanics 210, 475–500 (1990)

Jarrin, N., Uribe, J.-C., Prosser, R., Laurence, D.: Synthetic inflow boundary conditions for wall bounded flows. In: Peng, S.H., Haase, W. (eds.) Advances in Hybrid RANS–LES Modelling. Notes on Numerical Fluid Mechanics and Multidisciplinary Design, vol. 97, pp. 77–86. Springer, Heidelberg (2008)

Kok, J.C.: A high-order low-dispersion symmetry-preserving finite-volume method for compressible flow on curvilinear grids. Journal of Computational Physics 228, 6811–6832 (2009) (NLR-TP-2008-775)

Kok, J.C., Dol, H.S., Oskam, B., van der Ven, H.: Extra-large eddy simulation of massively separated flows. In: 42nd AIAA Aerospace Sciences Meeting, Reno, NV, January 5-8, AIAA paper 2004-264 (2004)

Kok, J.C., Soemarwoto, B.I., van der Ven, H.: X-LES simulations using a high-order finite-volume scheme. In: Peng, S.H., Haase, W. (eds.) Advances in Hybrid RANS–LES Modelling. Notes on Numerical Fluid Mechanics and Multidisciplinary Design, vol. 97, pp. 87–96. Springer, Heidelberg (2008)

Leith, C.E.: Stochastic backscatter in a subgrid-scale model: Plane shear mixing layer. Physics of Fluids A 2(3), 297–299 (1990)

Schlüter, J.U., Pitsch, H., Moin, P.: Large eddy simulation inflow conditions for coupling with Reynolds-averaged flow solvers. AIAA Journal 42(3), 478–484 (2004)

Schumann, U.: Stochastic backscatter of turbulence energy and scalar variance by random subgrid-scale fluxes. Proc. R. Soc. Lond. A 451, 293–318 (1995)

Spalart, P.R.: Detached-eddy simulation. Annual Review of Fluid Mechanics 41, 181–202 (2009)

Spalart, P.R., Jou, W.-H., Strelets, M., Allmaras, S.R.: Comments on the feasibility of LES for wings, and on a hybrid RANS/LES approach. In: Liu, C., Liu, Z. (eds.) Proc. 1st AFOSR Int. Conf. on DNS/LES, Advances in DNS/LES, Ruston (LA), USA. Greyden Press (1997)

Tenaud, C., Pellerin, S., Dulieu, A., Ta Phuoc, L.: Large eddy simulations of a spatially developing incompressible 3D mixing layer using the v–ω formulation. Computers & Fluids 34, 67–96 (2005)

Prediction of Transonic Buffet around a Wing with Flap

G. Barbut, M. Braza, Y. Hoarau, G. Barakos, A. Sévrain, and J.B. Vos

Abstract. The present study presents numerical simulations and turbulence modelling of the flow around a NACA0012 airfoil including a deflected aileron. The results are compared with experiments that have been performed in the N-3 wind tunnel of the Institute of Aviation (IoA), Warsaw. The experiment focused on unsteady flow characteristics and buffet phenomena arising as the result of the transonic shock wave / boundary layer interaction (SWBLI). The transonic buffet is a natural and self-sustaining oscillation of the shock wave and separated flow region, caused by pressure fluctuation. The first objective is to capture the transonic buffet unsteadiness by means of URANS and DES turbulence modelling approaches. Secondly, the periodic flap oscillation has been used to modify the oscillation amplitudes towards an outlook of attenuation of the transonic buffet.

1 Introduction

Within the UFAST (Unsteady Effects in Shock Wave induced separation) program, it has been shown through its main test-cases, that URANS approaches were not successful to capture unsteady shock wave boundary layer interaction, SWBLI for internal geometries, whereas they were more successful for external flow configurations. These facts are reported in the UFAST books, published also by Springer in 2010. The DES approaches were more successful than URANS for internal flow configurations, whereas they are comparable to URANS for external flow configurations (example: the INCAS-biconvex test-case), where also transonic buffet occurs. In the present study, the transonic buffet over the NACA0012 wing with deflected aileron is investigated. First, the natural unsteadiness is examined as a self-sustaining oscillation of the shock wave and of the separated flow region,

G. Barbut · M. Braza · A. Sévrain
Institut de Mécanique des Fluides de Toulouse, France
barbut@imft.fr

Y. Hoarau
Institut de Mécanique des Fluides et de Solides de Strasbourg, France

G. Barakos
University of Liverpool, UK

J.B. Vos
CFS Engineering, Switzerland

S.-H. Peng et al. (Eds.): Progress in Hybrid RANS-LES Modelling, NNFM 111, pp. 191–204.
springerlink.com

caused by pressure fluctuation (Seegmiller *et al*, 1978). A first objective of the present study is to capture the transonic buffet unsteadiness by means of URANS and DES turbulence modelling approaches. A second objective is to perform a sensitivity study concerning the effect of the periodic flap oscillation, that is used to modify the oscillation amplitudes. This goal is related to current and prospective studies for attenuation of the transonic buffet by deflected ailerons. In the present study, the URANS-Spalart-Allmaras (SA), the URANS-k-ω, the DES-SA and the DDES – SA turbulence modelling approaches have been used.

2 Test-Case Description

2.1 Experimental Set-Up

The NACA0012 airfoil model equipped with 22.6% of airfoil chord flap and modified gap as shown in Fig. 1 was tested. The model's chord is equal to 180 mm and flap length is 40.69 mm. The axis of rotation of the main part of the model is at 35% of the chord. There is a smaller, 0.3 mm gap between airfoil main part and flap leading edge. The flap leading edge radius is 4.69 mm. The axis of the aileron/ flap rotation is at 80% of the airfoil model chord. The span of the model is 596 mm.

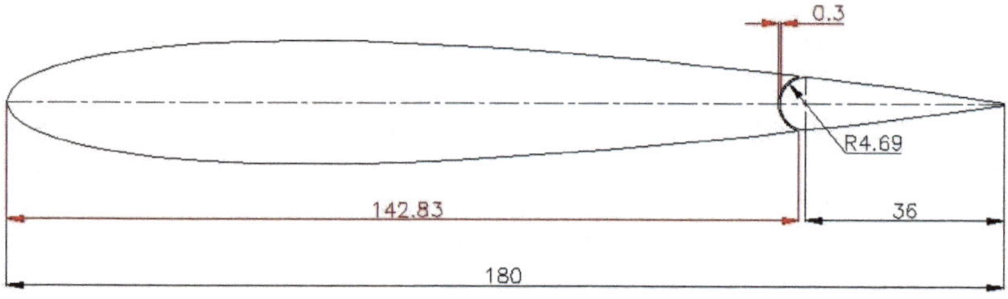

Fig. 1 The geometry of the NACA 0012 model equipped with the flap

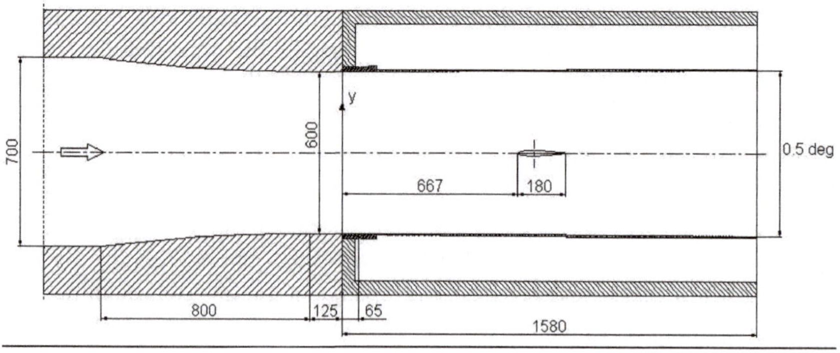

Fig. 2 Geometry of the test section, IoA N-3 wind tunnel with the NACA 0012 flapped model

The computations have been carried by using the real wind-tunnel 3D geometry (fig.2). The angle of attack is 5°. The Mach number is M = 0.73 and Reynolds number Re = $2.69 \cdot 10^6$. The amplitude of the flap deflection angle up to $\Delta\delta_\phi = \pm 2°$. The oscillation frequency of the flap, f_f = 100 Hz and 30 Hz (close to the buffet frequency and one third of the buffet frequency, respectively).

2.2 Advanced Signal Processing on Experimental Signal

The experimental signals display a strong chaotic character, interspersed by peaks of more organised events in respect to the buffet phenomenon (figure 3). This complex behaviour was enhanced by the fact that the upper and lower walls are not porous but solid, hence, fixed walls. IMFT has applied advanced signal processing techniques, as autoregressive modelling (AR), Burg algorithm as described in Marple (1987) and Morlet's wavelets analysis (Grossmann & Morlet-1984), to extract the organised part and to assess the contribution of the organised and random parts in the physical process.

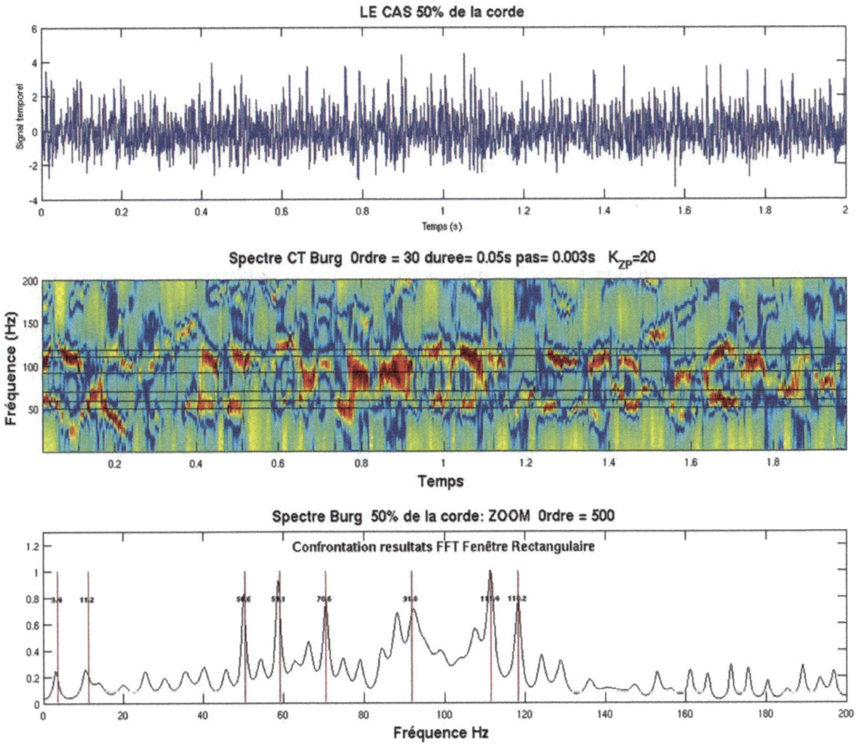

Fig. 3 Experimental signal, surface pressure at 50% of chord versus time, (up); autoregressive modeling (AR)-(middle), FFT spectrum (down), M=0.7, α=6.8, Re=2.63×10^6

The AR model indicates organisation of the flow according to a bi-modal pattern with a predominant frequency of order 95-100 Hz, corresponding to the buffet phenomenon and a first sub-harmonic of order 50 Hz. The third sub-figure in fig. 3 shows the FFT analysis that provides a multitude of frequencies in the present case. A similar behaviour is obtained from the Morlet's wavelet analysis, according to the previously mentioned bi-modal pattern. In case of forced oscillation of the flap by a frequency f=35 Hz, the organisation of the signal shown by AR and by Morlet's wavelets indicates a shift of the organised motion around the forcing frequency (35Hz), as well as the appearance of a first harmonic of this frequency, of order 70 Hz (Doerffer et al, UFAST research program, 2010).

3 Numerical Simulation of the IoA Test-Case

The transonic flow case of the IoA was studied for clean and flapped wing configurations. From the very start, the influence of the wind tunnel walls was identified as a key problem and a special mesh was put together to allow for modelling of the walls and easy comparison with free-stream conditions. At the time this report was compiled very few experimental data were available for comparisons and the exact flow case conditions were not specified. For this purpose, the University of Liverpool (ULIV) and IMFT focussed on a set of cases and comparisons with test data were made. The numerical grids are provided by ULIV (fig. 4 up) with 2.65 million cells including 18 blocks and by IMFT that performed refinement on the upper and lower walls as well as an optimised grid around the gap (fig.4 down). The refinement in the upper and lower walls was done in respect of the $y+$ value (less than 1). IMFT final grids are of order 3.3 million and 3.75 million cells. A special attention has been attributed to correctly mesh the gap region and to provide CFD on a computational domain with the real wind-tunnel walls. The grid spacing in the mid-span is of order 0.03. The DES computation has been active in the flow regions beyond the near-wall ones.

The resolution of the wing surface was higher than the side walls which were put at exactly the same location as in the experiment. Care was taken to allow adequate grid resolution near the front 20% of the NACA-0012 wing section. This would allow for shocks to be resolved and capture any separation due to shock/boundary-layer interaction. The work focussed at relatively high incidence angles (up to 6.8 degrees) for which buffet clearly appears and for this reason, the upper surface of the wing was also meshed at a finer level. In total 2.65 million cells were used and the baseline grid included 18 blocks at each span-wise station.

The time step size used was 10^{-7} for the URANS and DES computations. Statistics were evaluated following the elimination of initial transient phases.. The statistics obtained for the DES would need to be checked for convergence by using longer sample sizes in a future study.

The numerical code used by ULIV is the PMB code (Barakos, 2009) and by IMFT the NSMB, (Navier-Stokes MultiBlock code) (Vos et al, 1998) in which IMFT, being part of the NSMB consortium, has implemented higher-order turbulence models.

In the present study, the employed turbulence models concerning URANS are the Spalart Almaras (SA) (1992) and the k-ω model (Wilcox 1994). Concerning Detached Eddy Simulation, the DES-Spalart Allmaras (DES-SA) (1997), the DDES-Spalart-Allmaras (DDES-SA) (2006) have been used.

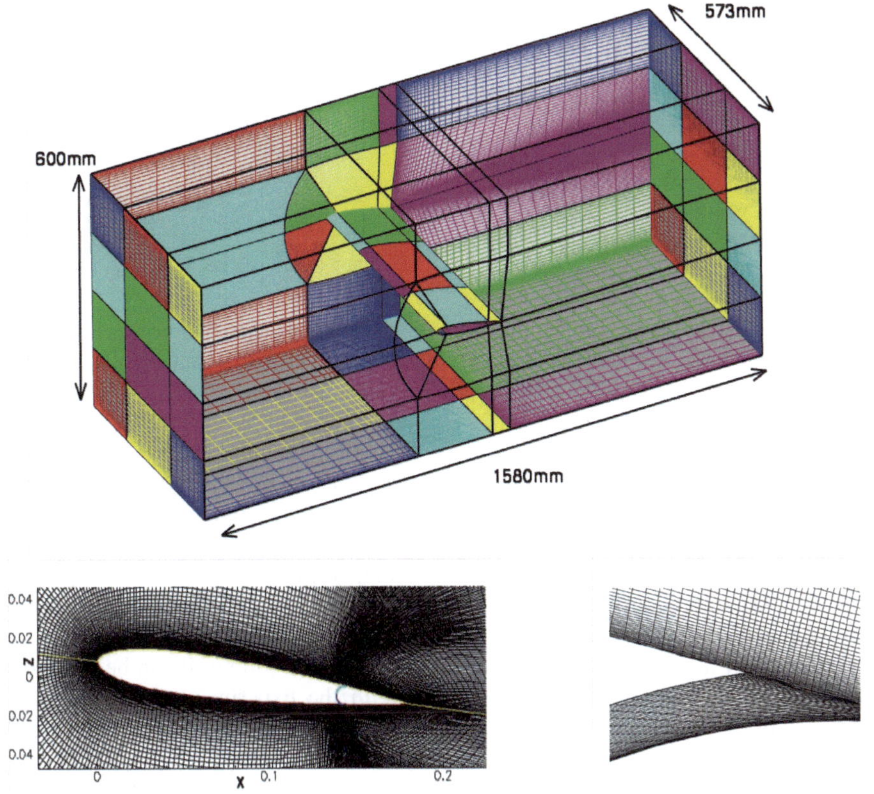

Fig. 4 View of the baseline mesh constructed by ULIV with ICEM-CFD Hexa. Up: overall domain configuration and block boundaries; down: adapted mesh by IMFT around the gap

3.1 Investigation of Flow Conditions

Figure 5 presents an overview of the flow configuration obtained for the wing with the tunnel walls present. The colour surface represents pressure that illustrates formation of a strong shock near the leading edge of the wing at a Mach number of 0.7 and 5 degrees of angle of attack. The strength of the shock is high enough to cause separation of the boundary layer and the flow near the side-walls of the tunnel is dominated by flow re-circulation regions. Near the mid-span of the wing, there is an extended region where the flow shows little spanwise variation. This suggests that a zero gradient condition could be imposed in this direction and a slice of the wing near the centre would be taken into account for the simulation. This approach would result in lower computational cost. However, the top and lower walls have a stronger effect and as was investigated, their effect must be accounted for. The flow reversal near the mid-plane of the wing and the walls is clearly visible in Figure 5, as well as the corner vortices.

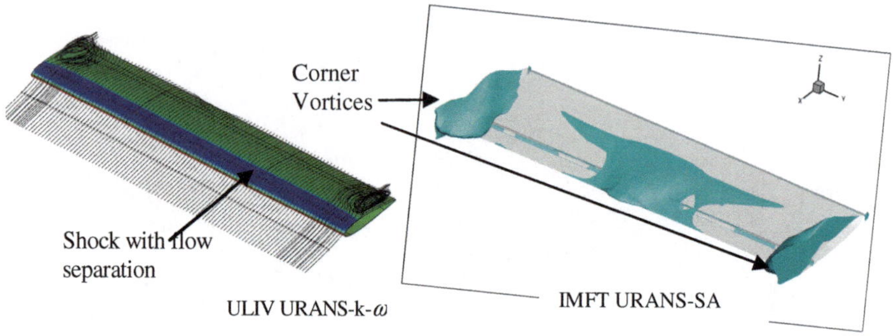

Fig. 5 Overview of flow configuration: flow recirculation and corner vortices near the side walls of the tunnel. Streamlines and iso contours of pressure coefficicient (left). Iso- *(U=0)* velocity contours (right). IMFT: URANS-SA

The effect of the tunnel walls was studied. Computations with the upper and lower walls and with free-stream conditions result in different surface pressure coefficients Overall, the presence of the walls resulted in higher levels of surface pressure coefficient comparing with the non-confined case. Without the presence of the walls, the buffet onset is shifted to higher Mach numbers and angle-of-attack. This significant influence suggested that the upper and lower wind tunnel walls must be taken into account for the flow simulation.

A study of the spanwise conditions is presented on Figure 6. It can be seen that the full 3D configuration compares better with the experimental data than the computation where spanwise symmetry was imposed. It is therefore considered necessary to represent also the lateral walls in the simulation. An interesting observation is that the flow near the flap may be influenced by the exact geometry of the gap between the main element and the flap of the wing. The slotted flap configuration has a small effect near the trailing edge, fig. 6 (right).

To enhance the possibility of using a simplified geometry for actuation of the flap, slotted and blended flap configurations were used. The conditions were Mach number of 0.75 and Reynolds number of 2.81 million. The angle of attach of the wing was changed between 0 and five degrees (where the onset of the buffet was close). At the same time the flap was deflected by three degrees. For all cases, blended and slotted flaps were used. Flow visualisation is shown in Fig. 7 for the flow region near the slot. The conditions were selected so that the pressure difference between the lower and upper wing surfaces drives the flow through the gap. As a result a small flow distortion is shown. This flow distortion is also evident in the surface pressure coefficient plots shown in Fig 6 (right).

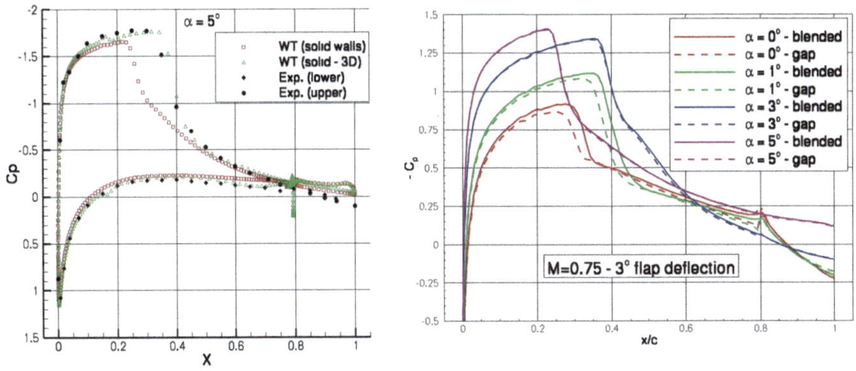

Fig. 6 Cp wall pressure coefficient compared with the experiment; axisymmetric assumption providing lower Cp, versus full 3D WT geometry computation. Slotted mesh including the gap between main and flap elements (left). Comparison between blended and slotted flap configurations (right), Re= 2.81 x 10^6

A zoom of the flow within the gap region is shown in figure 8, indicating the suction and pressure effects in the gap. Figure 9 shows IMFT computations concerning the time history of pressure on the wing surfaces, according to the refined mesh of fig. 4 (lower). The computations have been carried out during enough time to achieve several buffet cycles. The time-dependent pressure coefficients at two different chord positions (25% and 50%) are shown. In each figure, the pressure side and suction side positions are presented simultaneously. The suction side develops more irregular fluctuations due to the shock motion, especially at the 50% chord position. The pressure signals at each chord position are practically in phase opposition. Because of solving the phase-averaged Navier-Stokes equations in the URANS modelling, understood as a decomposition of the flow into coherent (resolved) and turbulent (modelled) motions, the high-frequency noise appearing in the experimental signals has been attenuated.

The buffet frequency obtained by the above IMFT results is of the order of 95 Hz, in good agreement with ULIV results and with the physical experiment. It is recalled that the experimental data processing in respect of the buffet frequency has been performed later than the CFD evaluation of this frequency. Therefore, the present test-case was a 'blind' one in respect of the predominant frequency value.

The results presented in Fig. 10 (left) show flow visualisation by means of instantaneous iso-*U* contours at the median plane. The SWBLI is clearly illustrated and compared with the experimental Schlieren visualisation (fig. 11 left) of the IoA. Figure 10 (right) shows *rms* of the pressure field at the suction side. This illustrates the region affected by the shock motion. Figure 11 (right) shows an instantaneous field of the ratio (*gradρ/ρ*), *ρ* being the density. This compares quite well with the instantaneous Schlieren visualisation.

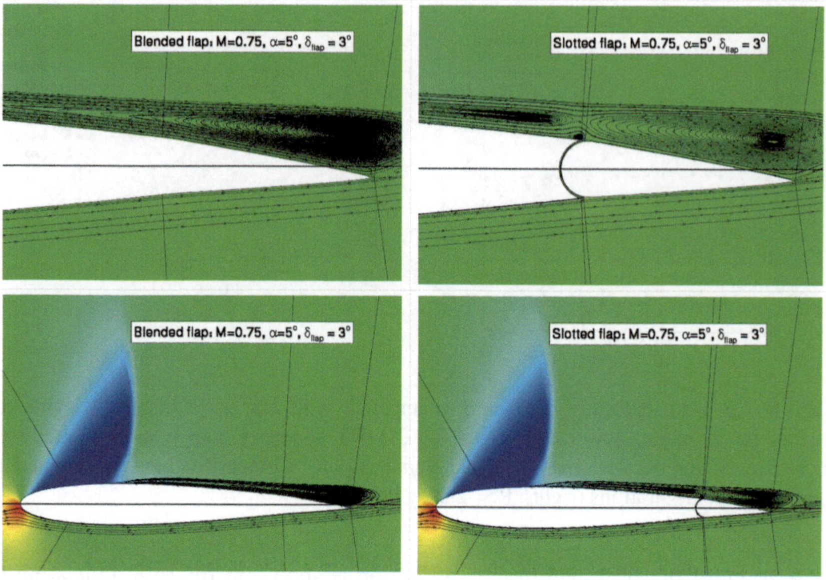

Fig. 7 Iso-streamlines and iso-Mach, blended (on the left) and slotted (on the right) flap configurations. Flap deflected by 3°, URANS-k-ω, (ULIV)

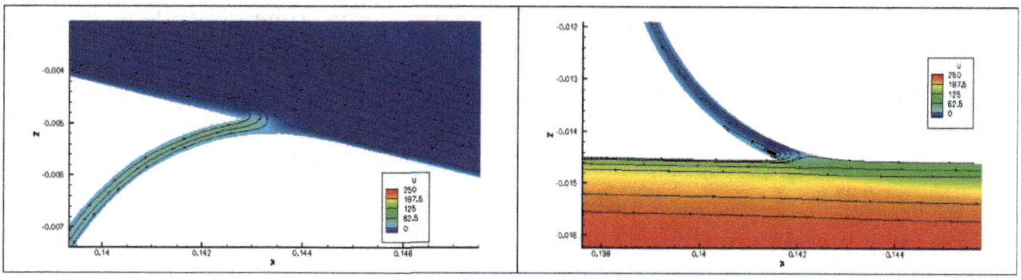

Fig. 8 Instantaneous iso-U contours in the gap region

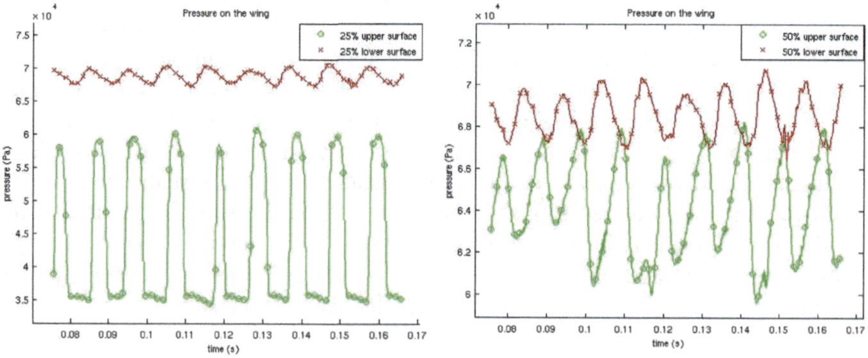

Fig. 9 Pressure signals, wing with gap, M= 0.73, a=5°, Re 2.65 x 10^6, URANS-SA

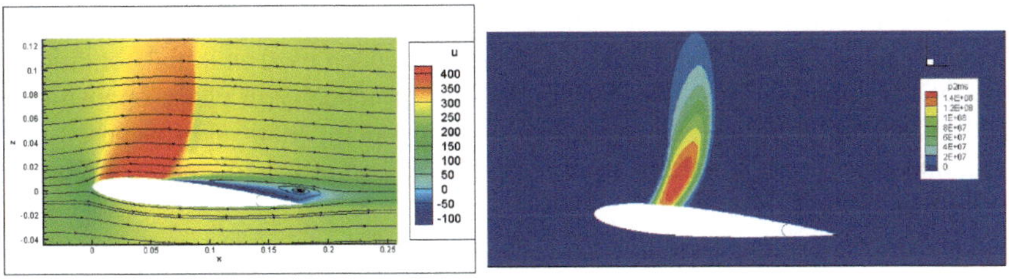

Fig. 10 Left: Instantaneous Iso-U contours at the medium span. Right: *rms* fluctuation of midspan pressure illustrating the compressibility effects in the shock region contours; URANS-SA

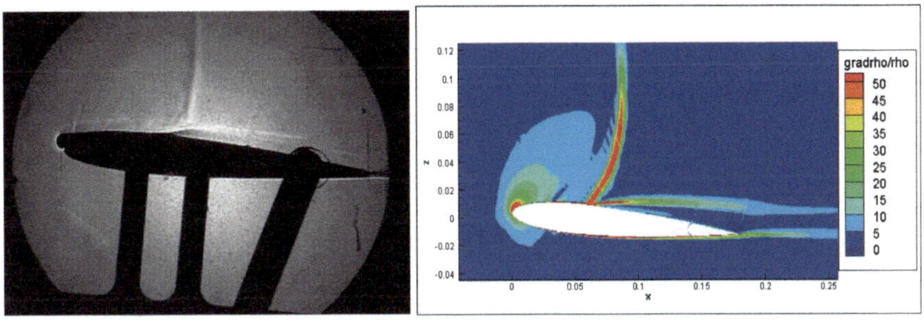

Fig. 11 Left : IoA shadowgraph showing SWBLI on the NACA 0012 airfoil with undeflected flap at $\alpha = 5°$, Mach number M = 0.70 and Reynolds number Re = $2.69{\cdot}10^6$. Right: Instantaneous iso-contours of the ratio ($grad\rho/\rho$), ρ being the density

Furthermore, comparisons of the CFD pressure coefficient with the experiment are presented on fig. 12. ULIV have used the DES-SA modelling, where IMFT has used the DDES-SA (Delayed DES). A quite good comparison of the mean wall pressure coefficient is achieved, especially by ULIV. The difference between the two DES could however be a consequence of the short preliminary statistical sample computed.

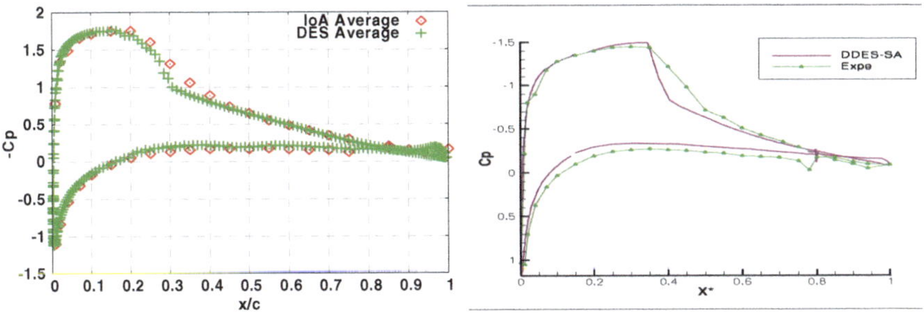

Fig. 12 Comparison of the mean wall pressure coefficient with the experiment: Left: ULIV; right IMFT

Figure 13 shows URANS-SA and DDES computations at 50% of chord, for the suction and pressure sides. The URANS field has been used as initial condition for the DDES computations. The two signals are practically in phase opposition by 180°. The buffet frequency is clearly pointed out by both approaches. Longer computational times should be used for the DDES approach to capture a more rich statistical content.

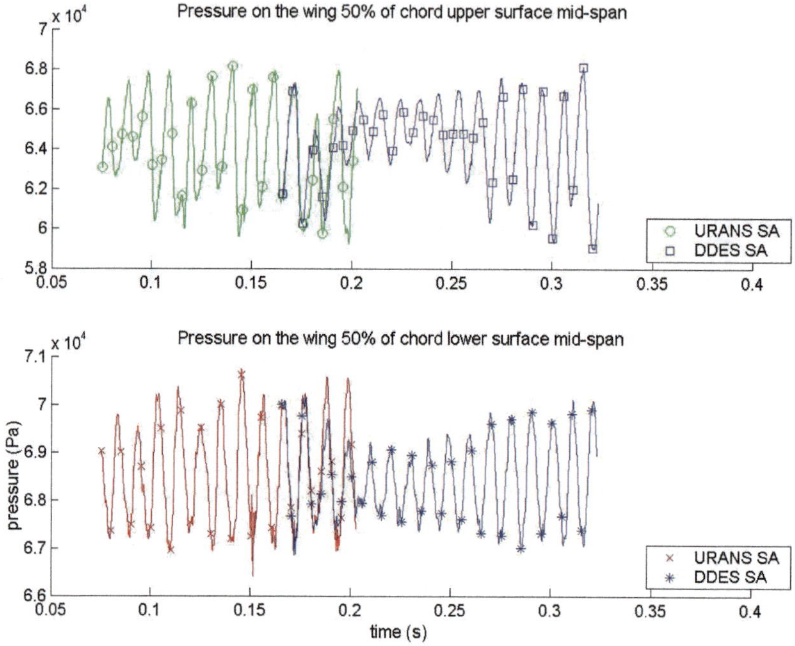

Fig. 13 Surface pressure signal at 50% of chord, URANS-SA and DDES-SA; up: suction side; down: pressure side

3.2 Flap Actuation for Flow Control

The effect of the flap actuation has been studied in two cases in the UFAST program, with actuation frequency of 30 Hz, close to one third of the buffet frequency and a frequency of 100 Hz, closed to the buffet frequency respectively. The flap was oscillating at an amplitude of 2 degrees.

After an initial time where the flow adapts from a partially-converged steady-state to buffet, the loads show an almost periodic response.

Figure 14 shows the lift and drag time-dependent coefficients with and without flap deflection. It is noticeable that the present actuation does not considerably attenuate the buffet frequency.

The results of Figure 15 present comparison of the controlled and uncontrolled cases with blended and slotted flaps. In the case of actuation with a frequency

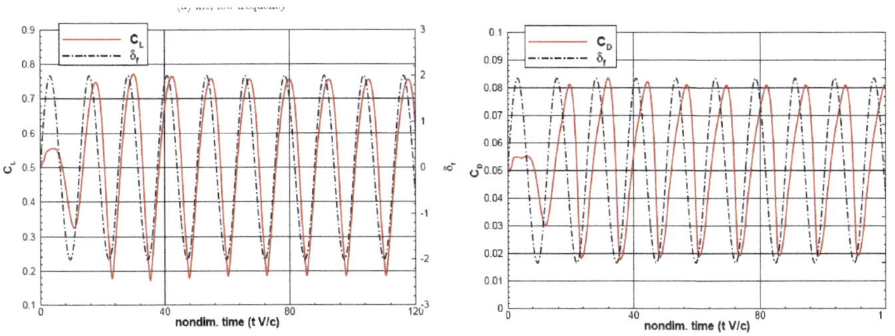

Fig. 14 Flap oscillation of 2 degrees at the buffet frequency; left: lift coefficient and flap angle, δ_f, versus time. Right: drag coefficient and δ_f. M= 0.7, Re= 2.63 x10^6 (ULIV)

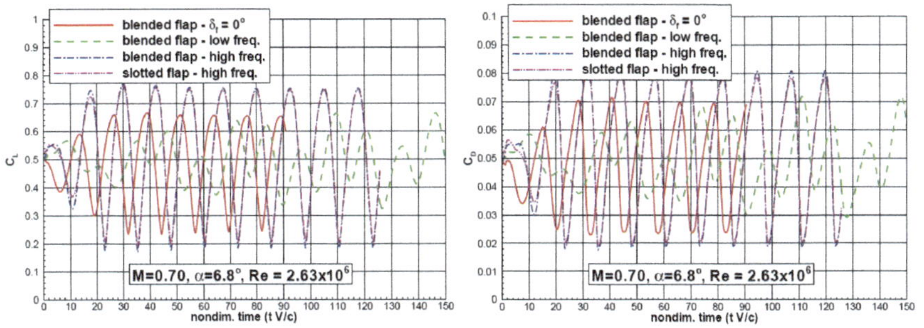

Fig. 15 Comparison of controlled and uncontrolled cases (flow parameters as in previous figure), ULIV, with two oscillation frequencies

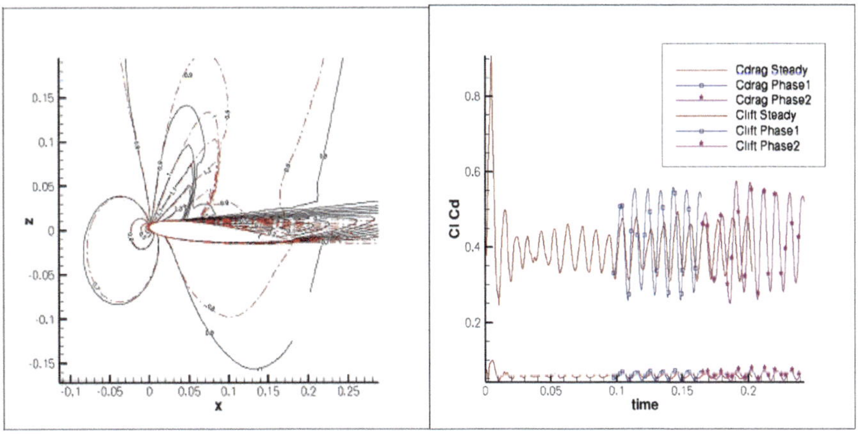

Fig. 16 Left: mid span iso-Mach lines at two different phases of the natural buffet oscillations (maximum and minimum of the Clift), corresponding to the starting of flap actuation. Right: Flap actuation at forcing frequency of 100 Hz and a deflection of ± 2°. In legend, "steady" means fixed flap. IMFT – URANS-SA

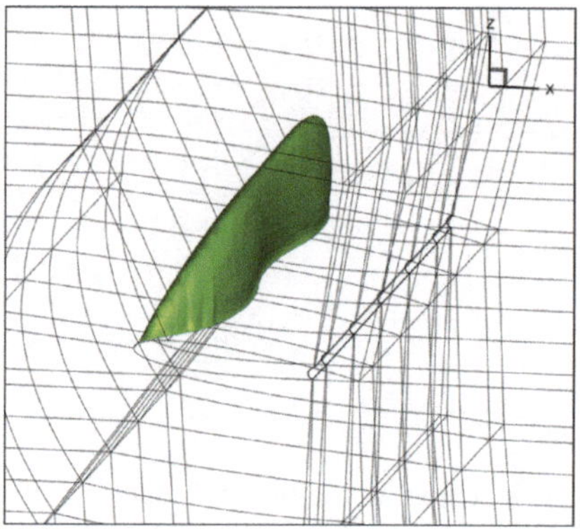

Fig. 17 Instantaneous iso-Mach surface (*M*=1), illustrating the supersonic 3D area along the span (right), M=0.73, a=5°, R=2.63 x 10^6

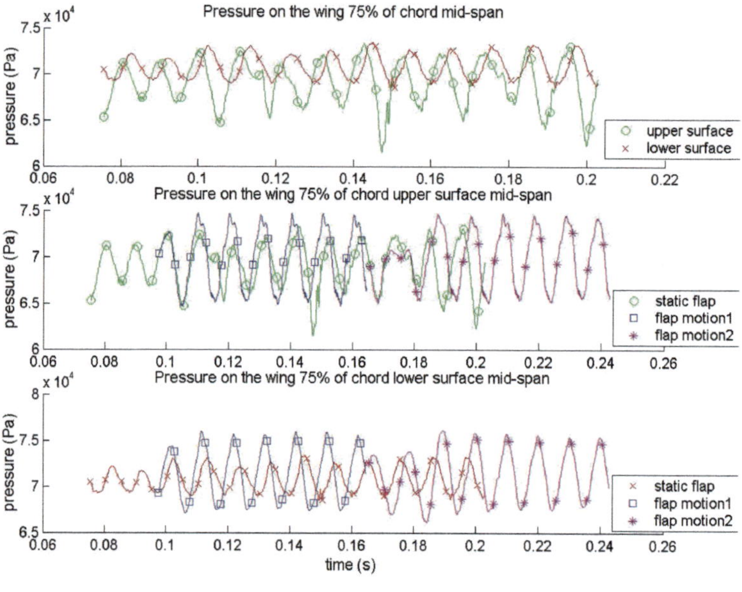

Fig. 18 Wall pressure coefficient oscillations as a function of time at 75% chord position and at mid-span distance. "Upper surface" means suction side. Motion 1=starting flap motion at minimum of Clift oscillation. Up: Static reference case. Middle: comparison of forced case with static case, suction side. Down: comparison of forced case with static case, pressure side, URANS-SA

very close to the buffet natural frequency, it is found that the amplitudes of the oscillation increase. In the case of actuation with a third of the buffet frequency, these amplitudes decrease. Furthermore, the flap motion appears introducing extra frequencies in the flow and induce loads modulation.

Figure 16 shows the influence of forcing at 100Hz (very close to the natural buffet frequency), actuated at two different phases (maximum and minimum of the Clift coefficient). The computations have been carried out by IMFT by URANS-SA. Both cases lead to an increase of lift coefficient amplitudes, as was pointed out by the computations of ULIV. This can be explained as a 'resonance' lock-in between the buffet frequency and the forcing frequency (100 Hz), created by the flap motion. Figure 17 shows the 3D modulation of the shock area along the span in the first case (flap oscillations starting at an instant of maximum Clift).

Figure 18 shows the wall pressure fluctuations at 75% of the chord, at mid section of the span in the case of forcing by 100Hz, (first phase case). It is shown that the selected flap motion leads to an increase of the amplitudes and of the absolute mean values of the pressure fluctuations.

Furthermore, this figure shows the influence of the phase concerning the starting of the flap motion, at minimum ("Phase 1") or at maximum ("Phase 2") of lift coefficient respectively. It is found that the buffet mechanism finally follows the phase of the flap motion oscillations.

4 Conclusions

A numerical study has been carried out to analyse the transonic interaction around a NACA0012 wing of constant section, including a trailing edge aileron (flap). First, the static case has been analysed and secondly, a forced flap motion case with frequencies equal to the natural buffet and to one third of the buffet frequency. The results are compared with the experimental results of the Institute of Aviation, IoA, provided in the UFAST research program.

A detailed signal processing of the experimental results has been carried out by IMFT by using especially wavelet analysis and Autoregressive (AR) modelling. This treatment was useful to extract the organised frequencies of the buffet phenomenon from the complex chaotic behaviour due to turbulence. A special attention has been attributed to examine the influence of a low amplitude (2°) flap motion.

The numerical simulations have been carried out *respecting the exact wind tunnel geometry*, by the University of Liverpool (ULIV) and the Institut de Mécanique des Fluides de Toulouse (IMFT). The approaches that have been used in the context of URANS and preliminary DES, show ability of capturing the buffet frequency, as well as the wall pressure coefficient, in a quite good agreement with the experiment. Comparisons were first carried out between the two sets of simulations without knowing the experimental values and afterwards, detailed comparisons with the experiment have been performed. The buffet frequency has been found around 95 Hz.

Concerning the flap motion, both simulations have shown an increase of the amplitudes at forcing equal to the natural buffet frequency. A slight attenuation of the aerodynamic coefficients has been achieved in the case of forcing by one third.

This study has shown that the phase response of the system is independent on the initial starting phase of flap motion and it opens a promising way towards a further significant control of the buffet, by using this kind of simulation approaches within an optimisation kernel. In this way, it may be possible to attenuate rather than amplify the oscillations arising from transonic buffet.

Acknowledgement. This work was sponsored by the UFAST research program, "Unsteady Effects in Shock Wave Induced separation", coordinated by IMP-PAN (P. Doerffer). The authors are grateful for the use of the French supercomputing facilities of CINES, IDRIS, GRID-5000.

References

Barakos, G.: UFAST Report, 48-month IoA Case Contribution (April 12, 2009)

Grossmann, A., Morlet, J.: Decomposition of Hardy functions into square integrable wavelets of constant shape. SIAM J. (1984)

Marple, S.L.: Digital Spectral Analysis. Prentice-Hall, Englewood Cliffs (1987)

Miller, M., Kania, V.: Parameters for aerofoil and aileron oscillations, UFAST Deliverable D3.1.3 (March 2009)

Seegmiller, H.L., Marvin, J.G., Levy, L.L.: Steady and unsteady transonic flow. AIAA J. 16(12), 1260–1270 (1978)

Spalart, P.R., Allmaras, S.R.: A One-Equation Turbulence Model for Aerodynamic Flows. AIAA Paper 92-0439 (1992)

Spalart, P.R., Jou, W.-H., Strelets, M., Allmaras, S.R.: Comments on the Feasibility of LES for Wings and on the Hybrid RANS/LES Approach. In: Advances in DNS/LES, Proceedings of the First AFOSR International Conference on DNS/LES (1997)

Spalart, P.R., Deck, S., Shur, M.L., Squires, K.D., Strelets, M., Travin, A.: A new version of detached eddy simulation, resistant to ambiguous grid densities. Theor. Comp. Fluid Dyn. 20, 181–195 (2006)

Vos, J., Chaput, E., Arlinger, B., Rizzi, A., Corjon, A.: Recent advances in aerodynamics inside the NSMB (Navier-Stokes Multi-Block) consortium. In: 36th Aerospace Sciences Meeting and Exhibit, AIAA Paper 1998-0802, Reno, USA (1998)

Wilcox, D.C.: Turbulence Modelling for CFD, DCW Industries, California, USA (1994)

RANS/LES Simulation of a Separated Flow in a 3D Curved Duct

I. Mary

1 Introduction

Prediction of internal flow over curved surface remains challenging like for air intake configuration. Indeed the boundary layers are subjected to curvature and pressure gradient effects, which often lead to a boundary layer separation. Despite the continuous improvement of modelisation, RANS can only predict the low frequency part of an unsteady signal, whereas it suffers from a lack of accuracy in case of separated flows. LES is relatively free of these problems, but it suffers from its computational cost. Therefore a zonal RANS/LES approach is retained for the simulation of an internal flow over a curved surface. Attached boundary layers are modelled with a Spalart-Almarras model, whereas the separation zones are taken into account by a LES model.

2 Zonal RANS/LES Coupling

2.1 Main Characteristics

In case of zonal RANS/LES coupling domains with different kinds of information communicate. For RANS the whole turbulence is modelled, whereas for LES only the very high frequency turbulence is modelled while the rest is resolved. So, to tranfer information from LES to RANS, the LES fields require a filtering to be a suitable RANS solution and, to transfer information from LES to RANS, high frequencies are added to RANS fields. Our zonal coupling is based on the use of two overlapping domains: a RANS domain (clear color in Fig. 1) is used to discretize the whole computational domain, whereas an overlapping LES domain (dark color in Fig. 1) is located where higher accuracy is needed. Since LES must be able to simulate fluctuations in space and time the latter domain has a finer mesh than the former. Besides each space direction has its own integer ratio between RANS and LES grids.

I. Mary
ONERA, Châtillon, 29 av. Division Leclerc, BP 72, F-92322, Châtillon Cedex
imary@onera.fr

S.-H. Peng et al. (Eds.): Progress in Hybrid RANS-LES Modelling, NNFM 111, pp. 205–211.
springerlink.com

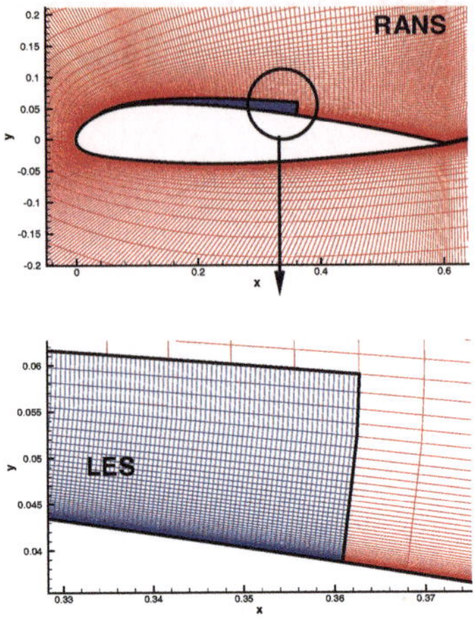

Fig. 1 Zonal RANS/LES coupling

2.2 Information Transfer from LES to RANS

The information transfer towards RANS domain divides into two steps. First the LES velocity and energy fields are filtered to remove the frequencies corresponding to the turbulence. Two different filters have been used. The first one is a space filter. At each timestep, the value of the conservative variables Q_{RANS} in the RANS domain in the overlapping zone are computed as follows:

$$\begin{aligned} \widehat{Q}_{LES}(x,t) &= \Omega_{RANS}^{-1} \int_{\Omega_{RANS}} Q_{LES}(x,t) d\Omega \\ Q_{RANS}(x,t) &= \widehat{Q}_{LES}(x,t) \end{aligned} \tag{1}$$

where Ω_{RANS} denotes the volume of the cell in the RANS domain and $\widehat{Q}_{LES}$ the space filtered LES conservative variables. The second filter employed is based on the time filtering applied to the space filtered variable obtained thanks to Eq. 1:

$$\begin{aligned} Q^{*}_{LES}(x,t^{n}) &= \tfrac{N-1}{N}\, Q^{*}_{LES}(x,t^{n-1}) + \tfrac{1}{N}\, \widehat{Q}_{LES}(x,t^{n}) \\ Q_{RANS}(x,t^{n}) &= Q^{*}_{LES}(x,t^{n}) \end{aligned} \tag{2}$$

where N specifies the number of time steps over which the fields are integrated and ^{n} indicates the time of the simulation ($t^{n} = n\,\Delta t$).

2.3 Information Transfer from RANS to LES

The main difficulty relies on the generation of unsteady inflow condition for LES, when the LES zone is located downstream a RANS region where the flow is turbulent. The information transfer between the RANS domain and the LES one requires some boundary conditions, which is achieved by the use of ghost cells. Three-linear interpolations are used as well as enrichment procedure Quéméré et al (2001). At an inflow interface within a turbulent flow field realistic turbulent fluctuations need to be generated in order for the LES to provide a realistic answer. The generation of fluctuations at this boundary can be performed with higher or lower perfection, depending on the flow investigated and the purpose of the simulation conducted. The method described in the following is a combination of a database technique and rescaling as used in the recycling technique for the generation of unsteady inflow data. More details can be found in Nolin et al (2006). At the upstream end of the LES domain, values of the mean conservative fields are known from the upstream RANS domain, so only the determination of the fluctuating part of these fields is needed. These fluctuations are extracted from a database, then scaled to fit the desired time and length scales and to have the proper magnitude and finally added to the average field at the inflow plane.

3 Air Intake Simulation

3.1 Flow Configuration Description

The test section of the ONERA S19 windtunnel represents in fact the air intake S19. It consists of a curved channel shown in Fig. 2. The shape of the vein in the upstream section is rectangular and the ratio between height (H = 150mm) and width (L = 300mm) is equal to 0.5. After the curve (below x = 0mm), the section is growing rapidly, becoming almost square ($H/W \geq 0.92$), then this value gradually decreases to stabilize to a value close to 0.66 at x = 1800mm. The 3 components of the velocity field have been measured by LDV and PIV technique for many sections of the vein. There is therefore a very complete experimental Geffroy and Sipp (2004) database for this geometric configuration. Concerning the flow conditions in the upstream section of the vein (H = 150mm), the velocity at the center is equal to 30ms^{-1} and the Mach number is 0.08.

3.2 Zonal RANS/LES Partitioning

To minimize the CPU cost of this computation, it seems worthwhile to place the LES zone only near areas of flow separation responsible for the highest dynamic distortion, as shown in Fig 2. A (U)RANS model is used in the dark zone to take into account accurately and inexpensively attached boundary layers, whereas a LES model is used in the clear zone where the flow is likely to separate from the wall and generate strong pressure fluctuations. At the beginning of the study, two-dimensional

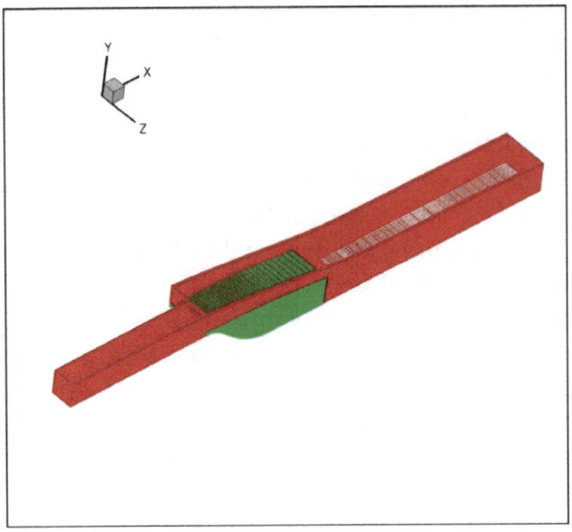

Fig. 2 S19 Air intake simulation: RANS (dark zone) and LES (clear zones)

geometry (with periodic condition in the spanwise direction) were considered to assess the influence of mesh resolution. Three meshes, M1, M2 and M3 have been created. In the direction normal to the wall, three refinements are introduced: the upper and lower walls, but also in the mixing layer created by the separation of the boundary layer. The mesh M1 respects the usual resolution to deal with a boundary layer: $\Delta x^+ < 50$ (axial direction), $\Delta y^+ < 2$ (direction normal to the wall),$\Delta z^+ < 15$ (transverse direction). For the M1 mesh, the spanwise extent, Lz, is equal to 0.5 H, where H is the height of the vein upstream of the bend. This size was originally chosen because it represents about two-thirds of the distance between the transverse vortices from the mixed layer. The influence of the spanwise extent Lz is studied using mesh M2, which has the same characteristics as M1, except that Lz was increased to 0.76 H. The influence of the transverse resolution was also evaluated. As the boundary layer is separated from the wall in the greater part of the LES zone, the standard criteria established for attached boundary layers are probably too restrictive. These are indeed guided by the need to resolve the streaks, with a spanwise size of about 100 wall units, whereas their influence on the dynamics of the boundary layer is certainly less in case of separation. Also the M3 mesh adopts the same characteristics as M1, except that its transverse resolution has been halved:$\Delta z^+ < 40$. For the simulation of the real 3D configuration, the resolution of the M3 mesh has been kept both for the lower and side walls. In the upstream channel (upstream the curved surface), only half of the channel width is computed and a symmetry condition is employed to limit the total number of the grid points. After the separation point, the instantaneous flow field can no longer be symmetric and the full channel must be discretized both in the LES and RANS domain. All these simulations required between 12 and 30 millions of cells. At the interfaces between RANS and LES zones, the specific conditions presented in the previous section are used to generate or filter turbulence. A slight modification of the LES inflow turbulence condition is introduced to take into account the corner formed by the lower and side walls for the 3D geometry.

3.3 Results

The time required to simulate 0.5 (0.2s for transient flow and 0.3s for the calculation of the statistics) is about 4000h CPU for the 3D geometry. The initial solution is obtained thanks to an interpolation of the instantaneous solution obtained on the M3 mesh with some damping function to take into account the side walls. The time step of the simulation is equals to 3 μs and the technique of Daude Daude et al (2006) is employed to optimize the number of Newton sub-iterations. The efficiency of the proposed inflow and outflow coupling method can be observed in Fig. 3, which represents a Schlieren view of the density fluctuation in three plans of the LES zone: one parallel to the lower wall and two parallel to the side wall. The right part of the figure shows a zoom near the LES inflow plan. A realistic turbulence is present immediately after this inflow plane. At the outflow of the LES domain, the treatment is also correct, because no damping of the turbulence is introduced by the boundary condition. However a failure of the LES tangential coupling can be observed in the right of Fig. 3 in the side wall boundary layer. The dark zone means that there is almost no unsteady perturbation in this boundary layer near the inflow. This is due to the fact that the streamlines come from the RANS zone at the upper face of this bloc. As the enrichment procedure is used at this interface, the fluctuations cannot be reconstructed accurately by this procedure. Despite the problem mentionned above at the interface between RANS and LES domain in the side wall boundary layer, the accuracy of the present 3D simulation is encouraging. The comparison of numerical results with experimental data in the Fig. 4 shows good agreement before and in the beginning of the detachment region (x $\leq$ 350mm) both for the mean velocity and TKE profiles. Further downstream, when the separation becomes more important, all the simulations differ significantly from the experience. This is particularly

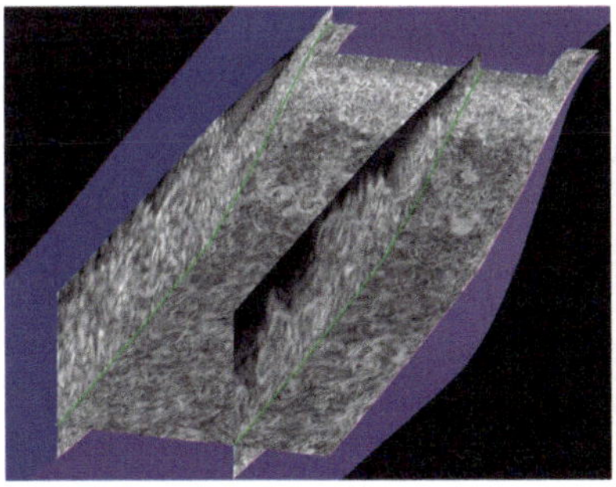

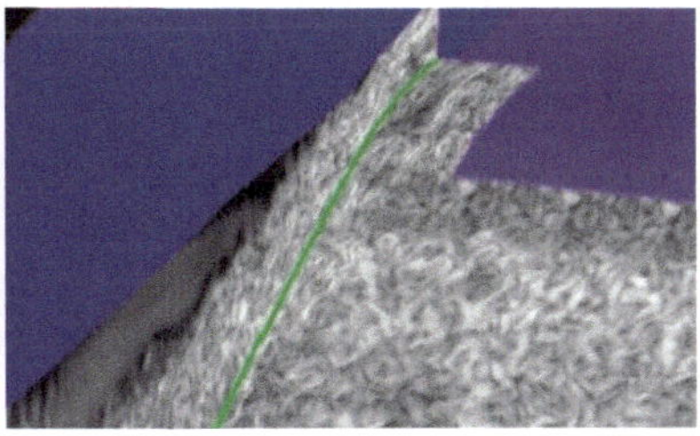

Fig. 3 Schlieren view of pressure fluctuation in the LES zone: general view (left), Zoom at the upstream corner near the RANS/LES interface (right)

the case for the simulation based on a periodic condition in the spanwise direction with the smallest spanwise extent (M1 mesh). These differences can have several origins. As experience shows a 3D character quite pronounced, this may explain the discrepencies in 2D geometry calculations. However no conclusion can be retained from the 3D geometry calculation. Indeed, the sampling time of 0.3s, which may appear very long in the academic context of turbulent boundary layers simulated by LES, is far too short (certainly an order of magnitude) to obtain a statistically converged field. Indeed a flapping movement of the separation zone in the spanwise direction was observed experimentally at a very low frequency of about 0.1 Hz. The current simulation, which is limited to a duration of 0.3 s for the calculation of statistics, can not take into account this flapping phenomenon, which would require at least a recording signal of 30-40s. Computation of such a long signal is not realistic because it would require around 40,000 to 50,000 CPU hours of NEC SX8.

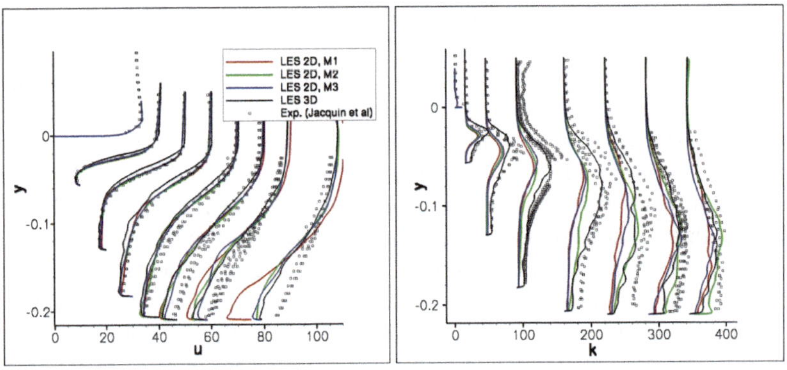

Fig. 4 Mean profile of the velocity field in the symmetry plan at x=0, 150, 250,...,650 and 750mm: streamwise velocity component (left), TKE (right)

4 Conclusion

The zonal RANS/LES coupling method has been applied to the simulation of the vein S19. Simulations of a simplified 2D geometry and of the true 3D geometry were performed. The method allows a strong coupling between the RANS and LES zones and is relatively flexible to use. When averaged in time, the continuity of the solution between the RANS and LES areas is satisfactory, either at the inflow or outflow of the LES area. The comparison of numerical results with experimental data shows good agreement before and in the beginning of the separated region. Further downstream, when the separation becomes more important, all the simulations differ significantly from the experience. Computation of the real geometry improve the quality of the results compared to those obtained with a periodic condition in the spanwise direction. The first explanation of the discrepencies between the 3D computation and the experimental results relies on the too short time signal of the computation, which cannot account for very low frequency flapping phenomenon observed in the experiment. Others parameters of the numerical simulation should also be improved. For instance the characteristics of the unsteady database is too far from those of the boundary layer that develops on the low wall of the vein (Reynolds and pressure gradient). The establishment of databases scanning a variety of settings, and taking into account the conditions of the corner formed by the intersection of the side walls and lower wall, seem to be necessary for a flexible use of the zonal coupling method. The resolution of the mesh was probably decreased too abruptly in some areas to limit the size of the mesh, as in the transverse direction just after the detachment of the boundary layer. Another way to improve the efficiency and accuracy of the coupling method for this type of configuration would be to use RANS model appropriate for junction flows, which cannot be modelled accurately by the Spalart-Allmarras model. The use of EARSM or RSM models could allow to significantly increase the size of the RANS zone and to limit the use of LES only in areas where accurate pressure fluctuations are seeked.

References

Daude, F., Mary, I., Comte, P.: Local optimization of the convergence rate of implicit time advancements for les of complex flows. AIAA Paper 06–3545 (2006)

Geffroy, P., Sipp, D.: Contrle des dcollements - etude exprimentale fondamentale s19ch d'une manche coude. NASA CP–3291, pp. 821–831 (2004)

Nolin, G., Mary, I., Ta, P.L.: Les inflow conditions for turbulent boundary layers of complex flows using database. AIAA Paper 06–2888 (2006)

Quéméré, P., Sagaut, P., Couailler, V.: A new multi-domain/multi-resolution method for large eddy simulation. Int. J. Numer. Meth. Fluids 36, 391–416 (2001)

A New Partially Integrated Transport Modelling (PITM) Method for Continuous Hybrid Non-zonal RANS/LES Simulations

B. Chaouat

Abstract. The partially integrated transport modelling (PITM) method viewed as a continuous approach for hybrid RANS/LES modelling allowing seamless coupling between the RANS and LES regions is presented. This method bridges two different levels of description in a consistent way and transposes the advantages of advanced RANS modelling such as Reynolds stress models (RSM) to LES simulations. In that sense, the PITM method fills up the gap between URANS and LES methodologies. This method is general and can be applied to almost all statistical models to derive its hybrid LES counterpart. In this work, the subfilter stress model is applied for simulating flows encountered in space propulsion and aeronautical applications that present a complex physics such as transition regime, separation, recirculation and reattachment.

1 Introduction

Advanced RANS models such as new generation Reynolds Stress Models (RSM) or Algebraic Reynolds Stress Models (ARSM) Speziale et al (1991) appear well suited for tackling engineering flows encountered in aeronautical applications Chaouat (2006); Gatski et al (2007) whereas LES methods using subgrid scale models such as for instance the dynamic model Germano et al (1992) or the structure model Lesieur and Metais (1996) are rather applied for simulating academic flows with emphasis on fundamental aspects and structural aspects. Although these methods are very useful, each of them has its own specific field of application. It appears for instance that statistical RANS models fail to reproduce the flow structures which evolve rapidly in time and space and that large eddy simulations are not accurate when performed on coarse grids. In practice, highly resolved large eddy simulations require very large computing time which is not affordable in industrial applications, even with the rapid increase in computer speed. For these reasons, hybrid RANS/LES zonal methods, capable of reproducing a RANS-type behaviour in the vicinity of a solid boundary and an LES-type behaviour far away from the wall boundary have been developed in the last decade. The detached eddy simulation (DES) Spalart (2000, 2009) where the model is switching from a RANS behaviour

Bruno Chaouat
ONERA, 92322 Châtillon, France
bruno.chaouat@onera.fr

S.-H. Peng et al. (Eds.): Progress in Hybrid RANS-LES Modelling, NNFM 111, pp. 213–224.
springerlink.com

to an LES behaviour depending on a criteria based on the turbulent length-scale is one of the most popular models applied in aeronautical industries.

Recently, the partially integrated transport modelling (PITM) method viewed as a continuous approach for hybrid RANS/LES modelling allowing seamless coupling between the RANS and LES regions has been developed by Chaouat and Schiestel (2005); Schiestel and Dejoan (2005); Chaouat and Schiestel (2007, 2009), especially for simulating unsteady flows on relatively coarse grids, allowing a saving of the computation time. This method bridges two different levels of description in a consistent way and transposes the advantages of advanced RANS modelling to LES simulations. Thus, it gains major interest on the fundamental point of view because it allows some unifying formalism that conciliates RANS and LES approaches. In that sense, the PITM method fills up the gap between URANS and LES computations. This method can be applied to almost all statistical models to derive its hybrid LES counterpart. Among these models, the subfilter scale stress model based on the transport equations for the subfilter stresses and the dissipation rate is one of the most sophisticated model for simulating unsteady flows with high resolution of the turbulence anisotropy. This paper presents the principle of the PITM method which finds its basic foundation in the spectral space Chaouat and Schiestel (2007). Then, we apply this subfilter stress model for simulating flows of complex physics encountered in aeronautical applications Apte and Yang (2003); Fröhlich et al (2005); Breuer et al (2009).

2 Governing Equations

Large eddy simulations assume that each flow variable ϕ can be decomposed into a filtered part including mean value and large-scale fluctuation $\bar{\phi}$ and a subgrid-scale fluctuating part ϕ' such that $\phi = \bar{\phi} + \phi'$ where the quantity $\bar{\phi}$ is defined by the filter function G_Δ as

$$\bar{\phi} = \int \prod_{i=1}^{3} G_{\Delta_i}(x_i, x_i')\, \phi(x_1', x_2', x_3')\, dx_1' dx_2' dx_3' \tag{1}$$

where Δ_i is the filter width in the i-direction. The Reynolds statistical average of ϕ is denoted $\langle \phi \rangle$ so that the large scale fluctuation is $\bar{\phi} - \langle \phi \rangle$. By applying the filter on the instantaneous momentum equation, one then derives the filtered momentum equation

$$\frac{\partial \bar{u}_i}{\partial t} + \frac{\partial}{\partial x_j}(\bar{u}_i \bar{u}_j) = -\frac{1}{\rho}\frac{\partial \bar{p}}{\partial x_i} + \nu \frac{\partial^2 \bar{u}_i}{\partial x_j \partial x_j} - \frac{\partial \tau(u_i, u_j)}{\partial x_j} \tag{2}$$

in which the subgrid-scale tensor is defined by the mathematical function

$$(\tau_{ij})_{sgs} = \tau(u_i, u_j) = \overline{u_i u_j} - \bar{u}_i \bar{u}_j \tag{3}$$

The statistical average of the subgrid-scale stress is the average of the small scale fluctuating velocities $\langle(\tau_{ij})_{sgs}\rangle = \langle u_i' u_i' \rangle$. The resolved scale stress is defined by the relation $(\tau_{ij})_{les} = \bar{u}_i \bar{u}_j - \langle u_i \rangle \langle u_j \rangle$. So that, the Reynolds stress tensor τ_{ij} is computed by the sum of the statistical average of subgrid and resolved stresses $\tau_{ij} = \langle(\tau_{ij})_{sgs}\rangle + \langle(\tau_{ij})_{les}\rangle$. Closure of equation (2) is then necessary for the subgrid-scale tensor. In the present approach, this is achieved by means of a transport equation in order to get a more general closure allowing a more realistic and precise description of the flow anisotropy than eddy viscosity models. This is particularly important when the cutoff goes into the non-universal range of the spectrum located before the inertial range.

3 Partial Integrated Transport Modelling Method

3.1 Principle of the Method

From a physical point of view, the PITM method finds its basic foundation in the spectral space by considering the Fourier transform of the two-point fluctuating velocity correlation equations in homogeneous turbulence Chaouat and Schiestel (2007). The extension to non-homogeneous turbulence is developed easily within the approximate framework of the tangent homogeneous space Chaouat and Schiestel (2007). Note that a formalism based on temporal filtering has been also proposed recently to handle non-homogeneous flows leading to the temporal partial integrated transport modelling (TPITM) method Fadai-Ghotbi et al (2009). When transposing the spectral equation in the physical space by inverse Fourier transform involving partial integration of the turbulence field in the range $[\kappa_c, \kappa_d[$ where $\kappa_c = \pi/\Delta$ is the cutoff wave number computed by the grid size width Δ, and κ_d is the dissipative wave number placed at the end of the inertial range of the spectrum completely after the transfer zone, one can derive a subgrid-scale model based on the transport equations for the subgrid scale stresses $(\tau_{ij})_{sgs}$ and the subgrid dissipation rate ε_{sgs}. These equations look formally like the corresponding RANS/RSM equations but the coefficients used in the model are no longer constants Chaouat and Schiestel (2007). They are now some functions of the dimensionless parameter η_c involving the cutoff wave number κ_c and of the turbulent length scale L_e built using the total turbulent kinetic energy k and the total dissipation rate ε including the subgrid dissipation rate ε_{sgs} and the macro-scale dissipation rate denoted $\varepsilon^{<}$.

$$\eta_c = \kappa_c L_e = \frac{\pi k^{3/2}}{(\Delta_1 \Delta_2 \Delta_3)^{1/3}\,\varepsilon} \tag{4}$$

As a result, the subfilter-scale stress model varies continuously with respect to the ratio of the turbulent length-scale to the grid-size L_e/Δ. For the limiting condition when the parameter η_c goes to zero, the subfilter-scale model behaves like a RANS/RSM model whereas when η_c goes to infinity, the computation switches to

DNS or under resolved DNS if the grid-size is not enough refined because the energy cannot be maintained. So that in regard with academic LES simulations which require that the spectral cutoff must be located within the inertial range, the present subfilter-scale stress model allows to perform flow simulations on relatively coarse grids since the cutoff wavenumber can be located almost anywhere inside the spectrum. In the case when the cutoff wave number is placed before the inertial range, an appreciable part of the turbulence fluctuations is then modelled while the large scales are resolved by the numerical scheme.

3.2 Modelling of the Subfilter-Scale Stress Transport Equation

By using the material derivative operator $D/Dt = \partial/\partial t + \bar{u}_k \partial/\partial x_k$, the transport equation of the subgrid stress tensor can be written in the simple compact form as

$$\frac{D(\tau_{ij})_{sgs}}{Dt} = P_{ij} + \Psi_{ij} + J_{ij} - (\varepsilon_{ij})_{sgs} \tag{5}$$

where the terms appearing in the right-hand side of this equation are identified as production, redistribution, diffusion and dissipation. The transport equation for the subgrid energy is obtained as the half trace of equation (5)

$$\frac{Dk_{sgs}}{Dt} = P + J - \varepsilon_{sgs} \tag{6}$$

where $P = P_{mm}/2$, $J = J_{mm}/2$, and $\varepsilon_{sgs} = (\varepsilon_{mm})_{sgs}/2$. The production term P_{ij} involves the interaction between the subgrid stress and the filtered gradient velocity

$$P_{ij} = -(\tau_{ik})_{sgs}\frac{\partial \bar{u}_j}{\partial x_k} - (\tau_{jk})_{sgs}\frac{\partial \bar{u}_i}{\partial x_k}, \tag{7}$$

Contrary to the production term P_{ij} which is exact, the redistribution, diffusion and dissipation terms need to be modelled in the range wave number $[\kappa_c, \kappa_d]$. In second-moment closures, the term Ψ_{ij} plays a pivotal role in reproducing the flow anisotropy by the redistribution of the turbulent energy among the stress components. The redistribution term Ψ_{ij} can be decomposed into a slow part Ψ^1_{ij} and a rapid part Ψ^2_{ij}. This term Ψ_{ij} is modelled assuming that the usual statistical Reynolds stress models must be recovered in the limit of vanishing cutoff wave number κ_c and considering also that the small scales return more rapidly to isotropy than the large scales before cascading into smaller scales by non-linear interactions

$$\Psi^1_{ij} = -c_{sgs_1}\frac{\varepsilon_{sgs}}{k_{sgs}}\left((\tau_{ij})_{sgs} - \frac{2}{3}k_{sgs}\,\delta_{ij}\right) \tag{8}$$

where c_{sgs_1} is an increasing function of the parameter η_c Chaouat and Schiestel (2007); Schiestel (1987). The second term Ψ^2_{ij} is modelled by

$$\Psi_{ij}^2 = -c_2 \left(P_{ij} - \frac{1}{3} P_{mm}\,\delta_{ij} \right) \tag{9}$$

where the coefficient c_2 remains the same than in statistical modelling Chaouat and Schiestel (2007); Schiestel (1987). The diffusion term J_{ij} is modelled assuming a gradient law hypothesis

$$J_{ij} = \frac{\partial}{\partial x_k} \left(\nu \frac{\partial (\tau_{ij})_{sgs}}{\partial x_k} + c_s \frac{k_{sgs}}{\varepsilon_{sgs}} (\tau_{kl})_{sgs} \frac{\partial (\tau_{ij})_{sgs}}{\partial x_l} \right) \tag{10}$$

where c_s is a numerical coefficient set to 0.22.

3.3 Modelling of the Subgrid-Scale Dissipation Rate Transport Equation

Closure of equation (5) needs to model the subgrid tensorial dissipation rate $(\varepsilon_{ij})_{sgs}$. It is decomposed into an isotropic part $2/3\varepsilon_{sgs}\delta_{ij}$ and an anisotropic part $(\varepsilon_{ij})_{sgs} - 2/3\varepsilon_{sgs}\delta_{ij}$. The modelling of dissipation-rate ε_{sgs} is made in the present case by means of its transport equation without referring to the grid information in the aim to obtain an accurate result, in particular in situation of non-equilibrium flows when the grid-size is no longer a good estimate of the characteristic turbulence length-scale. As a result of the theory developed in the spectral space by Chaouat and Schiestel (2007), the fluctuating modelled transport equation for the subgrid-scale dissipation-rate ε_{sgs} reads

$$\frac{D\varepsilon_{sgs}}{Dt} = c_{sgs\varepsilon_1} \frac{\varepsilon_{sgs}}{k_{sgs}} P - c_{sgs\varepsilon_2} \frac{\varepsilon_{sgs}^2}{k_{sgs}} + J_\varepsilon \tag{11}$$

where $c_{sgs\varepsilon_1}$ is a constant coefficient whereas the coefficient $c_{sgs\varepsilon_2}$ appearing in equation (11) is now a function of the ratio to the subgrid energy to the total energy $\langle k_{sgs} \rangle /k$ as follows Chaouat and Schiestel (2007)

$$c_{sgs\varepsilon_2} = c_{\varepsilon_1} + \frac{\langle k_{sgs} \rangle}{k} (c_{\varepsilon_2} - c_{\varepsilon_1}) \tag{12}$$

and where the coefficients c_{ε_1} and c_{ε_2} appearing in this equation denote the usual constants used in the statistical dissipation rate transport equation. The theory Chaouat and Schiestel (2007) shows that the coefficients of the production term remain the same for both RANS and LES dissipation-rate equations $c_{sgs\varepsilon_1} = c_{\varepsilon_1} = 3/2$. In the present case, the values retained are $c_{\varepsilon_1} = 1.45$ and $c_{\varepsilon_2} = 1.9$. Equation (11) using the relation (12) constitutes the main feature of the PITM approach where only the part of the spectrum for $\kappa > \kappa_c$ is modelled. The ratio $\langle k_{sgs} \rangle /k$ is evaluated by means of an accurate energy spectrum $E(\kappa)$ inspired from a Von Kármán like spectrum valid on the entire range of wavenumbers leading to the result

$$c_{sgs\varepsilon_2}(\eta_c) = c_{\varepsilon_1} + \frac{c_{\varepsilon_2} - c_{\varepsilon_1}}{[1 + \beta_\eta\, \eta_c^3]^{2/9}} \tag{13}$$

where β is a constant coefficient calibrated by the decay of homogeneous isotropic turbulence Chaouat and Schiestel (2009). Equation (13) indicates that the function $c_{sgs\varepsilon_2}$ acts like a dynamical parameter which controls the spectral distribution of turbulence. Note that this model is basically different from an URANS (unsteady Reynolds averaged Navier-Stokes) approach, although it is fully compatible with it at the limit of vanishing cutoff. The diffusion term J_ε appearing on the left hand side of equation (11) is modelled assuming a well-known gradient law hypothesis

$$J_\varepsilon = \frac{\partial}{\partial x_j}\left(\nu \frac{\partial \varepsilon_{sgs}}{\partial x_j} + c_\varepsilon \frac{k_{sgs}}{\varepsilon_{sgs}} (\tau_{jm})_{sgs} \frac{\partial \varepsilon_{sgs}}{\partial x_m}\right) \tag{14}$$

where the coefficient c_ε is set to 0.18.

4 Hybrid RANS/LES Simulations

Numerical simulations of engineering flows encountered in space propulsion and aeronautical applications by using the subfilter-scale stress model are presented for illustrating the potentials of the PITM method. The first case is concerned with a channel flow with an appreciable fluid injection through a permeable wall which simulates the propellant burning in solid rocket motors. This type of flow is difficult to reproduce numerically because it differs appreciably from a standard duct flow bounded by impermeable walls. The geometry is a plane channel where the values of the length, height and width of the channel are $L_1 = 58.1$ cm, $L_2 = 6$ cm and $\delta = 1.03$ cm, respectively. The PITM simulation is performed on a coarse mesh composed of $400 \times 44 \times 80$ grid points and the results are compared with the highly resolved LES performed by Apte and Yang (2003) using a dynamical Smagorinsky model (DSM) Germano et al (1992) on a refined mesh (see table 1).

Table 1 Flow with wall injection

Authors	Turbulence model	Grid points	Processors	Computer
Apte and Yang (2003)	DSM	$8.4\ 10^6$	18	Cray T3E
Chaouat and Schiestel (2005)	PITM	$1.4\ 10^6$	1	NEC-SX5

Figure 1 shows the isosurfaces of the instantaneous spanwise filtered vorticity $\tilde{\omega}_i = \varepsilon_{ijk}\partial \tilde{u}_k/\partial x_j$ and reveals the detail of the flow structures subjected to mass injection as well as the location of the transition. In the downstream transition location, the flow is then characterised by the presence of roll-up vortex structures of large magnitude of vorticity in the spanwise direction.

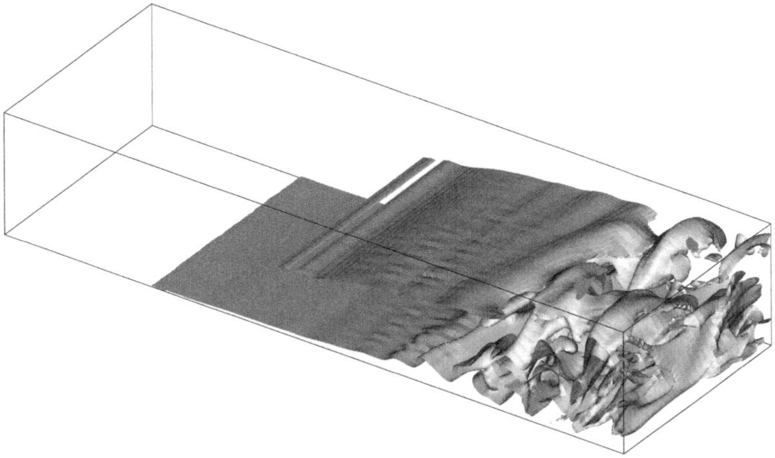

Fig. 1 Isosurface of instantaneous filtered vorticity in the spanwise direction

Figure 2 shows the streamwise rms of the stresses $\langle \tau_{11} \rangle^{\frac{1}{2}} / u_m$ normalised by the bulk velocity at two different locations close to the exit of the channel. As a result of interest, one can see that the computed profiles agree well with the the experimental data Avalon et al (1998) except in the immediate wall region.

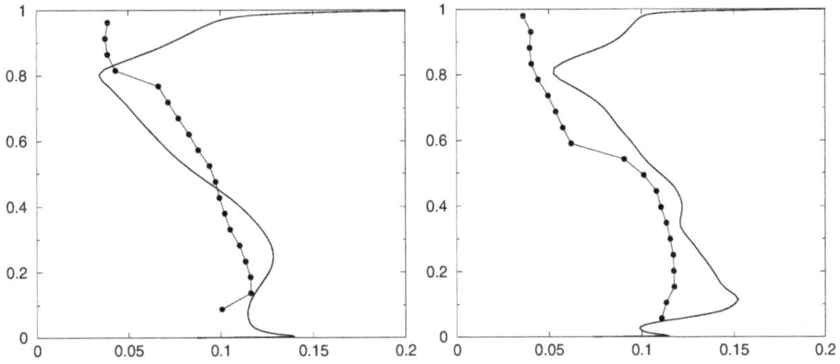

(a) $\langle \tau_{11} \rangle^{\frac{1}{2}} / u_m \; x_1 = 50$ cm (b) x_1=57 cm

Fig. 2 Streamwise turbulent stresses $< (\tau_{11})^{\frac{1}{2}} > / u_m$ in different cross sections (a) 50 cm; (b) 57 cm. + : LES ; •: experimental data

The second case is concerned with a channel flow with streamwise periodic constrictions. The hills constricts the channel by about one third of its height and are spaced at a distance of about 9 hill heights. The dimension of the computational domain are $L_1 = 9h$, $L_2 = 4.5h$ and $L_3 = 3.036h$, where h denotes the hill height, respectively in the streamwise, spanwise and normal directions. Note that an other team Jakirlic et al (2009) also performed this case using a variant PITM model. A no-slip and impermeability boundary conditions are used at the wall whereas periodic conditions are applied in the streamwise and spanwise directions. The Reynolds number, based on the hill height and the bulk velocity about the crest is $R = U_b h/\nu = 10595$. Both PITM and RSM Chaouat (2006) simulations of the contracted channel flow are performed on a coarse curvilinear mesh $80 \times 30 \times 100$ and the results are compared with the highly resolved LES performed by Breuer et al (2009) using a dynamical Smagorinsky model (DSM) on a refined mesh (see table 2).

Table 2 Flow over periodic hills

Authors	Turbulence model	Grid points	Processors	Computer
Breuer *et al.* (2009)	DSM	$13.1\ 10^6$	64	Hitachi SR8000-F1
Chaouat and Schiestel (2009a)	PITM	$2.4\ 10^5$	8	NEC-SX8

Figure 3 shows the isosurfaces of the instantaneous filtered vorticity modulus and demonstrates clearly the three dimensional nature of the flow. A strong turbulence activity is visible near the lower wall.

Figures 4 and 5 describe the streamline contours of the instantaneous and averaged flow field, respectively, in a mid-plane plan and reveal the presence of the recirculation zone caused by the strong streamwise curvature of the lower wall. The flow separates at the hill crest at $x_1/h \approx 0.2$ and reattaches about 5 hill heights. These flow characteristics including the detachment and reattachment of the boundary layer are quit close to those provided by the highly resolved large-eddy simulations performed on refined meshes Fröhlich et al (2005); Breuer et al (2009).

Figure 6 displays the velocity profiles at four locations $x/h = 0.05, 2, 4, 8$ for both the PITM and RSM simulations. One can see that the PITM profiles agree very well with the reference data whereas the RSM profiles present some discrepancies with the data, particularly in the near wall region. This result seems indicated that the flow predicted by the RSM model reattaches too early in the downstream channel. This difference should probably attributed to the three dimensional flow effects that cannot be reproduced by a RANS computation because of the 2d-geometry. Figure 7 shows the profiles of the turbulent energy at the locations $x/h = 0.05, 2, 4, 8$. One can see that the turbulent intensity of the flow is qualitatively reproduced by both the RSM and PITM simulations. The PITM simulation, however, provides more accurate results than the RSM computation. But the turbulent peak of the turbulent energy is not well captured in the immediate vicinity of the lower wall. We assume that the mesh is not enough refined in the near wall region for properly reproducing the turbulent boundary layer.

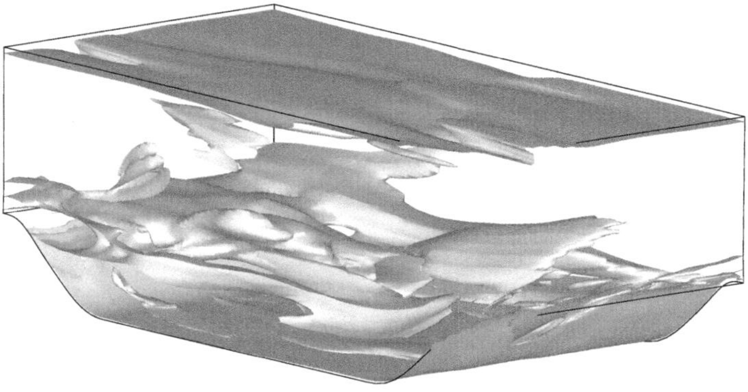

Fig. 3 Isosurfaces of the instantaneous filtered vorticity modulus

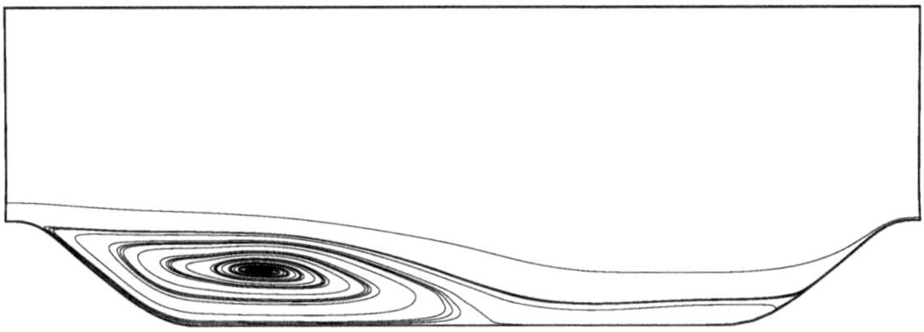

Fig. 4 Streamlines of instantaneous filtered flowfield

Fig. 5 Streamlines of the averaged filtered flowfield

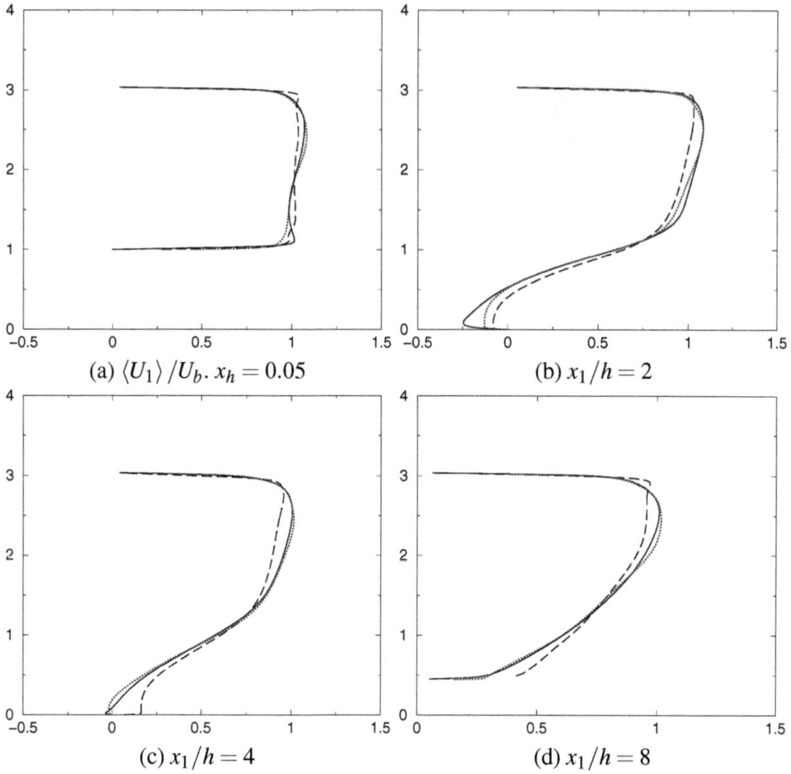

Fig. 6 Streamwise velocity profiles $\langle U_1/U_b \rangle$ at different locations (x/h=0.05, 2, 4, 8.) LES (Breuer *et al.*, 2009) —; PITM $\cdots$; RSM - -.

5 Conclusion

The PITM method viewed as a continuous approach for continuous hybrid non-zonal RANS/LES simulations has been applied to derive a subfilter scale stress model. This model has been used for simulating flows encountered in engineering applications. As a result, it is found that the model is able to reproduce fairly well these flows of complex physics. This hybrid method can be useful in industries because it allows a drastic saving of the computation time, due to the fact that the simulations can be performed on coarse meshes. Note that the present PITM model can be easily implemented in RANS codes, provided the numerical schemes are precise enough.

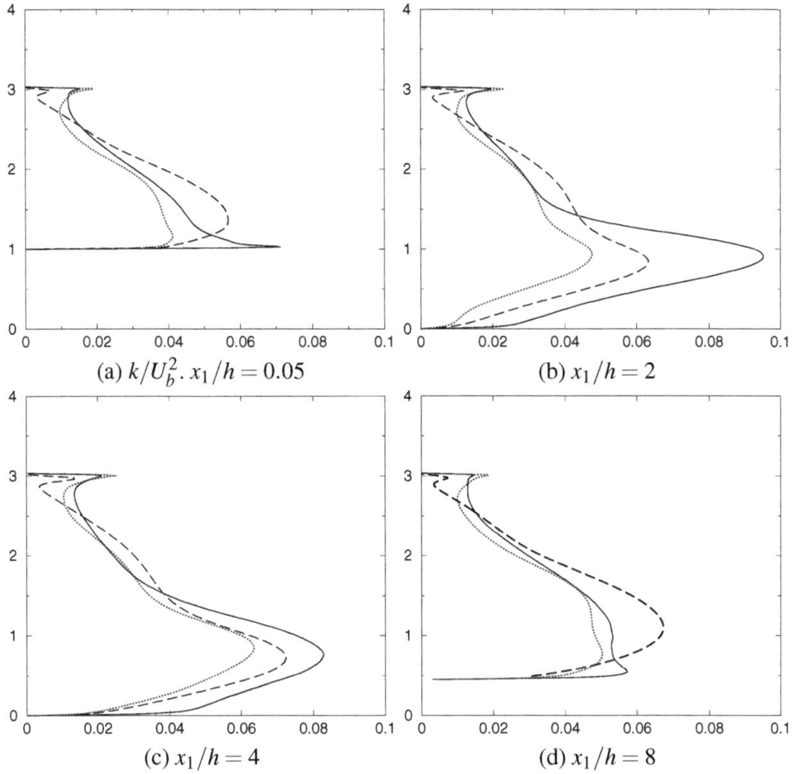

Fig. 7 Turbulent energy profiles k/U_b^2 at different locations (x/h=0.05, 2, 4, 8.) LES (Breuer *et al.*, 2009) —; PITM ···; RSM - -.

References

Apte, S.A., Yang, V.: A large-eddy simulation study of transition and flow instability in a porous-walled chamber with mass injection. Journal of Fluid Mechanics 477, 215–225 (2003)

Avalon, G., Casalis, G., Griffond, J.: Flow instabilities and acoustic resonance of channels with wall injection. AIAA Paper 98–3218 (1998)

Breuer, M., Peller, N., Rapp, C., Manhart, M.: Flow over periodic hills. Numerical and experimental study in a wide range of reynolds numbers. Computers and Fluids 38, 433–457 (2009)

Chaouat, B.: Reynolds stress transport modeling for high-lift airfoil flows. AIAA Journal 44(10), 2390–2403 (2006)

Chaouat, B., Schiestel, R.: A new partially integrated transport model for subgrid-scale stresses and dissipation rate for turbulent developing flows. Physics of Fluids 17(6) (2005)

Chaouat, B., Schiestel, R.: From single-scale turbulence models to multiple-scale and subgrid-scale models by Fourier transform. Theoretical Computational Fluid Dynamics 21(3), 201–229 (2007)

Chaouat, B., Schiestel, R.: Progress in subgrid-scale transport modelling for continuous hybrid non-zonal RANS/LES simulations. International Journal of Heat and Fluid Flow 30, 602–616 (2009)

Fadai-Ghotbi, A., Friess, C., Manceau, R., Gatski, T., Borée, J.: Toward a consistent formalism for seamless hybrid RANS-LES modelling in inhomogeneous turbulence based on temporal filtering. In: Proceeding 6nd Symposium on Turbulent Shear Flow Phenomena, vol. 3, pp. 1051–1056 (2009)

Fröhlich, J., Mellen, C., Rodi, W., Temmerman, L., Leschziner, M.: Highly resolved large-eddy simulation of separated flow in a channel with streamwise periodic constriction. Journal of Fluid Mechanics 526, 19–66 (2005)

Gatski, T.B., Rumsey, C.L., Manceau, R.: Current trends in modelling research for turbulent aerodynamic flows. Phil. Trans. Royal Society 365, 2389–2418 (2007)

Germano, M., Piomelli, U., Moin, P., Cabot, W.H.: A dynamic subgrid-scale eddy-viscosity model. Physics of Fluids 3(7), 1760–1765 (1992)

Jakirlic, S., Saric, S., Kadavelil, G., Sirbubalo, E., Basara, B., Chaouat, B.: SGS modelling in LES of wall-bounded flows using transport RANS models: from a zonal to a seamless hybrid LES/RANS method. In: Proceeding 6nd Symposium on Turbulent Shear Flow Phenomena, vol. 3, pp. 1057–1062 (2009)

Lesieur, M., Metais, O.: New trends in large-eddy simulations of turbulence. Ann. Rev. Journal of Fluid Mechanics 28, 45–82 (1996)

Schiestel, R.: Multiple-time scale modeling of turbulent flows in one point closures. Physics of Fluids 30(3), 722–731 (1987)

Schiestel, R., Dejoan, A.: Towards a new partially integrated transport model for coarse grid and unsteady turbulent flow simulations. Theoretical Computational Fluid Dynamics 18, 443–468 (2005)

Spalart, P.R.: Strategies for turbulence modelling and simulations. International Journal of Heat and Fluid Flow 21, 252–263 (2000)

Spalart, P.R.: Detached-eddy simulation. Annual Reviews Fluid Mechanics 41, 181–202 (2009)

Speziale, C.G., Sarkar, S., Gatski, T.B.: Modelling the pressure-strain correlation of turbulence: an invariant dynamical systems approach. Journal of Fluid Mechanics 227, 245–272 (1991)

A Hybrid RANS–LES Model Based on Temporal Filtering

A. Fadai-Ghotbi, C. Friess, R. Manceau, T.B. Gatski, and J. Borée

1 Introduction

Since large eddy simulation methods (LES) are too CPU-demanding for many complex industrial applications, a multitude of unsteady low-cost strategies have gained prominence over the last decade Fröhlich and von Terzi (2008); Sagaut et al (2006). Some of these models can be described as *seamless hybrid* RANS–LES models since the transition between RANS and LES occurs in a continuous manner so that there is no need to define explicit boundaries between RANS and LES regions. In statistically homogeneous flows, such a model can be seen as an LES with a filter width Δ_S continuously going to infinity – a limit that corresponds formally to a RANS approach. However, the majority of flows of practical relevance are inhomogeneous, and in that case such models suffer from an important conceptual weakness due to the fact that the LES yields spatially filtered fields and the RANS yields long-time averaged fields.

In order to develop a consistent seamless hybrid RANS–LES model in inhomogeneous flows, an Eulerian temporal filtering approach Pruett (2000); Pruett et al (2003) within the LES formalism can be used. The discussion is then focused within the context of a TLES (*Temporal Large-Eddy Simulation*) approach. This new approach, the so-called *Temporal Partially-Integrated Transport Model* (TPITM), is a transposition to temporal filtering of the PITM approach Schiestel and Dejoan (2005); Chaouat and Schiestel (2005), and enables the adaptation of any RANS model to a hybrid RANS–TLES model. In addition, it is then possible to provide a consistent interpretation of the usual eddy-resolving methods, such as DES, SAS, OES, URANS, ..., within a temporal filtering framework.

The work presented here is a summary of the detailed development of a hybrid RANS–TLES model developed in Fadai-Ghotbi et al (2009a,b).

A. Fadai-Ghotbi · C. Friess · R. Manceau · T.B. Gatski · J. Borée
Laboratoire d'Études Aérodynamiques (LEA), Université de Poitiers, ENSMA, CNRS, 86962 Futuroscope Chasseneuil Cedex, France
remi.manceau@lea.univ-poitiers.fr

T.B. Gatski
Center for Coastal Physical Oceanography and Ocean, Earth and Atmospheric Sciences, Old Dominion University, Norfolk, Virginia 23529 USA

S.-H. Peng et al. (Eds.): Progress in Hybrid RANS-LES Modelling, NNFM 111, pp. 225–234.
springerlink.com

2 Developing a Consistent Formalism

In formulating a seamless hybrid temporally filtered methodology that is capable of spanning the range of solution methodologies from DNS to RANS, it is necessary to insure that the resulting equations retain the same invariance properties as the Navier-Stokes and Reynolds averaged equations. A Galilean invariant, time filtering operation can be defined and applied in the partitioning of the flow variables into resolved and unresolved (subfilter) parts. In order to decompose the instantaneous velocity $\mathbf{u}^*$ into filtered (resolved) $\mathbf{U} = <\mathbf{u}^*>$ and residual $\mathbf{u}''$ parts, the resolved part would be defined as

$$<\mathbf{u}^*>(\mathbf{x},t) = \iint \mathscr{G}(\mathbf{x}' - \xi(\mathbf{x},t'-t),t'-t)\, \mathbf{u}^*(\mathbf{x}',t')\, \mathrm{d}\mathbf{x}'\mathrm{d}t', \tag{1}$$

with a filter of the form

$$\mathscr{G}(\mathbf{x}' - \xi(\mathbf{x},t'-t),t'-t) = \delta(\mathbf{x}' - \xi(\mathbf{x},t'-t))\, G_{\Delta_T}(t'-t) \tag{2}$$

with Δ_T the temporal filter width. The introduction of $\xi(\mathbf{x},t'-t)$, with $\xi(\mathbf{x},0) = \mathbf{x}$, in the definition of the filter aims at ensuring that the filtering operation preserves the translational, or Galilean, invariance, of the field Pruett (2000). In order to preserve the Galilean invariance, the filter $\mathscr{G}$ must be defined using $\xi(\mathbf{x},\tau) = \mathbf{x} + \mathbf{V}_{\mathrm{ref}}\,\tau$, i.e., the temporal kernel G_{Δ_T} of the filter is necessarily applied at a location moving with some reference velocity. To further unambiguously define the filter, this $\mathbf{V}_{\mathrm{ref}}$ reference velocity must be related to the flow configuration, e.g., the boundary conditions. With such definitions for $\mathbf{V}_{\mathrm{ref}}$, the Galilean invariance of the filtering process can be retained.

2.1 Consistency with the Reynolds Average

A consistent hybrid RANS–TLES method requires that the respective operators, that is the Reynolds averaging and temporal filtering operators, are mutually compatible in all inertial frames. In the case of a stationary flow, Reynolds averaged quantities are obtained as the limit of filtered quantities when the temporal width of the filter goes to infinity. For a Galilean invariant top-hat filter for example,

$$\mathscr{G}(\mathbf{x}' - \xi(\mathbf{x},t'-t),t'-t) = \delta(\mathbf{x}' - \mathbf{x} - \mathbf{V}_{\mathrm{ref}}(t'-t))\, \frac{1}{\Delta_T} H(t'-t+\Delta_T) H(t-t')\,, \tag{3}$$

the filtering operation is equivalent to a running average from $t - \Delta_T$ to t, denoted by $[\cdot]_{\{\Delta_T;\mathbf{V}_{\mathrm{ref}}\}}$,

$$<\mathbf{u}^*>(\mathbf{x},t) = \frac{1}{\Delta_T}\int_{t-\Delta_T}^{t} \mathbf{u}^*(\xi,t')\,\mathrm{d}t' = [\mathbf{u}^*]_{\{\Delta_T;\mathbf{V}_{\mathrm{ref}}\}}(\mathbf{x},t)\,, \tag{4}$$

and the generalized long-time averaging, defined by

$$[\mathbf{u}^*]_{\{\infty;\mathbf{V}_{\text{ref}}\}}(\mathbf{x},t) = \lim_{\Delta_T \to \infty} <\mathbf{u}^*> (\mathbf{x},t) \;, \tag{5}$$

inherits the Galilean invariance from the filter. This property ensures that, in any reference frame, the generalized long-time average (5) is equivalent to the Reynolds average. The definition of the filter (1), and the associated generalized long-time average (5), thus provide the consistent formalism for seamless hybrid RANS–TLES for a particular class of flows: the flows that are stationary in a particular reference frame.

2.2 Consistency between Averaged and Filtered Equations

If the temporal filter width Δ_T is constant, the temporal filter commutes with the differential operators, and it can be shown that the filtered field satisfies both continuity and momentum equations that are form-invariant with their Reynolds-averaged counterparts. Similarly, the subfilter stress tensor that now appears in the momentum equation correspondingly satisfies the same form-invariant equation as its Reynold-averaged counterpart – the Reynolds stress tensor.

One of the advantages of Eulerian temporal filtering in the context of stationary flows is that the following property is satisfied,

$$\overline{<\mathbf{u}^*>} = \overline{\mathbf{u}^*} \;, \tag{6}$$

so that the total Reynolds stress $\tau_{ij}(=\overline{u_i u_j})$ is exactly decomposed Germano (1992) into a resolved part and a modeled part given by

$$\begin{aligned}\tau_{ij} &= \tau_{ijr} + \tau_{ijm} \\ &= \left(\overline{\tilde{U}_i \tilde{U}_j} - U_i U_j\right) + \overline{\tau_{ij_{SFS}}}.\end{aligned} \tag{7}$$

The last term, $\overline{\tau_{ij_{SFS}}}$, in this equation is generally not accounted for in standard LES, since the cutoff wavenumber is located well inside the inertial region. In the case of hybrid methods, however, the contribution of the subfilter scales can be dominant (e.g. 100 % of the energy in RANS regions). Consequently, an accurate evaluation of this part is necessary, in general through transport equations for the subgrid energy k_{SFS} or, as in Chaouat and Schiestel (2005, 2009), by directly solving transport equations for the subfilter stress.

The main issue in this framework is the modeling of the subfilter stress, which must be compatible with all the possible locations of the characteristic cutoff frequency ω_c in the turbulent spectrum. The form-invariance of the TLES transport equation for the subfilter stress and the standard RANS equation for the Reynolds stress, suggests that the form of the models used in TLES and RANS regions can be identical, and that a sensitization of the model to the temporal filter width can provide the appropriate adaptation of the level of subfilter stresses in the momentum equation to account for the transition from a TLES to a RANS behavior.

3 Subfilter Models

Two approaches to developing a subfilter model for a hybrid RANS-TLES method have been investigated. One has been an extension of the subfilter stress model with the TLES framework and the other has been adapting a spatial hybrid methodology to the temporal domain. In the remainder of this section, aspects of utilizing either of these approaches are summarized.

3.1 TLES Models

The first possibility to be discussed in developing a hybrid RANS–TLES model is to extend an existing TLES subfilter stress model so that it becomes compatible with the RANS limit. For this purpose, the behavior of the models in the limit $\Delta_T \to \infty$ must be investigated. Since in Smagorinsky-type models a length scale characterizing the filter is needed to determine the subfilter viscosity, these models cannot be directly adapted to temporal filtering.

Models for the subfilter stresses for TLES based on Eulerian temporal filtering have been proposed by Pruett (2000); Pruett et al (2003); Tejada-Martinez et al (2007). In these models the filter, such as a scale-similarity model or a deconvolution model, is explicitly used. In the scale-similarity model, termed a Temporal Scale-Similarity Model (TSSM), the subfilter stress is approximated by

$$\tau_{ij_{SFS}} \approx << u_i^* >< u_j^* >> - << u_i^* >><< u_j^* >> \ . \tag{8}$$

Unfortunately, this TSSM model vanishes in the limit $\Delta_T \to \infty$, so that the model is not compatible with the RANS limit. The better candidate for an extension to the hybrid methodology is the Temporal Approximate Deconvolution Model (TADM) proposed by Pruett et al (2003), as an adaptation to time filtering of the ADM model Stolz et al (2001). This model is based on a spectral partitioning in the frequency domain

$$\omega \in [0,\infty] = \underbrace{[0,\omega_{\Delta_T}]}_{\substack{\text{Resolved}\\ \text{scales}}} \cup \underbrace{[\omega_{\Delta_T},\omega_{\Delta t}] \cup [\omega_{\Delta t},\infty]}_{\substack{\text{Subfilter}\\ \text{scales}}} \ , \tag{9}$$

where ω_{Δ_T} and $\omega_{\Delta t}$ are the frequencies associated with the temporal width of the filter Δ_T and with the time step Δt, respectively. Two parts are distinguished in the subfilter scales: the resolved subfilter scales (RSFS), in the range $[\omega_{\Delta_T},\omega_{\Delta t}]$, and the unresolved subfilter scales (USFS) in the range $[\omega_{\Delta t},\infty]$.

The RSFS are reconstructed using the temporal approximate deconvolution method

$$\tau_{ij_{RSFS}} \approx <v_i v_j> - <v_i><v_j> , \qquad (10)$$

where v_i is an approximation of the instantaneous (unfiltered) velocity u_i^* obtained by applying the approximate inverse filter $G^{-1}_{\Delta_T a}$ to the resolved velocity. The remaining contribution to the subfilter stress, the USFS, is obtained by a regularization procedure

$$\frac{\partial \tau_{ij_{USFS}}}{\partial x_j} = \chi(I - G_{\Delta_T} * G^{-1}_{\Delta_T a}) * <u_i^*>, \qquad (11)$$

which is purely dissipative. The behavior of the TADM model for large values of the temporal filter width Δ_T was investigated by Pruett et al (2003), who showed that the approximate deconvolution procedure is able to accurately reproduce the exact filtered velocity, provided that the order q of the approximate inverse is sufficiently high. Unfortunately, they also found that in the limit $\Delta_T \longrightarrow \infty$, the model is not valid, because the deconvolution procedure assumes that the filter is invertible, which is not the case in this limit: the frequency ω_c characterizing the width of the Fourier transform of the filter $\widehat{G}_{\Delta_T}$ goes to zero, such that $\lim_{\Delta_T \to \infty} \widehat{G}_{\Delta_T}(\omega) = 0$.

The current TLES models are thus not adaptable to a hybrid strategy since they suffer from the incompatibility between the requirement that the filter must tend to the Reynolds average in the limit $\Delta_T \longrightarrow \infty$ to ensure consistency, and the properties of the Reynolds average which result in the degeneracy of both the scale-similarity and approximate deconvolution models.

3.2 Temporal PITM (TPITM)

The second possibly to be explored is an adaptation/extension of the PITM model Schiestel and Dejoan (2005); Chaouat and Schiestel (2005) to the temporal filtering domain. This new subfilter stress model now proposed is the Temporal Partially Integrated Transport Model (TPITM). The TPITM model is based on a spectral analysis in the frequency domain, in order to guarantee compatibility between the two methodologies – RANS and TLES. In the TPITM formulation, the energy partition between the filtered and residual scales is cast in terms of the frequency domain. The transport equation for the Eulerian temporal energy spectrum $E_T(\mathbf{x}, \omega)$ is first derived Fadai-Ghotbi et al (2009a,b) in order to obtain the transport equation for the subfilter energy k_m,

$$k_m(\mathbf{x}) = \int_0^\infty \left[1 - \widehat{G}_{\Delta_T}(\omega)\widehat{G}^*_{\Delta_T}(\omega)\right] E_T(\mathbf{x}, \omega)\, d\omega \, . \qquad (12)$$

This equation is then the basis for the sensitization of a standard RANS dissipation rate equation to the temporal filter width.

In seamless hybrid RANS–LES methods, the amount of resolved energy is to be controlled by making the equations of the model dependent on the filter width. This can be achieved by modifying the scale variable, the subfilter dissipation rate ε_{SFS}, through a RANS-type transport equation. A detailed analysis is given in Fadai-Ghotbi et al (2009a,b) and the resulting transport equation for ε_{SFS} is given by

$$\frac{D\varepsilon_{SFS}}{Dt} = C_{\varepsilon_1}\frac{\varepsilon_{SFS}}{k_m}P_m - \underbrace{\left[C_{\varepsilon_1} + r(C_{\varepsilon_2} - C_{\varepsilon_1})\right]}_{C^*_{\varepsilon_2}}\frac{\varepsilon^2_{SFS}}{k_m} + D_{\varepsilon_{SFS}}\,, \tag{13}$$

where $r = k_m/k$ is the ratio of the subfilter energy to the total energy and $D_{\varepsilon_{SFS}}$ is a diffusion term. The RANS–TLES transition is controlled by the parameter r, with the RANS limit corresponding to $r = 1$. During the computation, the value of this parameter must be prescribed as a function of the filter width.

In TLES, the choice of the temporal filter width Δ_T, that is the frequency $\omega_c = \pi/\Delta_T$, is not as obvious as in spatial LES. As a first approach, it would seem straightforward to link Δ_T to the time step used in the computation, Δt; however, since Δt is necessarily the same everywhere in the domain, this relation does not allow local variations of the filter width, which is in contradiction to the objective of hybrid models. Another possibility is to link the temporal filter width to the local mesh refinement. For this purpose, ω_c must be related to the wavenumber associated with the mesh size $\kappa_c = \pi(\Delta x\Delta y\Delta z)^{-1/3}$. The main difficulty then lies in the fact that a dispersion relation $\omega = f(\kappa)$ is required to evaluate the frequency ω_c, and this relation is not known in general and combines complex effects. However, since the Eulerian frequency spectrum $E_T(\omega)$ and the wavenumber spectrum $E(\kappa)$ are related by

$$\mathrm{d}k = E(\kappa)\,\mathrm{d}\kappa = E_T(\omega)\,\mathrm{d}\omega, \tag{14}$$

the energy ratio $r = k_m/k$ can be evaluated either in the frequency domain or in the wavenumber domain. If a spectral cutoff filter is considered, the ratio r is given by

$$r = \frac{1}{k}\int_{\omega_c}^{\infty} E_T(\omega)\,\mathrm{d}\omega = \frac{1}{k}\int_{\kappa_c}^{\infty} E(\kappa)\,\mathrm{d}\kappa\,, \tag{15}$$

and when coupled with the Kolmogorov spectrum

$$E(\kappa) = C_\kappa \varepsilon^{2/3}\kappa^{-5/3}\,, \tag{16}$$

provides the relation between r and the cutoff wavenumber

$$r = \frac{3}{2}C_\kappa\left(\kappa_c\frac{k^{3/2}}{\varepsilon}\right)^{-\frac{2}{3}}. \tag{17}$$

The transport equations for the subfilter stresses are then indirectly sensitized to the filter width through the associated dissipation rate equation; the temporal filter implicitly enters the latter via the energy ratio r, and the above analysis shows that r can be related to the local grid step.

Up to this point, the discussion has focused on the development of a spectrally sensitive closure methodology for the subfilter field. Just as in the case of applying high Reynolds number RANS methodologies to practical wall-bounded flows, various near-wall modifications need to be made. One such near-wall modification is the elliptic blending methodology which can be adapted to the hybrid formulation presented here, and which has been validated against channel flow direct numerical simulation data at $Re_{\tau} = 395$ Fadai-Ghotbi et al (2009a,b).

4 Extension to Other Methodologies

Numerous approaches have been proposed in recent years to achieve the same goals as described here, that is, to provide a level of description of turbulence intermediate between LES and RANS. These approaches are collectively termed as *eddy-resolving methods* by Fröhlich and von Terzi (2008), and those approaches that ensure a continuous transition from a RANS to an LES model, depending on the local grid size, are called seamless, global Sagaut et al (2006), unified Fröhlich and von Terzi (2008) or non-zonal hybrid RANS–LES models. Other methods, that do not rely explicitly on the grid step, are denoted by *second-generation URANS models* by Fröhlich and von Terzi (2008).

One of the main features shared by these models is an ambiguous definition of the resolved variables. In the regions where the model works in LES mode, the spatially-filtered momentum equations are solved; in the RANS region, the Reynolds average is used, which is a statistical average. Contrary to zonal models where the LES and RANS regions do not overlap, a fundamental inconsistency appears in seamless methods, since the transition from RANS to LES is continuous. Indeed, the two approaches do not continuously match, except in the case of homogeneous flows. Although the presentation here has focused on the introduction of temporal filtering to provide a consistent framework for a new hybrid methodology for inhomogeneous, stationary flows, it can be argued that many, if not all, seamless and second-generation URANS methods should be regarded as RANS–TLES hybrid methods.

Recall that this newly proposed TPITM model as well as the seamless and second-generation URANS methods share the common objective of controlling the level of turbulent stress in the momentum equations to ensure a transition from a RANS to a (T)LES behavior where necessary. In many of these other hybrid methods, a *subfilter viscosity* ν_t is proposed and is related to the eddy-viscosity ν_{tRANS} of a RANS model. The same reduction of the level of stresses in the momentum equation is achieved by a direct damping of the eddy-viscosity or, equivalently, a reduction of the length scale. For instance, the most widespread hybrid method, DES Spalart et al (2009), in its original formulation, directly reduces the level of eddy-viscosity in detached regions, by increasing its destruction term. In SAS Menter et al (2003), the same effect is achieved by a decrease of the production term in the eddy-viscosity transport equation. Other methods, such as OES Bourguet et al (2007, 2008), XLES Kok et al (2004) or the method of Fan et al (2004), directly modify the Boussinesq relation, either by redefining the coefficient C_{μ} or by blending the RANS and LES formulae. In LNS Batten et al (2004) and FSM Fasel et al (2002), the stresses in the momentum equations are directly premultiplied by a damping factor, as originally proposed by Speziale (1998), which is equivalent to modifying the eddy-viscosity. In the case of PANS Girimaji (2006), which controls the ratio modeled energy/total energy through the C_{ε_2} coefficient, the similarity with PITM is obvious, despite a completely different argumentation and significant differences in the practical details.

In many of these other hybrid methods, the length scale entering the model coincides with the grid step, in order to be equivalent to LES. However, it can be shown Fadai-Ghotbi et al (2009b) that a relation exists between the wavenumber and the frequency characterizing the energy partition, which reflects the fact that a spatial filtering implicitly induces a temporal filtering, and vice versa. The Fourier transforms of the two filters are related by

$$\widehat{G}_{\Delta_S} = \widehat{G}_{\Delta_T} \circ f \ , \tag{18}$$

where $\omega = f(\kappa)$ is the dispersion relation. Therefore, a hybrid formulation based on a filter width related to the grid step can be viewed as a temporally-filtered approach.

Some additional comments pertaining to URANS in stationary flows is in order. In such flows, the Reynolds average is equivalent to a long-time average, so that the solution of the RANS equations is expected to be independent of time. However, in practice, the solutions are often found to be time-dependent, and the method is then denoted by URANS. Although the RANS equations are recovered by making the temporal filter width go to infinity, the time scale entering the RANS models is the integral time scale; therefore, the equations of the model are formally identical to the equations of a subfilter stress model with a filter width equal to the integral time scale.

5 Concluding Remarks

Temporal filtering provides a consistent formalism for a class of seamless hybrid RANS–LES methodologies for inhomogeneous and stationary turbulent flows. In such flows, the generalized temporal filter consistently bridges the two approaches, RANS, based on Reynolds averaging, and TLES, based on temporal filtering. The form-invariance between the RANS and TLES equations provides the basis for the use of similar models for the subfilter stresses, with a continuous variation of the coefficients of the model to transition between the two methodologies.

In order to seamlessly transition from RANS to TLES, the hybrid methodology must be able to control the amount of resolved energy. Within this framework, the TPITM, the temporal version of the PITM, has been developed. It was also shown through an analysis in the frequency domain that any RANS model using a transport equation for a scale variable, such as the dissipation rate, can be sensitized to the cutoff frequency introduced by the filter. Consequently, the temporal filter width can be implicitly fixed by the grid step, which enables local refinement.

An important consequence of this analysis is the fact that a temporally filtered approach does not necessarily need to use explicitly the temporal filter width, but can rather be based on a corresponding spatial filter width. It can then be concluded that most of the existing seamless hybrid methodologies, such as the widely used DES, or the so-called *second-generation URANS models* Fröhlich and von Terzi (2008), such as SAS, and even the basic URANS, can be regarded as temporally filtered approaches. This observation then reconciles an apparent inconsistency in the formalisms adopted in many previously proposed hybrid methods.

References

Batten, P., Goldberg, U., Chakravarthy, S.: Interfacing statistical turbulence closures with large-eddy simulation. AIAA J. 42, 485–492 (2004)

Bourguet, R., Braza, M., Perrin, R., Harran, G.: Anisotropic eddy-viscosity concept for strongly detached unsteady flows. AIAA J. 45, 1145–1149 (2007)

Bourguet, R., Braza, M., Harran, G., Akoury, R.E.: Anisotropic Organised Eddy Simulation for the prediction of non-equilibrium turbulent flows around bodies. J. Fluid Struct. 24, 1240–1251 (2008)

Chaouat, B., Schiestel, R.: A new partially integrated transport model for subgrid-scale stresses and dissipation rate for turbulent developping flows. Phys Fluids 17, 1–19 (2005)

Chaouat, B., Schiestel, R.: Progress in subgrid-scale transport modelling for continuous hybrid non-zonal RANS/LES simulations. Int. J. Heat Fluid Flow 30, 602–616 (2009)

Fadai-Ghotbi, A., Friess, C., Manceau, R., Gatski, T.B., Borée, J.: Toward a consistent formalism for seamless hybrid rans–les modelling in inhomogeneous turbulence based on temporal filtering. In: Kasagi, N., Eaton, J.K., Friedrich, R., Humphrey, J.A.C., Johansson, A., Sung, H.J. (eds.) Proceedings Sixth International Symposium on Turbulent Shear Flow Phenomena (2009a)

Fadai-Ghotbi, A., Friess, C., Manceau, R., Gatski, T.B., Borée, J.: Toward a consistent formalism for seamless hybrid rans–les modelling in inhomogeneous turbulence based on temporal filtering. Int. J. Heat Fluid Flow (2009b) (Submitted for Publication)

Fan, C.C., Xiao, X., Edwards, J.R., Hassan, H.A., Baurle, R.A.: Hybrid large-eddy/Reynolds-averaged Navier-Stokes simulations of shock-separated flows. J. Spacecraft Rockets 41, 897–906 (2004)

Fasel, H.F., Seidel, J., Wernz, S.: A methodology for simulations of complex turbulent flows. J. Fluid Eng-T ASME 124, 933–942 (2002)

Fröhlich, J., von Terzi, D.: Hybrid LES/RANS methods for the simulation of turbulent flows. Prog. Aersp. Sci. 44, 349–377 (2008)

Germano, M.: Turbulence: the filtering approach. J. Fluid Mech. 238, 325–336 (1992)

Girimaji, S.S.: Partially-averaged Navier-Stokes model for turbulence: A Reynolds-averaged Navier-Stokes to direct numerical simulation bridging method. J. APPL. Mech-T ASME 73, 413–421 (2006)

Kok, J., Dol, H., Oskam, H., van der Ven, H.: Extra-large eddy simulation of massively separated flows. AIAA Paper 2004-0264 (2004)

Menter, F., Kuntz, M., Bender, R.: A scale-adaptive simulation model for turbulent flow predictions. AIAA Paper 2003-0767 (2003)

Pruett, C.D.: Eulerian time-domain filtering for spatial large-eddy simulation. AIAA J. 38, 1634–1642 (2000)

Pruett, C.D., Gatski, T., Grosch, C.E., Thacker, W.D.: The temporally filtered navier-stokes equations: properties of the residual stress. Phys. Fluids 15, 2127–2140 (2003)

Sagaut, P., Deck, S., Terracol, M.: Multiscale and multiresolution approaches in turbulence. Imperial College Press, London (2006)

Schiestel, R., Dejoan, A.: Towards a new partially integrated transport model for coarse grid and unsteady turbulent flow simulations. Theoret Comput Fluid Dynamics 18, 443–468 (2005)

Spalart, P.R., Jou, W.H., Strelets, M., Allmaras, S.R.: Comments on the feasibility of LES for wings, and on a hybrid RANS/LES approach. In: Liu, C., Liu, Z. (eds.) Advances in DNS/LES. Greyden Press, Columbus (2009)

Speziale, C.G.: Turbulence modeling for time-dependent $_{RANS}$ and VLES: a review. AIAA J. 36, 173–184 (1998)

Stolz, S., Adams, N.A., Kleiser, L.: An approximate deconvolution model for large-eddy simulation with application to incompressible wall-bounded. Phys. Fluids 13, 997–1015 (2001)

Tejada-Martinez, A.E., Grosch, C.E., Gatski, T.B.: Temporal large eddy simulation of unstratified and stably stratified turbulent channel flows. Int. J. Heat Fluid Flow 28, 1244–1261 (2007)

Scale-Adaptive Simulation with Artificial Forcing

F.R. Menter, A. Garbaruk, P. Smirnov, D. Cokljat, and F. Mathey

Abstract. The present paper describes first efforts carried out to extend the range of Scale-Adaptive Simulation (SAS) methods into the area of RANS-stable flows. Instead of relying on inherent flow instabilities for generating resolved turbulence in unstable regimes, the current approach employs forcing terms for transferring modelled turbulence energy into resolved energy in pre-specified regions of the simulation domain.

1 Introduction

In recent years there was a strong drive for including unsteady elements into established RANS turbulence models. The most prominent example is the Detached Eddy Simulation (DES) approach of Spalart and co-workers (Spalart 2000) in its multiple forms. The underlying assumption of DES was to avoid the high resolution demands of Large Eddy Simulation (LES) in wall bounded flows by keeping established RANS models active there, while at the same time switching to LES-like formulations in large separation ("detached") zones. The switching is achieved by comparing the RANS and (a suitably defined) LES length scale (or the eddy-viscosity) and selecting the smaller of both. In addition, shielding functions are introduced for preserving the RANS model inside the boundary layer. The idea behind the DES methodology was essentially that RANS (or more precisely URANS) models are not capable of resolving turbulent structures in separated flow regions. This is based on the observation that standard RANS models provide only very large scale unsteady structures in unstable flows like flow past a cylinder. This behaviour is often associated with the Reynolds averaging technique underlying SAS model formulations.

F.R. Menter
ANSYS Germany GmbH, Staudenfeldweg12, 83624 Otterfing, Germany
florian.menter@ansys.con

A. Garbaruk · P. Smirnow
NTS, Russia

D. Cokljat
ANSYS, U.K.

F. Mathey
Air Liquide, France (formerly ANSYS France)

S.-H. Peng et al. (Eds.): Progress in Hybrid RANS-LES Modelling, NNFM 111, pp. 235–246.
springerlink.com

In a series of publications, Menter and Egorov (2004, 2005a, 2005b, 2005c, 2006) have demonstrated that this behaviour of RANS models is not a result of the Reynolds averaging, but of the specifics of model formulations within the RANS concept. It was shown that URANS models show entirely different unsteady characteristics when basing the scale-equation on an exact transport equation for the turbulent length scale (Rotta, 1972). This introduces the von Karman length scale into the system, which allows the model to adjust its behaviour to resolved scales. The resulting modelling concept was termed Scale-Adaptive Simulation (SAS) and it has been demonstrated that SAS models behave in many flows similarly to DES, but with less explicit impact of the grid spacing on the model formulation. Industrial experience (e.g. Grahs and Othmer, 2006, Widenhorn et al. 2009) over the last few years has shown that numerous inherently unstable flows can be handled well with the SAS formulation.

In order to produce unsteady structures, both DES and SAS rely in their classical formulation on "strong enough" flow instabilities. Such instabilities are typically observed for flows around blunt obstacles (cylinders, buildings, stalled airfoils/wings, mixers, ...) or flows with a strong swirl instability (combustion chambers etc.). However, there are also many flows, where the instability is not strong enough and no, or "insufficient" unsteadiness is produced (Davidson, 2006). For DES models, this results in the well known "grey areas", where the model switches to LES, but the flow is not sufficiently resolved and the model behaviour is undefined. In SAS, the result is a steady state RANS, or a low-fidelity URANS solution.

In order to extend the SAS approach to flows which do not exhibit a sufficiently strong instability for generating LES content, an explicit transfer of turbulent kinetic energy, *k*, is needed. This can be achieved with forcing terms in the momentum equations, which convert modelled *k* into resolved *k*.

The idea of forcing consists in introducing a volume stochastic source term (velocity fluctuations based on a modified Random Flow Generator (RFG) of Smirnov et al., 2001, Batten et al., 2004) in the momentum equations and a corresponding sink term in the *k*-transport equation of the SST-based SAS model. These terms are introduced in a confined user-specified flow region ("forcing zone"), where switching from RANS to LES mode is desired. In this zone, RFG generates unsteady structures which the SAS can then detect. The SAS model thereby automatically reduces its length scale (and eddy-viscosity) to these structures resulting in an LES-like formulation. The RFG, in turn, adapts its synthetic length sales to these smaller structures, which ultimately results in an automatic effective deactivation of the RFG in the most of the specified region. As a result, the RFG remains active only in a close vicinity of the user-specified boundary of the forcing zone. The forced SAS model will be termed SAS-F model.

2 Model Formulation

2.1 SST-SAS Model

The underlying model is the SST-SAS model which consists of the original SST model (Menter, 1994) plus an additional term in the ω–equation:

$$Q_{SAS} = \max\left[\rho\zeta_2 S^2\left(\frac{L}{L_{vK}}\right)^2 - C_{SAS}\frac{2\rho k}{\sigma_\Phi}\max\left(\frac{1}{k^2}\frac{\partial k}{\partial x_j}\frac{\partial k}{\partial x_j},\frac{1}{\omega^2}\frac{\partial \omega}{\partial x_j}\frac{\partial \omega}{\partial x_j}\right),0\right] \quad (1)$$

with $\zeta_2=1.47$, $\sigma_\Phi=2/3$ and $C_{SAS}=2$ and:

$$L_{vK} = \kappa\left|\frac{U'}{U''}\right|;\quad L = c_\mu^{-1/4}\frac{\sqrt{k}}{\omega};\quad U'' = \sqrt{\frac{\partial^2 U_i}{\partial x_k^2}\frac{\partial^2 U_i}{\partial x_j^2}};\quad U' = \sqrt{2\cdot S_{ij}S_{ij}} \quad (2)$$

In addition, a high wave number damping is used in the model to avoid accumulation of energy at the smallest scales. This is achieved by imposing a lower limit of the von Karman length scale:

$$L_{vK} = \max\left(\kappa\left|\frac{U'}{U''}\right|; C_S\sqrt{\frac{\kappa\zeta_2}{\beta/c_\mu-\alpha}}\Delta\right);\quad \Delta = \sqrt[3]{V} \quad (3)$$

with $C_s=0.11$, $\kappa=0.41$, $c_\mu=0.09$ and $\alpha=0.44$, $\beta=0.0828$ are the k-ε related coefficients of the SST model.

2.2 Forcing Terms for the SAS-F Model

Forcing terms are introduced into the momentum equations for modelling the transfer of modelled to resolved turbulent kinetic energy. In addition, a sink term is defined for the equation of the modelled turbulent kinetic energy, k, in order to preserve the energy between the resolved and modelled portions. The source terms read:

$$F_{mom,i} = \frac{\rho u_{f,i}}{\Delta t};\qquad F_k = -0.5\frac{\rho u_{f,i}^2}{\Delta t} \quad (4)$$

where ρ is the density, Δt the time step and u_{fi} the velocity fluctuation to be transferred. The velocity fluctuations are produced by using the Random Flow Generator (RFG) as proposed by Kraichnan (1969) and including modification of Batten et al. (2004) for the frequency distribution. The formulation is local and requires as input only the length and time scale of the RANS turbulence model:

$$u_{f,i} = \sqrt{\frac{2}{3}k}\cdot\sqrt{\frac{2}{N}}\sum_{n=1}^{N}\left[p_i^n\cdot\cos(\arg^n) + q_i^n\cdot\sin(\arg^n)\right] \quad (5)$$

$$p_i^n = \varepsilon_{ijk}\eta_j^n d_k^n,\quad q_i^n = \varepsilon_{ijk}\xi_j^n d_k^n$$

$$\arg^n = 2\pi\left(\frac{d_i^n x_i}{L_t} + \frac{\omega^n t}{\tau_t}\right)$$

where $L_t = C_L \frac{\sqrt{k}}{C_\mu \omega}$, $(C_L = 0.5)$ is the length scale of turbulence and $\tau_t = \frac{L_t}{\sqrt{k}}$ the time scale.

$$\eta_i^n = N(0,1),\quad \xi_i^n = N(0,1),\quad d_i^n = N(0,0.5),\quad \omega^n = N(1,1) \tag{6}$$

N(φ, ψ) is a normally distributed random variable with mean *φ* and standard deviation *ψ*. Numerical experiments have shown that typically *N~200* modes are sufficient for the required energy transfer. Furthermore, it is not sensible to transfer energy of modes which cannot be resolved with a given numerical grid and time step. A Nyquist limiter is therefore employed:

$$\frac{\tau_t}{\omega^n} \geq 2\cdot\Delta t;\quad \frac{L_t}{\left|d^n\right|} \geq 2\cdot\Delta h \tag{7}$$

where $\left|d^n\right| = \sqrt{d_i^n d_i^n}$ is the magnitude of the wave number, $\Delta h = \max(\Delta_x, \Delta_y, \Delta_z)$ is the characteristic cell size. Note that an isotropy assumption is used instead of the Cholesky decomposition in the formulation of the RFG. This is deemed sufficient for these first tests. It should also be noted that there is a dynamic interaction between the SAS turbulence model and the RFG, as the SAS model adjust automatically to the smallest resolved scales. As these scales are typically in the LES-limit, the assumption of isotropic turbulence seems justified.

3 Numerical Method

The simulations have been carried out with the general purpose CFD code ANSYS Fluent 12.0. The RFG model has been implemented via a User Defined Function (UDF). The code is based on a cell-centred finite volume method for unstructured hybrid grids. In the current simulations, the pressure-based incompressible option in the solver has been selected. For the spatial discretization of the momentum equations, a pure central differencing scheme of second order accuracy was applied, whereas for the turbulence equations a second order upwind scheme was used. A second-order implicit scheme was chosen for time integration, and a non-iterative time-advancement (NITA) scheme (see, for ex., Armsfield & Street, 1999) was utilized to solve the equations at each time step.

4 Testcases

4.1 Decaying Isotropic Turbulence (DIT)

DIT is typically a test used for the calibration of LES models. In this mode, an initial resolved velocity field is generated in a box of dimension 2π×2π×2π and the decay of the resolved kinetic energy inside the box is tracked in time. In addition, spectral information is extracted through FFT and compared with experimental data. It is important to note that with the SAS model, the DIT case can be run in two modes. The first is the RANS mode, where only 2 values are specified for k and ω (or L_t and τ) and their variation in time is computed (from the k- and ω-equation). However, as the SAS model can adjust to resolved scales, the model can also be run in "LES" mode. In this mode, an initial resolved velocity field is specified and the turbulence model is run on that frozen field to convergence in order to generate initial conditions for k and ω. Once the turbulence level are established, the SAS model is run in combination with the momentum equations to produce a dynamically changing and decaying velocity distribution, like in the LES case. The SAS model is calibrated for achieving correct spectral information in this mode (see Figure 1).

In the current application, another mode is of interest, whereby no resolved velocity field is specified initially, but only two variables for k and ω are given as in the RANS simulation. However, using the RFG, the goal is the conversion of the unresolved turbulent kinetic energy into a resolved part and a remaining modelled portion (subgrid). In other words, the specified values of k and ω will be used in RFG for switching from RANS to "LES" mode. The RFG formulation is such that this transfer takes place essentially in the first time step. While the RFG forcing remains active also for the remainder of the simulation, its effect is substantially reduced due to the resulting small length scale and the activation of the Nyquist limiter.

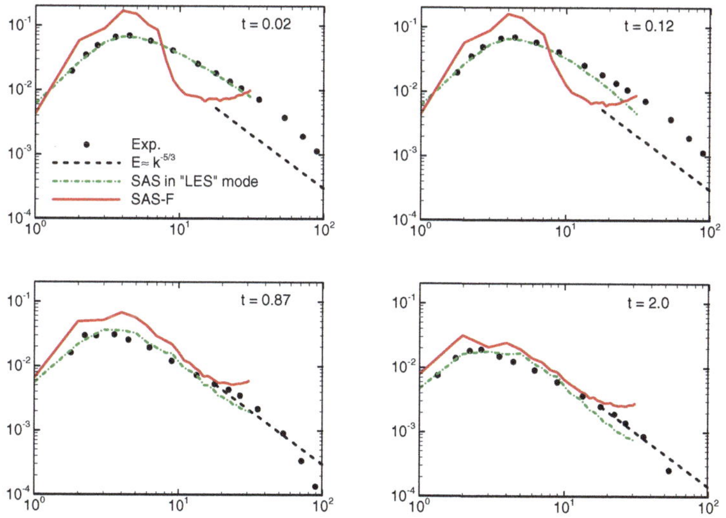

Fig. 1 Spectral development for Decaying Isotropic Turbulence (DIT) test case

Figure 1 shows the spectral development at 4 non-dimensional time values for the Compte-Bellot & Corrsin, (1971) experiment for a 64×64×64 grid. A time step of $\Delta t=0.005$ was used. Again, it has to be emphasized that the simulation is started with only two numbers: $k=0.81$, $\omega=3.75$ and the entire spectral content is generated through the forcing terms and the RFG.

As expected, the initial spectrum generated by the RFG is not in perfect agreement when compared with the experimental data. Also, the behaviour at the high wave numbers is currently not fully understood. Nevertheless, the energy is successfully transferred from the modelled to the resolved part and the spectral distribution improves with time once the process is controlled by the momentum equations and the eddy-viscosity provided by the SAS model. Note that the turn-up of the energy at high wave numbers is a result of the RFG as can be seen in comparison with the SAS simulations starting from a resolved field ("LES" mode).

Figure 2 shows the strength of the RFG terms in the *k*-equation for the duration of the simulation. It is clear that the transfer of energy takes place essentially at the first time step, as designed (note logarithmic scale).

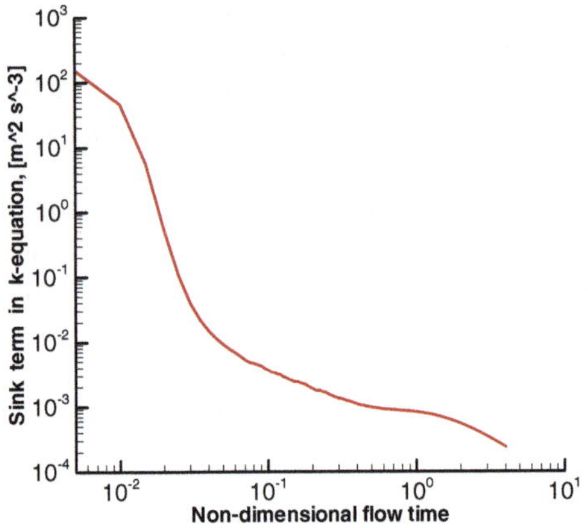

Fig. 2 Evolution of RFG sink term in the k-equation for the DIT test case

4.2 2D Channel Flow

The goal of the current formulation is to enable the SAS model to operate in unsteady mode even for flows which are URANS-stable. This applies to essentially all wall bounded flows (boundary layers, channel flows, ...). While it is often desirable that the SAS model returns a RANS solution in such portions of the flow domain, there can be situations, where the switch to an unsteady formulation is desirable (for increased accuracy, for capturing unsteady effects further downstream, for acoustics simulations, etc.). The most widely used testcases for LES models in wall bounded flows are periodic turbulent channel flows at different Reynolds numbers. These flows reveal the essential weakness of LES models requiring substantial increases in grid resolution with Reynolds number. The current application of the SAS-F model to this flow is a first test for investigating the general behaviour of this concept and exploring the Reynolds number scaling on under-resolved grids. Four different Reynolds numbers (Re_τ=*395, 950, 2000* and *20 000*) have been computed without any adjustment/optimization of the model formulation. Table 1 shows the non-dimensional resolution for the four cases and also the number of nodes in streamwise, wall normal and spanwise direction. The time step used was $\Delta t=0.02$ and in the averaging procedure over 10 000 time steps were used.

Table 1 Non-dimensional grid resolution and grid dimensions for the channel flows.

Re_τ	*395*	*950*	*2000*	*20 000*
N_x	73	81	81	81
N_y	75	97	97	141
N_z	73	61	61	61
Δx^+	35	100	200	2000
Δy^+	30	50	100	1000
Δz^+	17	50	100	1000

Figure 3 shows the turbulent structures computed with the SAS-F model for the four Reynolds numbers depicting iso-surfaces of $Q=S^2-W^2$ ($Q_{iso}=0.3$).

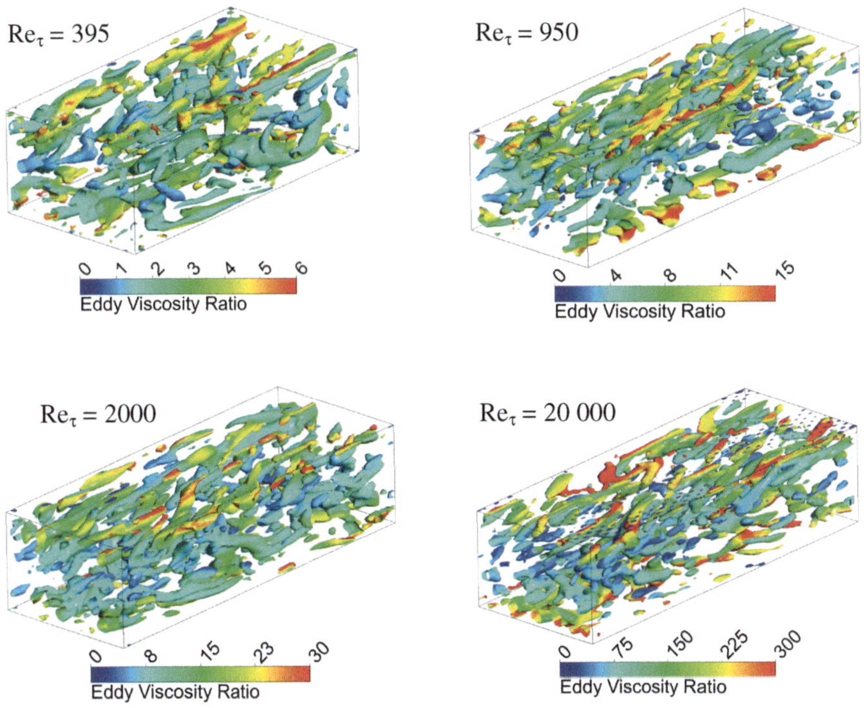

Fig. 3 Turbulent structures in developed channel flow computed with the SAS-F model for the four Reynolds numbers. Iso-surfaces correspond to the value of Q-invariant equal to 0.3 for all the cases, coloured with (eddy viscosity/molecular viscosity) ratio

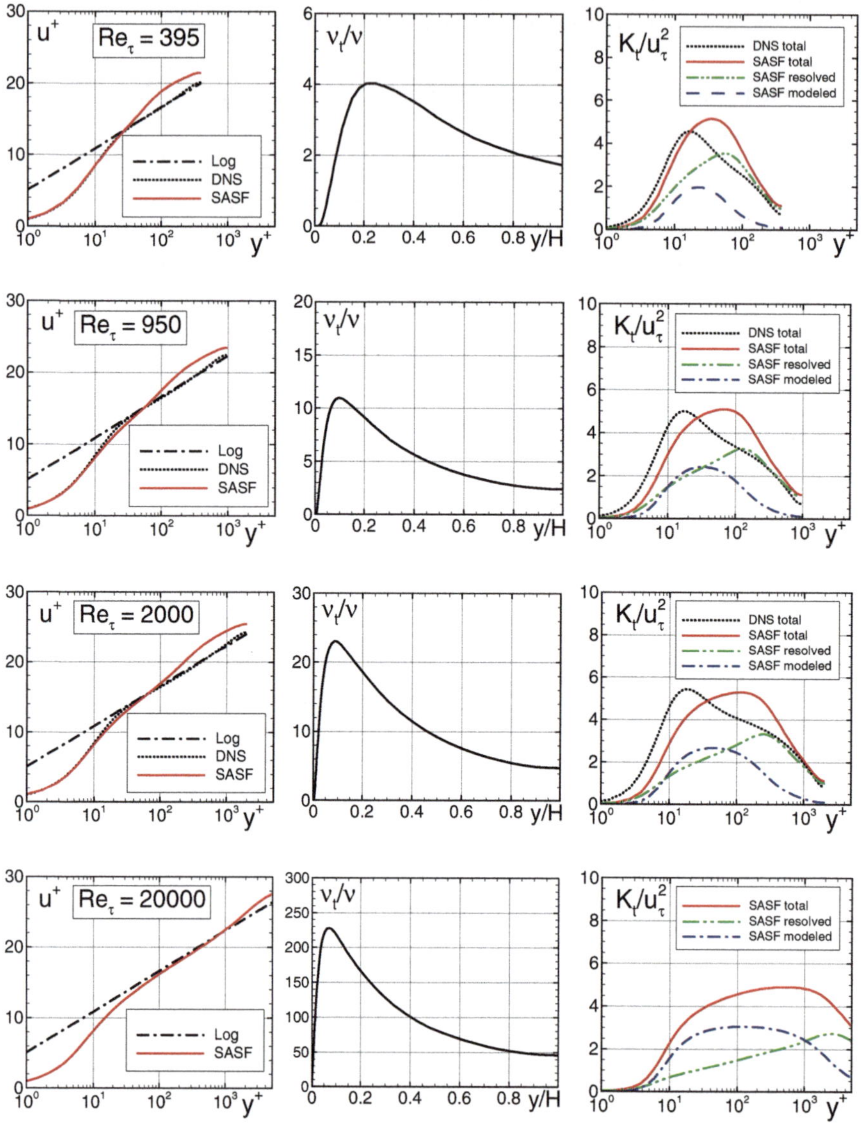

Fig. 4 Velocity profiles in wall-coordinates and eddy viscosity ratio across the channel for the four Reynolds numbers computed using the SAS-F model

Figure 4 shows the non-dimensional velocity profiles along with the corresponding profiles of the eddy-viscosity. It can be seen that the velocity profiles lay above the DNS and the desired Log profile. In addition, the model shows a double layer structure for the high Re number (and thereby under-resolved) cases similar to the profiles obtained with DES-hybrid models if no additional modifications are applied. While this is not an optimal behaviour, it is considered an encouraging starting point for the formulation of a more sophisticated model for such flows. It also appears that the simulations actually improve with higher Reynolds number. This behaviour and possible enhancements of the model will be explored in the future.

4.3 Backward Facing Step Flow

The backstep flow is an interesting practical candidate for the SAS-F model. Again the flow is stable when run with the SST-SAS model. It is however anticipated that the flow can be converted to unsteady mode with less impact of the forcing terms than for the channel flow. The reason lies in the shear layer emanating from the step. This layer is less stable than an attached boundary layer and will be able to quickly dominate the approximations of the RFG method, once unsteady structures appear. The situation for the backstep is actually more relevant for industrial flow simulations, than the channel flow. In engineering flows, it is often more desirable to resolve parts of the turbulent spectrum in separated flows, than to resolve the actual wall boundary layer.

In the current test, the experiment of Vogel & Eaton, 1985 is computed. The Reynolds number based on the step height is 28 000. Inlet profiles are specified from the experiment. The simulation is carried out on a grid of 2,309,715 nodes. A fragment illustrating the grid in the XY plane is shown in Figure 5. In the spanwise direction, the grid has 81 nodes and the domain size is Z/H = 4. The time step was $\Delta t=0.02$ and 5 000 time steps where accumulated in the averaging of the flow. The forcing terms are activated only downstream of the step.

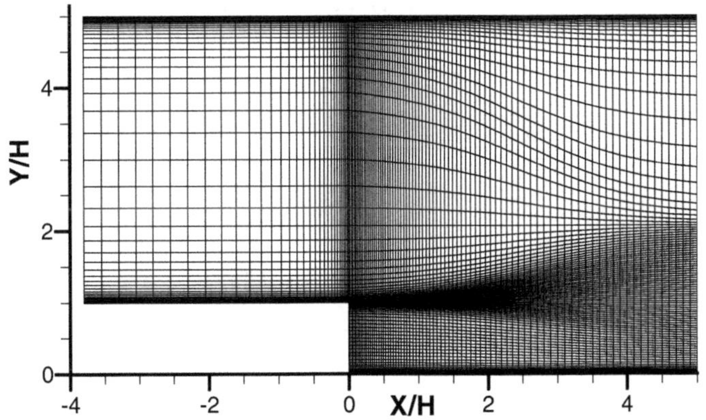

Fig. 5 Fragment of the grid for the backward-facing step in the XY plane

Figure 6 shows the turbulent structures again using an iso-surface of the Q-criterion ($Q_{iso}=0.5$). One can clearly see the flow instability emanating from the step. The effect of the RFG is actually relatively small in this flow, and is quickly dominated by the flow instability (note that RFG generates isotropic turbulent structures, contrary to the observed structures in Figure 6).

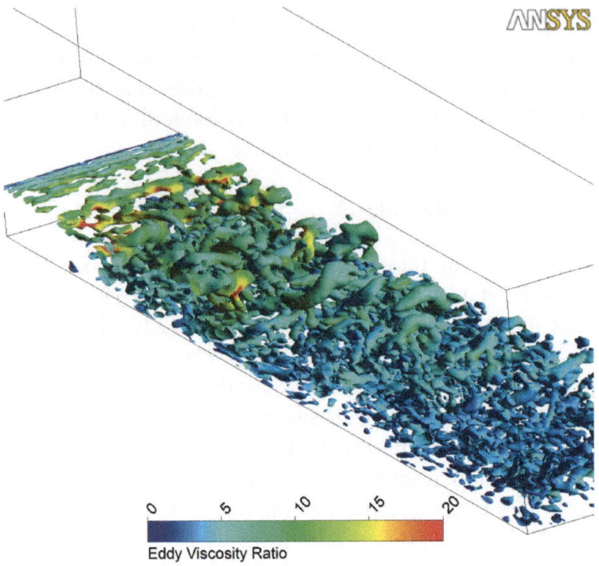

Fig. 6 Turbulent structures in the backward-facing step flow computed with the SAS-F model. Iso-surface of the Q-invariant = 0.5 colored with eddy viscosity ratio

Figure 7 shows the wall shear stress distribution downstream of the step. It is in good agreement with the experimental data, giving the correct flow reversal and reattachment location. The C_f in the recovery region downstream of reattachment is still low compared to the data. This was however also observed in pure LES simulations. An alternative set of experimental data indicates lower values – this will have to be explored in more detail.

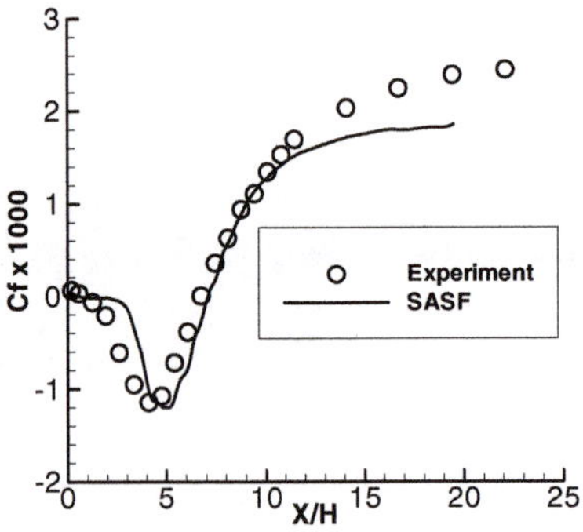

Fig. 7 Distribution of skin friction coefficient downstream of the backward-facing step. Result from the SAS-F model is averaged in time and in spanwise direction

Figure 8 shows a comparison of the velocity profiles. Again, the agreement with the data is quite good considering the early development status of the model.

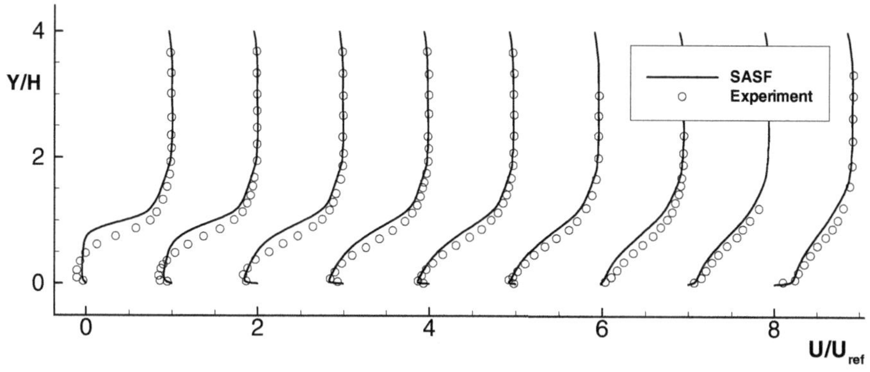

Fig. 8 Mean profiles of the streamwise velocity downstream of the backward-facing step. Results from the SAS-F model are averaged in time and in spanwise direction

5 Summary

Forcing terms for the SST-SAS model have been proposed. They allow the model to switch to unsteady mode even if the flow is stable for the conventional SST-SAS model. The forcing terms can be activated in limited regions of the domain and allow a convenient selection of the unsteady flow regime. The SST-SAS-F model has been applied in this first iteration to three testcases and has shown an overall good performance, considering that no optimization of the formulation has been attempted to capture additional effects as e.g. in wall bounded flows. The model will be developed further within the European Research project ATAAC.

Acknowledgement. The original SST-SAS model has been developed partly within the EU research project DESIDER (Detached Eddy Simulation for Industrial Aerodynamics) under contract No. AST3-CT-200-502842 (http://cfd.mace.manchester.ac.uk/desider/)

References

Armsfield, S., Street, R.: The Fractional-Step Method for the Navier-Stokes Equations on Staggered Grids: Accuracy of Three Variations. Journal of Computational Physics 153, 660–665 (1999)

Batten, P., Goldberg, U., Chakravarty, S.: Interfacing statistical turbulence closures with large-eddy simulation. AIAA Journal 42, 485–492 (2004)

Compte-Bellot, G., Corrsin, S.: Simple Eulerian time correlation of full – and narrow band velocity signals in grid-generated, "isotropic" turbulence. J. Fluid Mech. 48, 273–337 (1971)

Davidson, L.: Evaluation of the SST-SAS Model: Channel Flow, Asymmetric Diffuser and Axi-symmetric Hill. In: Proc. ECCOMAS CFD 2006 (2006)

Grahs, T., Othmer, C.: Evaluation of Aerodynamic Noise Generation: Parameter Study of a Generic Side Mirror Evaluating the Aeroacoustic Source Strength. In: Proc. ECCOMAS CFD (2006)

Kraichnan, R.H.: Diffusion by a random velocity field. Phys. of Fluids 13(1), 22–31 (1969)

Menter, F.R.: Two-Equation Eddy-Viscosity Turbulence Models for Engineering Applications. AIAA-Journal 32(8), 269–289 (1994)

Menter, F.R., Egorov, Y.: Re-visiting the turbulent scale equation. In: Proc. IUTAM Symp. One Hundred Years of Boundary Layer Research, Göttingen. Springer, Heidelberg (2004/2006)

Menter, F.R., Egorov, Y.: A Scale-Adaptive Simulation Model using Two-Equation Models, AIAA Paper 2005-1095, Reno/NV (2005a)

Menter, F.R., Egorov, Y.: Turbulence Models based on the Length-Scale Equation. In: Fourth International Symposium on Turbulent Shear Flow Phenomena, Williamsburg, 2005 – Paper TSFP4-268 (2005b)

Menter, F.R., Egorov, Y.: SAS Turbulence Modelling of Technical Flows. In: DLES 6 - 6th ERCOFTAC Workshop on Direct and Large Eddy Simulation September, Poitiers (2005c)

Menter, F.R., Egorov, Y., Rusch, D.: Steady and Unsteady Flow Modelling Using the $k-\sqrt{k}L$ Model. In: Hanjalic, K., Nagano, Y., Jakirlic, S. (eds.) Proc. Turbulence, Heat and Mass Transfer, vol. 5 (2006)

Rotta, J.C.: Turbulente Strömumgen. BG Teubner Stuttgart (1972)

Smirnov, A., Shi, S., Celik, I.B.: Random flow generation technique for large-eddy simulations and particle-dynamics modelling. J. of Fluids Engineering 123, 359–371 (2001)

Spalart, P.R.: Strategies for turbulence modelling and simulations. Int. J. Heat Fluid Flow 21, 252–263 (2000)

Vogel, J.C., Eaton, J.K.: Combined heat transfer and fluid dynamics measurements downstream of a backward-facing step. ASME J. Heat Transfer 107, 922–929 (1985)

Widenhorn, A., Noll, B., Aigner, M.: Numerical study of a non-reacting turbulent flow in a gas turbine model combustor, AIAA Paper, AIAA 2009-647, Orlando (2009)

RDDES for Strong Shock-Wave/Boundary Layer Interaction

A. Shams and P. Comte

Abstract. Detached eddy simulation (DES) is a suitable method for the simulation of complex turbulent flows, providing an access to the resolved turbulent scales at relatively low computational cost. The near wall region is solved by (U) RANS whereas the large scale detached region by large eddy simulation (LES). The RANS equations can give significant error when applied to flow involving shock-waves and their interaction with adverse pressure gradient boundary layers. Turbulence models based on eddy viscosity assumption yield very high amplification of the turbulent kinetic energy (TKE) under such non-equilibrium conditions and violate the realizability constraint. The main problem arises in shock-induced separated flows in rocket nozzles which are very sensitive to separation point prediction, and may completely change the flow physics with standard eddy viscosity models. In the present study a new version of DES has been proposed which deals fairly well with such type of flow configurations. Furthermore, emphasis is put on the problems includes (i) modelled stress depletion (MSD) and grid induced separation (ii) realizability issue and (iii) model sensitivity in supersonic separated flow regimes.

1 Introduction

Hybrid modelling methods, combining RANS and LES techniques, have been an area of interest since the last decade. Various hybrid techniques have been proposed in this regard for the better predictions of complex turbulent flows. The first idea of Detached Eddy Simulation was first proposed by Spalart et al. Spalart et al (1997) in order to overcome the deficiencies of RANS models for predicting the massively separated flows. This technique has recently received an increasing attention in the recent to fill up the gap between (U) RANS and LES. This original DES formulation is based on a modification to the Spalart-Allmaras RANS model Spalart and Allmaras (1992) such that the model reduces to RANS formulation near solid surfaces and to a sub-grid scale (SGS) model away from the wall. Later on, DES implementation based on two-equation model was proposed by Travin et al. Travin et al (2002) for Menter-SST model. The main reason why these models

Afaque Shams · Pierre Comte
LEA, Université de Poitiers, ENSMA, CNRS UMR-6609, CEAT,
43 route de l'Aérodrome, F-86036 Poitiers, France
afaque.shams@lea.univ-poitiers.fr,
Pierre.Comte@lea.univ-poitiers.fr

S.-H. Peng et al. (Eds.): Progress in Hybrid RANS-LES Modelling, NNFM 111, pp. 247–260.
springerlink.com

have been selected as the underlying RANS models lies in their improved separation prediction capabilities, but under certain conditions these models over-predict the extent of separation Menter et al (2003). This issue of separation point prediction becomes much more severe when we deal with shock-induced separated flows. One such example is supersonic separation flow configurations which occur in over-expanded rocket nozzles. This type of highly accelerated flows exhibit: high Reynolds number and strong shock-wave / boundary layer interactions Shams et al (2008).

Depending upon the nozzle contour and pressure ratio these free separated flows reattach back to the nozzle wall and form completely different flow configuration, characterized by special shock pattern, called cap-shock. Successive interactions of shock and expansion waves inside this annular supersonic jet cause the secondary separation. Correct prediction of the Mach stem, separation point, reattachment point and the pressure distribution downstream of it is very crucial. Recently Shams et al. Shams and Comte (6-9 July, 2009) has shown that a small change in the nozzle pressure ratio (NPR) leads to a significant change in the flow structures. Variety of turbulence models (Spalart-Allmaras (SA), $k-\omega$, Menter-SST, $k-\varepsilon$, Limited Numerical Scales (LNS) and Detached-Eddy Simulation (DES)) has been used by different researchers individually to predict these type of complex flows Shams et al (2008). So far Menter-SST model have shown some improved results as compare to the other models and their DES extensions. Numerical study performed by Cyril et al. Cyril and Abderrahmane (2004) for free shock separated flow in thrust ideal contour (TIC) nozzle has shown that for relatively lower NPR's (i.e. NPR<30) SST model over-predicts the separation point as compare to the experimental data. In the range of high NPR it under-predicts, and shows about 20% disagreement for NPR=75.9. The reason of such disagreement lies in the fact that in these eddy viscosity models Reynolds stresses are linearly related to the mean strain rate via eddy viscosity. In such type of highly non-equilibrium flows, time scale of distortion is significantly smaller than that of turbulence and the equilibrium concept of eddy viscosity breaks down. Usually models with eddy viscosity yield unrealistically high values of the Reynolds stresses, which in turn high production of turbulent kinetic energy. The amount of turbulence in a boundary layer affects the flow separation in an adverse pressure gradient case.

In this paper a new version of DES based on Wilcox $k-\omega$ model has been proposed. The main idea of this DES has been taken from Yan et al (2005), in which different length scales for DES have been proposed for $k-\omega$ model Wilcox (1988). The proposed DES is based on two steps, in the first step a modification based on realizability constraint applied to the $k-\omega$ model and in the second one extension of this realizable model has been performed and is denoted as "Realizable Delayed Detached Eddy Simulation (RDDES)". Simulations are performed on a thrust optimized contour nozzle (TOC) experimentally investigated at the Laboratoire d'Etudes Aérodynamiques (LEA) and is denoted as LEATOC nozzle Nguyen (2003). The paper is organized as follows: First the formulation of RDDES, followed by the flow configuration and numerical tools. Finally in the results section following issues are discussed: (i) modelled stress depletion and grid-induced separation (ii) realizability Issue (iii) model sensitivity & (iv) side-load calculations and spectral analysis.

2 RDDES Formulation

Realizable delayed detached eddy simulation formulation is based on $k-\omega$ model due to its well known capability of separation point prediction, especially in the strong of adverse pressure gradient flows. In much more complex flow configurations when we deal with supersonic separated flows these two-equation linear models based on Boussinesq assumption violate the realizability constraints. This also affects the separation point prediction capability of these standard models. Hence to achieve improved and realistic results this new version of DES based on $k-\omega$ model is used along with realizability correction. The formation of RDDES is given below:

2.1 (U) RANS Mode: $k-\omega$ Model

Transport equations for Wilcox standard $k-\omega$ model are given below:

$$\frac{\partial \rho k}{\partial t}+\frac{\partial}{\partial x_j}(\rho k U_j)=P_k-\rho C_\mu k\omega+\frac{\partial}{\partial x_j}\left((\mu+\frac{\mu_t}{\sigma_k})\frac{\partial k}{\partial x_j}\right) \tag{1}$$

$$\frac{\partial \rho \omega}{\partial t}+\frac{\partial}{\partial x_j}(\rho \omega U_j)=C_{\omega 1}\frac{\omega}{k}P_k-\rho C_{\omega 2}\omega^2+\frac{\partial}{\partial x_j}\left((\mu+\frac{\mu_t}{\sigma_\omega})\frac{\partial \omega}{\partial x_j}\right) \tag{2}$$

Where the production term and the strain tensor are:

$$P_k=\mu_t S^{*2},\;\; S^*=\sqrt{2S_{ij}S_{ij}},\;\; S_{ij}=\frac{1}{2}(\frac{\partial U_i}{\partial x_j}+\frac{\partial U_j}{\partial x_i})$$

The standard model constants are retained and are given by:

$$C_{\omega 1}=5/9,\quad C_{\omega 2}=3/40,\quad C_\mu=0.09,\quad \sigma_k=\sigma_\omega=2$$

2.2 (U) RANS Mode: Realizability Constraint

Schumann Schumann (1977) and Lumley Lumely (1978) defined the weak version of realizability: the variances of the fluctuating velocity components must be positive and the cross-correlations should be bounded by Schwartz inequality.

$$\tilde{u^2}\geq 0,\quad \tilde{v^2}\geq 0,\quad \tilde{w^2}\geq 0 \tag{3}$$

$$\widetilde{uv}^2\leq \tilde{u^2}\tilde{v^2},\quad \widetilde{uw}^2\leq \tilde{u^2}\tilde{w^2},\quad \widetilde{vw}^2\leq \tilde{v^2}\tilde{w^2} \tag{4}$$

These constraints are not fulfilled by the linear models based on Boussinesq assumption, and they can be infringed in the presence of strong shock wave and boundary layer interactions (shock-induced separation). To avoid these unrealistic results Moore et al. Moore and Moore (1999) proposed a correction based on realizability constraint. The general idea of this correction is based on observation that, realizability can be ensured by properly decreasing the level of turbulent eddy viscosity, and is given as:

$$\mu_t = \alpha_v C_\mu \overline{\rho} k/\omega \tag{5}$$

where

$$\alpha_v = min(1, \overline{\alpha}_v) \quad \& \quad \frac{1}{\overline{\alpha}_v C_\mu} = A_o + A_s(\overline{s}^2 + A_r \overline{\omega}^2)^{1/2}$$

Here s is the dimensionless mean strain rate S/ω with $\widetilde{S}^2 = 2\widetilde{S}_{ij}\widetilde{S}_{ij} - \frac{1}{2}S_{kk}^2$, and $\overline{\omega}$ is the dimensionless vorticity invariant $\sqrt{2\widetilde{\Omega}_{ij}\widetilde{\Omega}_{ij}}/\omega$.

$$\widetilde{S}_{ij} = \frac{1}{2}\left(\frac{\partial \widetilde{U}_i}{\partial x_j} + \frac{\partial \widetilde{U}_j}{\partial x_i}\right); \quad \widetilde{\Omega}_{ij} = \frac{1}{2}\left(\frac{\partial \widetilde{U}_i}{\partial x_j} - \frac{\partial \widetilde{U}_j}{\partial x_i}\right) \tag{6}$$

With $A_o = 2.85$; $A_s = 1.77$; $A_r = 1$.

2.3 DES Extension

Yan et al. Yan et al (2005) have investigated different length scale substitutions in DES based on Wilcox $k-\omega$ model. Following this, we propose the implementation of DES based on the Wilcox $k-\omega$ model with realizability correction. In order to understand this DES implementation equation (1) is rewritten as follows:

$$\frac{\partial \rho k}{\partial t} + \frac{\partial}{\partial x_j}(\rho k U_j) = P_k - \frac{\rho k^{3/2}}{L_k}\omega + \frac{\partial}{\partial x_j}\left((\mu + \frac{\mu_t}{\sigma_k})\frac{\partial k}{\partial x_j}\right) \tag{7}$$

In the above equation L_k have no independent meaning, but denote the turbulent length scale L_t which is defined as:

$$L_t = \frac{k^{1/2}}{C_\mu \omega} \tag{8}$$

Standard DES length scale is given as:

$$L_{DES} = min(L_t, C_{DES}\Delta) \tag{9}$$

Where C_{DES} is 0.7 Yan et al (2005) and $\Delta = max(\Delta_x, \Delta_y, \Delta_z)$ is the computational mesh size. L_{DES} is a continuous function, given by the minimum of the two length scales. For DES implementation, the turbulent length scale L_t in the dissipation term of k transport equation is replaced by the DES length scale L_{DES}. This substitution leads to calculate the eddy viscosity like in the standard RANS model with the realizability correction as discussed earlier.

2.3.1 DDES Part

Menter et al. Menter and Kuntz (2002.) use shielding functions F_1 & F_2 of the SST model to protect the "RANS mode" in the boundary layer from early switching to "LES mode". In this framework we have used F_1 function and the modified DDES length scale for the present DES is given below:

$$L_{DDES} = L_t - f_{DDES} \times max(0, L_t - C_{DES}\Delta) \tag{10}$$

Where $f_{DDES} = 1 - F_1$, when this function is 0: RANS mode activates near wall region and turns into LES mode when it is equal to 1.

3 Numerical Tools

The flow configuration considered in the present study is a convergent-divergent Thrust-Optimized Contour (TOC) nozzle, experimentally investigated at the Laboratoire d'Etudes Aérodynamiques (LEA), Poitiers / France, and is denoted as LEATOC nozzle. Numerical simulations on this LEATOC nozzle has been performed by using the TGNS3D code developed at CEAT/LEA (Poitiers, France). This code solves the three dimensional unsteady compressible Navier-Stokes equations on multiblock structured grid. The Navier-Stokes equations are discretized in space by using a cell centered finite volume method.

In the case free shock separation, Roe's flux difference splitting scheme is used to obtain the flux at the cell interface. The monotone upstream-centered schemes for conservation laws (MUSCL) approach is used to extend the spatial accuracy up to second order, based on the use of the minmod limiter to prevent the numerical instabilities in shock regions. Restricted shock separated flow is a very complex flow regime containing a number of flow discontinuities and require high order numerical scheme with low dissipation for flow analysis. Hence a fifth order Monotonicity Preserving Weighted Essentially Non-Oscillatory (MPWENO) scheme Balsara and Shu (2000.). The MPWENO schemes have high phase accuracy and high order of accuracy. The high order members of this family are almost spectrally accurate for smooth problems.

A second order implicit scheme has been implemented for time discretization. Diagonally dominant alternating direction implicit (DDADI) approach has been used for the inversion of the large sparse matrix system resulting from the implicit scheme.

Fig. 1 Mesh of axi-symmetric nozzle and the exterior domain

The computational domain consists of two zones (i) nozzle and (ii) the extended exterior in axial and radial direction, shown in Fig. 1. The exterior domain is 8 times the nozzle exit radius in the longitudinal direction, and 3 times in the radial direction. The computational domain is composed of 4032000 grid points with 3D configuration: 72 nodes in the azimuthal direction, and 1.73 million nodes inside the nozzle.

The dimensionless distance r^+, which is a Reynolds number based on the friction velocity u_τ, is given as

$$r^+ = (\sqrt{\rho_w|\tau_w|/\mu_w})r = (ru_\tau/\nu_w) \tag{11}$$

The value of r^+ at the first mesh point on the nozzle wall for these flow configurations was calculated and remains less than 0.8. This value of r^+ is quite sufficient for the correct calculation of the viscous stresses in the vicinity of the wall. In the case of free shock separation flow regime at NPR=15.5, separation line is located near to the nozzle throat. The corresponding nozzle radius in that attached region $\approx 2 * r_t$: where r_t is the throat radius. The maximum value regarding to the mesh resolution in the azimuthal direction $\Delta\theta^+ \approx 35$, and past the grid-induced separation. Qualification of RDDES based on the modelled stress depletion (MSD) and grid-induced separation is given in Fig. 3.

4 Results and Discussion

Three dimensional numerical investigations of supersonic separated flows is performed on a thrust optimized contour (TOC) nozzle. Experimentally and most recently by numerical calculations it has been found that in this LEATOC nozzle FSS flow regime appears at low nozzle pressure ratio (i.e. < 24). When NPR reaches this critical NPR value, 24, flow transition from FSS to RSS takes place. Depending upon flow regimes two different NPR are selected for the current research work.

In the present study, the simulations are run during a time $T = 80 * L_d/U_t$, where L_d is the length of the computational domain in the longitudinal direction and U_t is the velocity at throat of the nozzle. The physical time step was about 1 μs and the results were sampled every 10 steps corresponding to a physical sampling frequency of 100 KHz for a total physical sample time of 0.2 s. The length of this physical time sample is sufficient enough to obtain the range of minimum possible frequency of about ≈ 5 Hz for the spectral analysis (for details see section 7.).

Figure 2, presents successfully reproduced free and restricted shock separated flows at NPR=15.5 & 41.6, respectively. Large scale structure appears downstream of the nozzle are shown with the help of Q-criterion coloured with the thermodynamic (static) pressure. In addition, bottom and side wise projections of these flow fields are also shown. To identify flow discontinuities in these complex flow regimes shock function is used. This function represents the Mach number of the velocity component in the direction of the pressure gradient with bright zones indicating shocks or compression regions and dark ones highlights the expansion region.

In the case of free shock separation, the over-expanded flow fully separates from the nozzle wall. The resulting stream wise mean wall pressure evolution is mainly governed by the physics of the shock-wave / boundary layer interaction. The separation and the subsequent formation of the recirculation zone give rise to an oblique shock wave near the wall as depicted in Fig. 2 (left).

Appearance of restricted shock separated flow regime depends upon the nozzle contour in a defined range of NPR. This special type of separated flow exists only in the nozzles which contain an internal shock: appears near the throat region. RSS flow regime is characterized by a small recirculation pocket followed by reattachment of free separated jet on the nozzle wall. The main flow involves a special shock pattern called cap-shock followed by a trapped vortex surrounded by an annular supersonic jet. This annular supersonic flow is subjected to expansion and compression waves reflected between the nozzle wall and the mixing layer separating the high speed region from the central trapped vortex. These reflected waves give an oscillatory wall pressure distribution which can lead in some case to secondary separation as shown by Nguyen Nguyen (2003).

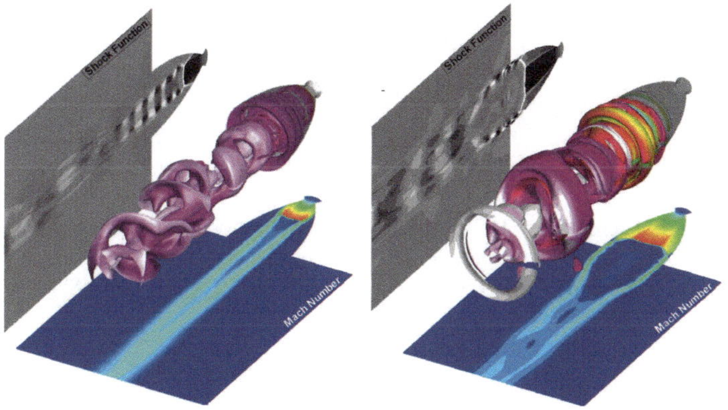

Fig. 2 (Left) Free & (right) restricted shock separation configurations: Iso-surface of Q-criterion coloured with static pressure, iso-contours of Mach number and shock function $f(x) \equiv \frac{\mathbf{U}}{c} \cdot \frac{\mathtt{grad}\,p}{|\mathtt{grad}\,p|}$ at bottom and side wise projections respectively

4.1 Modelled Stress Depletion and Grid-Induced Separation

As a result of numerous studies of the original DES method, several key shortcomings were identified: for grids with fine wall-tangential resolution, the grid-dependent switch between RANS and LES mode may encroach inside the boundary layer resulting in a depletion of the modeled stresses and subsequent strong under-production of the skin friction. In strong cases this may cause a spurious prediction of the skin friction of the separation known as grid-induced separation Menter and Kuntz (2002.). To avoid the problem of unwanted switch of LES mode inside the attached turbulent boundary layer, the DDES shield function is used here Spalart et al (2006). Evolution of this shielding function along the velocity vectors in the divergent part of the nozzle is shown in Fig. 3 (left). One can notice that along with the nozzle wall RANS mode is protected by this function inside the boundary layer. A comparison of $k-\omega$ model with RDEES has been done in a boundary layer with the grid spacing of about 1/30th of the boundary layer thickness, see Fig. 3 (right). It is depicted that the velocity profile for RDDES inside the boundary layer are almost in agreement with the RANS model. Besides DDES shielding function (f_{DDES}) safely rise from 0 in the outer part of the boundary layer and RDDES fully preserves the turbulent eddy-viscosity.

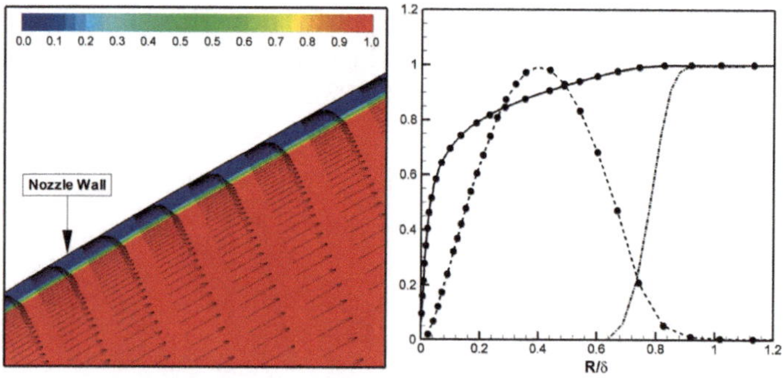

Fig. 3 (Left) Iso-contours of DDES shielding function along with velocity vectors for the attached flow in the divergent part of LEATOC nozzle & (right) Comparison of velocity profile and eddy viscosity for RDDES and k-ω model: –·–, f_{DDES}; RANS: —, U/U_∞, - - -, 0.017 μ_t/μ ; •, RDDES

5 Realizability Issue

In this section emphasis are given on the fact how eddy-viscosity models in RANS mode of DES yields unrealistic results. Aforementioned flow configurations exhibit a number of flow discontinuities and undergoes strong shock-wave / boundary layer interaction. The strength of this interaction increase with the increase in NPR.

Turbulence models based on Boussinesq assumption, Reynolds stresses are linear with respect to the strain rate S_{ij} via eddy viscosity, and is given as:

$$\frac{u_i u_j}{k} = -2\frac{\nu_t}{k}S_{ij} + \frac{2}{3}\delta_{ij} \tag{12}$$

Turbulent viscosity is related to the turbulent kinetic energy.

$$\nu_t = C_\mu k^2/\varepsilon \tag{13}$$

To understand the infringement of the realizability constraint, dimensionless strain invariant s is introduced:

$$s = \frac{k}{\varepsilon}S \quad with \quad S = \sqrt{2S_{ij}S_{ij}} \tag{14}$$

By using these relations for these flow configurations: which are quasi axi-symmetric, the realizability constraints becomes equivalent to:

$$s \leq \frac{2}{3C_\mu} \tag{15}$$

This shows that how large value of s (i.e. strong velocity gradients) leads to an infringement of the realizability constraints. Three different locations (R/δ=0.23, 0.44 & 0.67) inside the boundary layer are selected to point out the zones where this realizability constraint is violated.

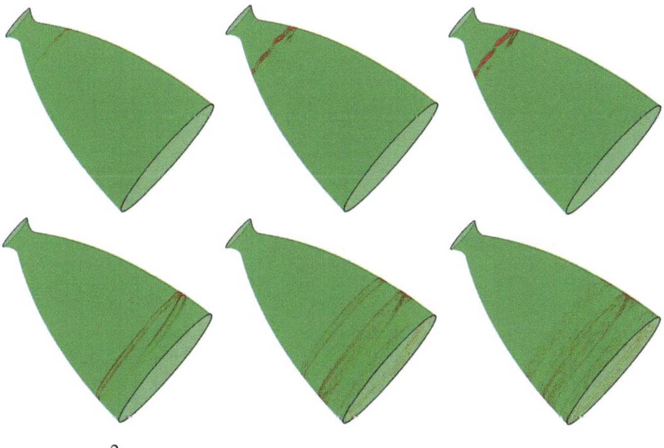

Fig. 4 Traces of $s \geq \frac{2}{3C_\mu}$ at R = $0.23/\delta$ (left), $0.44/\delta$ (middle) & $0.67/\delta$ (right): FSS (top) & RSS (bottom) regimes

Numerical solution obtained by $k-\omega$ based DDES for FSS and RSS flow regimes is shown in Fig. 4. One can notice the traces of realizability infringement

($s \leq \frac{2}{3C_\mu}$) appear in the region of shock-wave / boundary layer interaction. This zone increase as we move further in the outer region of boundary layer. On the other hand due to relatively strong shock-wave / boundary layer interaction, flow reattachment and successive shock and expansion waves interaction in the supersonic jet attached to the nozzle wall, RSS flow regimes shows considerably a large region undergoes the situation where relation (15) is not fulfilled. Whereas with the introduction of realizability correction no such traces has been observed and some improvement in the results has also been observed (see section. 6).

6 Model Sensitivity

Currently DES has been used in combination with Spalart-Allmaras and $k-\omega$ models and there popular invariants. The main reason why these models have been selected as the underlying RANS models lies in their improved separation prediction capability. A number of numerical studies in the past have been performed on such type of flow configurations, and SST has shown some improved results as compare to the other RANS model. In Fig. 5, numerically obtained mean wall pressure is compared with the available experimental data. At low NPR (i.e. 15.5) RDDES have shown better agreement. Whereas, in the case of DDES-SST, slight disagreement in the separation point location and the pressure downstream of it is due to higher level of turbulence prediction during shock-wave / boundary interaction. As we have seen in the previous section that this zone of realizability infringement in the outer parts part of the boundary layer and for higher NPR's. Moore et al. Moore and Moore (1999) pointed out that, when we go from two to three dimensions, Menter has substituted $\sqrt{S^2+\Omega^2}$ to $\frac{\partial U}{\partial y}$ instead of Ω. Only this point is missing to ensure that the SST model is realizable.

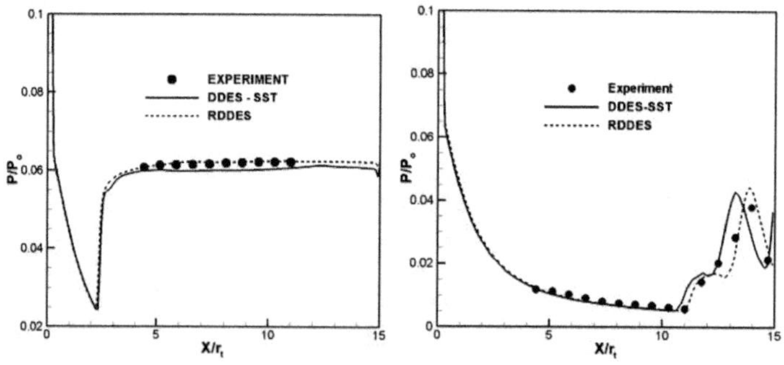

Fig. 5 Evolution of mean wall pressure along nozzle wall at NPR=15.5 & 41.6

It is well known that amount of turbulence in a boundary layer directly affects the flow separation in adverse pressure gradient. From Fig. 5, we can see that at high NPR (i.e. 41.6) DDES-SST under-predicts the mean separation point location.

This in turns seems to be the reason for incorrect flow recovery prediction downstream of the reattachment. This under-prediction of the separation point also leads to secondary separation caused by the interaction of shock and expansion waves in the annular supersonic jet. On other hand separation point location is well predicted by RDDES, but shows a slight disagreement for flow reattachment and downstream pressure. However, the static pressure upstream of the separation seems to be a bit higher in experiments.

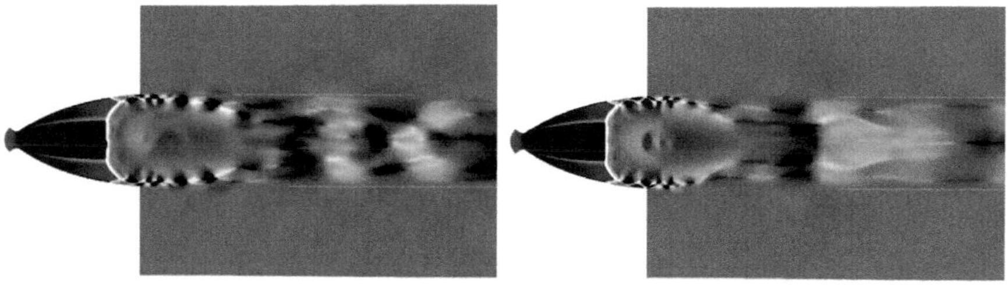

Fig. 6 Iso-contours of shock function at NPR=41.6 for (left) RDDES & (right) DDES-SST

At high NPR condensation of oxygen occurs in the experiments de Roquefort (Tokyo, Japan, 2002) which give two effects:

- An increase of the inviscid static pressure upstream of separation: this can be interpreted as a higher equivalent NPR.
- A decrease of the inviscid Mach number upstream of the shocks.

This can be interpreted as higher equivalent NPR upstream of the separation. Consequently, the plateau predicted by RDDES and DDES-SST is more pronounced than what the experimental data suggest, see Fig. 5(right).

Numerical solution obtained by RDDES and DDES-SST in shown in Fig. 6, it is evident that even a small under-prediction in the separation point location leads to secondary separation which eventually cause a large difference in flow structures downstream of the nozzle. As a consequence, it under-predicts the size of the trapped vortex appears downstream of the cap-shock pattern. This shock system is a quite complicated one and still unknown. Shams et al. Shams and Comte (6-9 July, 2009) in his detailed study about RSS flow regime has shown that there is closed loop between cap-shock, trapped vortex, separation line, reattachment line and a small recirculation bubble appears between them, and how the size of recirculation zones affect the flow structure.

7 Side-Loads and Spectral Analysis

In this section, the spectral properties of the wall pressure fluctuations and the resulting pressure loads are investigated. The unsteady and three-dimensional pressure field induces unsteady asymmetrical loads acting on the nozzle. These loads can be obtained by integrating the unsteady pressure field on the nozzle extension inner surface during the calculation.

Figure 7 (Left), shows a typical polar plot of the side-loads for FSS configuration, i.e. evolution of $F_y(t)$ component of the load vs. $F_z(t)$. One can notice that side-loads are centered at zero since for an averaged axi-symmetric flow the buffet load is absent, that is $F_y = F_z = 0$, and buffet load seems to be a rotating vector which is uniformly distributed. This simply means that side-load components are normally distributed and it is interesting to take into account the PDF of these side-loads.

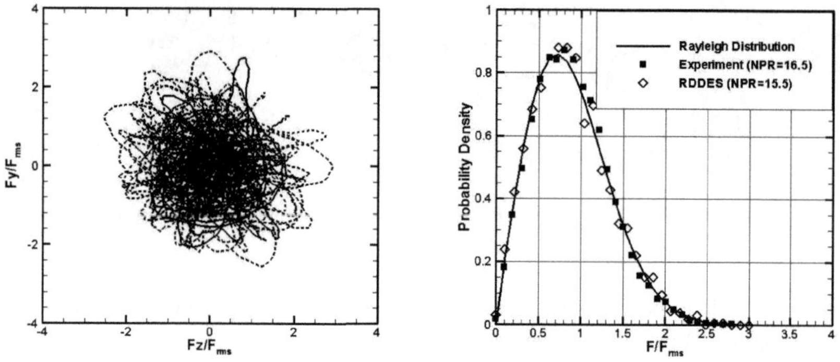

Fig. 7 (Left) Polar plot of the computed buffet load & (Right) PDF of side-load amplitude

The computed side-force probability density function is compared with experimental results in Fig. 7 (Right). Both values indicate that the distribution of side-load amplitude is a Rayleigh distribution. These results could be quite expected for an axi-symmetrical geometry because the Rayleigh distribution is the particular case of a χ^2 distribution with two degrees of freedom, which correlates with normally distributed side-load components.

Figure 8 (Left), presents the comparison of numerical computed one-sided PSD function $G(f)$ of FSS flow regime with the experiments. It is worthwhile to note that in the FSS regime the side loads are dominated by very low frequencies. Slight difference in PSD level is due to the fact that flow regime at NPR=16.5 exhibit relatively high unsteady and energetic character with respect to NPR=15.5 (see Nguyen (2003) for details). In the case of RSS configuration pressure spectra are computed at the station near to the nozzle exit ($x/r_t = 14.7$) and compared with the experiments, see Fig 8 (Right). The PSD displays a broadband aspect at low frequencies which are attributed to the large-scale motion of shock and the successive interaction of shock and expansion waves in the attached annular supersonic jet.

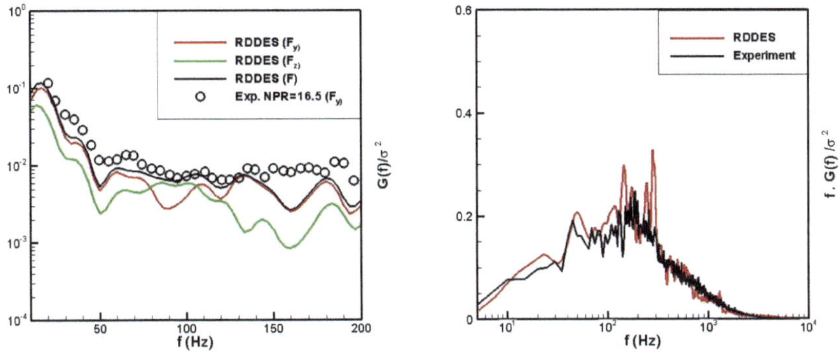

Fig. 8 (Left) PSD of side load in FSS regime at low frequency range ($\leq$ 200 Hz) & (Right) PSD of pressure spectra : RSS configuration

8 Conclusion

In the present study a new version of detached-eddy simulation has been proposed for shock-induced separated flows. Based on the three dimensional numerical investigations on supersonic separated flows in LEATOC nozzle different issues and problems were studied. Thanks to the shielding function which has safely protected boundary layer in the RANS mode to avoid onset separation and major MSD problem. It has been observed that in the RANS mode eddy-viscosity models across strong shock-wave / boundary layer interactions violates the realizability constraints and leads to an unrealistic increase in the level of turbulent kinetic energy (TKE). This behaviour of TKE production also affect the separation point prediction capabilities of the model, and in complicated and/or flow configurations which are sensitive to separation point location (like restricted shock separation) produce misleading solution. The unsteady and non-symmetric behaviour of FSS and RSS flow regimes has been studied and detailed comparison has been made with the available experimental data. Numerically obtained solution by RDDES is in excellent agreement with the experiments, which suggests that newly proposed RDDES is capable to predict the unsteady low frequency shock motion in these complex shock-induced separated flow configurations.

Acknowledgments. The Higher Education Commission (HEC), Government of Pakistan, is acknowledged for the research grant.

References

Balsara, D.S., Shu, C.W.: Monotonicity preserving weighted essentially non-oscillatory schemes with increasing high order of accuracy. Journal of Computational Physics 160, 405–452 (2000)

Cyril, P., Abderrahmane, N.: Flow separation in truncated ideal contour nozzle. Journal of Turbulence 5(1) (2004)

Lumely, J.L.: Computational modelling of turbulent flows. In: Advance in Applied Mechanics, vol. 18. Academic Press, London (1978)

Menter, F.R., Kuntz, M.: Adaptation of eddy-viscosity turbulence models to unsteady separated flows behind vehicles. In: The aerodynamics of heavy vehicles: Trucks, busses and trains. Springer, Asilomer (2002)

Menter, F.R., Kuntz, M., Langtry, R.: Ten years of industrial experience with the sst turbulence model. Turbulence, Heat and Mass Transfer (2003)

Moore, J.G., Moore, J.: Realizability in two equation models. AIAA Paper 99-3779 (1999)

Nguyen, A.T.: Découlement instationaire et charges latérales dans les tuyéres propulsives. Ph.D Thesis, Université de Poitiers (2003)

de Roquefort, T.A.: Low frequency fluctuations in separated turbulent compressible flows. In: Symposium on Advanced Fluid Information, Tokyo, Japan (2002)

Schumann, U.: Realizability of reynolds stress turbulence models. Physics of Fluids 20, 721–725 (1977)

Shams, A., Comte, P.: Formation of cap-shock pattern. In: 3rd European Conference for Aero-Space Sciences (EUCASS), Versailles-Paris, France, July 6-9 (2009)

Shams, A., Comte, P., Girard, S., Lehnasch, G., Shahab, M.F.: 3d unsteady numerical investigation of an overexpanded thrust optimized contour nozzle. In: Sixth European Symposium on Aerothermodynamics for Space Vehicles, Versaille, France (2008)

Spalart, P.R., Allmaras, S.R.: A one-equation turbulence model for aerodynamic flows. AIAA Paper 92-0439 (1992)

Spalart, P.R., Jou, W.H., Strelets, M., Allmaras, S.R.: Comments on the Feasibility of LES for Wings, and on a Hybrid RANS/LES Approach. In: Advances in DNS/LES, 1st AFOSR Int. Conf. on DNS/LES, pp. 137–148 (1997)

Spalart, P.R., Deck, S., Shur, M., Squires, K.D., Strelets, M., Travin, A.: A new version of detached-eddy simulation, resistant to ambiguous grid densities. Theoretical Computational Fluid Dynamics 20, 181–195 (2006)

Travin, A., Shur, M., Strelets, M., Spalart, P.R.: Physical and numerical upgrades in the detached-eddy simulation of complex turbulent flows. In: Friederich, R., Rodi, W. (eds.) Advances in LES of Complex Flows, Proceedings of EUROMECH Colloquium 412. Fluid Mech. Applic., vol. 65, pp. 239–254 (2002)

Wilcox, D.C.: Reassessment of the scale deterring equation for advanced turbulence models. AIAA Journal 26, 1299–1310 (1988)

Yan, J., Mockett, C., Thiele, F.: Investigation of alternative length scale substitutions in detached-eddy simulation. Flow, Turbulence and Combustion 74, 85–102 (2005)

Hybrid RANS/LES of Plane Impinging Jets with k-ω Based Models

S. Kubacki and E. Dick

Abstract. Plane impinging jets with nozzle-plate distances H/B=10 at Re=13500 and H/B=9.2 at Re=20000 are simulated with k-ω based hybrid RANS/LES models. Three ways of substitution of the turbulent length scale by the local grid size in the LES mode of the hybrid RANS/LES models are tested. The results show that the hybrid models give much better prediction of wall shear stress and heat transfer rate along the impingement plate than the RANS model. The good performance of the hybrid models is due to their ability to resolve the evolution and break-up of the vortices in the shear layer of the jet, which strongly affects the turbulent flow and convective heat transfer in the stagnation region and the developing wall-jet region.

1 Introduction

The newest version of the k-ω model of Wilcox (2008) is applied as a "state-of-the-art" turbulence model. Since the RANS model provides poor description of the flow physics for free jet flows, hybrid RANS/LES models are constructed in order to resolve the evolution of large scale structures in flow regions where the grid density is fine enough, replacing the turbulence length scale by the local grid size. In the near wall regions, the model switches to RANS mode which is known to be adequate for modelling fine scale structures. Three different ways of substitution of the turbulent length scale by the local grid scale are tested.

Results of simulations of plane impinging jets with nozzle-plate distances H/B=10, 9.2, and Re=13500, 20000 (Reynolds number based on slot width B and centreline velocity V_0) are presented and compared to experimental data and LES results. It is demonstrated that the hybrid RANS/LES models are able to reproduce

Slawomir Kubacki
Department of Flow, Heat and Combustion Mechanics, Ghent University,
St.–Pietersnieuwstraat 41, 9000 Gent, Belgium
Institute to Aeronautics and Applied Mechanics, Warsaw University of Technology,
Nowowiejska 24, 00-665 Warsaw, Poland
Slawomir.Kubacki@UGent.be

Erik Dick
Department of Flow, Heat and Combustion Mechanics, Ghent University,
St.–Pietersnieuwstraat 41, 9000 Gent, Belgium
Erik.Dick@UGent.be

S.-H. Peng et al. (Eds.): Progress in Hybrid RANS-LES Modelling, NNFM 111, pp. 261–270.
springerlink.com

the flow unsteadiness in the shear layers of the jet and give realistic turbulent flow prediction along the impingement plate. Overall, very good results are obtained.

2 The k-ω Model for Hybrid Formulation

The new k-ω model of Wilcox (2008) reads

$$\frac{Dk}{Dt}=\tau_{ij}\frac{\partial U_i}{\partial x_j}-\beta^* k\omega+\frac{\partial}{\partial x_j}\left[\left(\nu+\sigma^*\frac{k}{\omega}\right)\frac{\partial k}{\partial x_j}\right], \tag{1}$$

$$\frac{D\omega}{Dt}=\alpha\frac{\omega}{k}\tau_{ij}\frac{\partial U_i}{\partial x_j}-\beta\omega^2+\frac{\partial}{\partial x_j}\left[\left(\nu+\sigma\frac{k}{\omega}\right)\frac{\partial \omega}{\partial x_j}\right]+\frac{\sigma_d}{\omega}\frac{\partial k}{\partial x_j}\frac{\partial \omega}{\partial x_j}. \tag{2}$$

The components of the modelled stress tensor are given by $\tau_{ij}=2\nu_t S_{ij}-2/3k\delta_{ij}$. The turbulent viscosity ν_t is defined by

$$\nu_t=\frac{k}{\tilde{\omega}} \quad \tilde{\omega}=\max\left(\omega, C_{\lim}\sqrt{\frac{2S_{ij}S_{ij}}{\beta^*}}\right), \tag{3}$$

with $\beta^*=0.09$, $C_{lim}=7/8$. The components of the rate of strain tensor are $S_{ij}=1/2(\partial U_i/\partial x_j+\partial U_j/\partial x_i)$. The closure coefficients and auxiliary relations are

$$\alpha=\frac{13}{25},\quad \beta=\beta_0 f_\beta,\quad \beta_0=0.0708,\quad \sigma=\frac{1}{2},\quad \sigma^*=\frac{3}{5},\quad \sigma_{do}=\frac{1}{8}, \tag{4}$$

$$\sigma_d=\begin{cases}0, & \dfrac{\partial k}{\partial x_j}\dfrac{\partial \omega}{\partial x_j}\le 0\\ \sigma_{do}, & \dfrac{\partial k}{\partial x_j}\dfrac{\partial \omega}{\partial x_j}>0\end{cases},\quad f_\beta=\frac{1+85\chi_\omega}{1+100\chi_\omega},\quad \chi_\omega=\left|\frac{\Omega_{ij}\Omega_{jk}S_{ki}}{(\beta^*\omega)^3}\right|, \tag{5}$$

where $\Omega_{ij}=1/2(\partial U_i/\partial x_j-\partial U_j/\partial x_i)$.

The model contains a stress limiter through Eq. (3). In flow regions characterized by large levels of strain S (e.g. in impingement flow regions), the term $C_{lim}(2S_{ij}S_{ij}/\beta^*)^{1/2}$ becomes higher than ω in Eq. (3), which makes the production term in Eq. (1) proportional to S instead of proportional to S^2. Our tests showed that for impinging jet simulation with hybrid RANS/LES models, the importance of the stress-limiter is small. Since the LES mode is active in the free shear layers and in the stagnation flow region, the model is capable of resolving the flow physics there, without necessity of introducing further damping to the modelled shear stress, which would be necessary with a RANS model. Therefore, the stress limiter is omitted in the hybrid model. Notice that it is omitted everywhere, so also in the RANS mode, since it has negligible effect on the results in near-wall regions of wall parallel boundary layer flows.

3 Length Scale Substitution for Hybrid Formulation

Three different ways are studied for the substitution of the turbulent length scale $L_t=k^{1/2}/(\beta^*\omega)$ by the local grid size Δ when the hybrid RANS/LES functions in LES mode. The local grid size Δ is defined by $\Delta=\max(\Delta_x, \Delta_y, \Delta_z)$ where Δ_x, Δ_y, Δ_z denote the distances between the cell faces in x, y and z directions.

In the first model, the DES-type formulation of Strelets (2001) is implemented by modifying the destruction term in Eq. (1):

$$\beta^* k\omega = \varepsilon \rightarrow \max\left(\beta^* k\omega, \frac{k^{3/2}}{C_{DES}\Delta}\right). \tag{6}$$

The constant C_{DES} has been taken from Kok et al. (2004), namely, $C_{DES}=0.67$. A similar value of the constant C_{DES} has been determined by Yan et al. (2005) through simulation of the decay of isotropic turbulence using the k-ω based DES model, namely $C_{DES}=0.7$. Hereafter, this model is called the M1-model.

The second model is the model of Kok et al. (2004), where the underlying k-ω model has been used for construction of the "X-LES" approach. In Kok et al. (2004), the turbulent length scale is replaced by the grid size in both

$$\nu_t = \min\left(\frac{k}{\omega}, \beta^* C_{DES}\sqrt{k}\Delta\right) \tag{7}$$

and the destruction term of the k-equation (Eq. 6). This formulation is referred as the M2-model.

The third model, M3-model, is according to the latency factor concept of Batten et al. (2004), based on inclusion of the local grid size only in the definition of ν_t. An LES subgrid viscosity can be expressed by

$$\nu_t = \beta^* \Delta^{4/3} \varepsilon^{1/3}. \tag{8}$$

A RANS-type eddy viscosity can be written as

$$\nu_t^{RANS} = \beta^* \frac{k^2}{\varepsilon} = \beta^* L_t^{4/3} \varepsilon^{1/3}. \tag{9}$$

According to the kinetic energy cascade, turbulent kinetic energy is transferred from the largest scales of motion towards smaller and smaller scales approaching that of Kolmogorov size without loss. As a result, a local equilibrium assumption can be formulated equating the dissipation at the small scales to the rate at which the turbulent kinetic energy is transferred from large to small scales. With this assumption, Eqs. (8) and (9) lead to

$$\nu_t = \nu_t^{RANS} \left(\frac{\Delta}{L_t} \right)^{4/3} . \tag{10}$$

The term $(\Delta/L_t)^{4/3}$ in Eq. (10) can be interpreted as a damping function which bridges the RANS and LES zones. In order to recover the RANS turbulent viscosity in Eq. (10) for $L_t < \Delta$, the function $(\Delta/L_t)^{4/3}$ is limited by unity.

4 Computational Set Up

Fig. 1 (left) shows a scheme of the rectangular box computational domain. Length L, height H and width W and the number of cells N_x, N_y, N_z in x, y and z directions are

Table 1 Case parameters

Case	*L/B*	*H/B*	*W/B*	N_x	N_y	N_z
H/B=10, *Re*=13500	50	10	π	140	150	44
H/B=9.2, *Re*=20000	80	9.2	π	140	165	44

At the inlet to the computational domain (jet exit) an almost flat mean velocity profile was specified according to Tsubokura et al. (2003):

$$V(x,0,z) = -V_0 \left(1 - (2x/B)^{14} \right). \tag{11}$$

As shown in Fig. 1 (right), the profile defined by Eq. (11) agrees well with the experimental data of Maurel and Solliec (2001). The same definition of the inlet mean velocity profile was used for both cases.

As noted by Maurel and Solliec (2001), a uniform profile of the fluctuating velocity component can be assumed at the jet exit. In the experiments, the nozzle exit turbulence intensity varied in the range Tu=1.6-2.8% for the case H/B=10 and Re=13500. For the case H/B=9.2 and Re=20000, the turbulence intensity was Tu=1% (Ashforth-Frost et al., 1997, Zhe and Modi, 2001). The turbulence length scale was not measured. In the present computations, constant values of k and ω are specified at the inlet to the computational domain (for 2D RANS and hybrid RANS/LES) with Tu=2.5% for H/B=10 and Re=13500 and Tu=1.0% for H/B=9.2 and Re=20000 while $l_{t,inlet}$=0.015B (according to Jaramillo et al., 2008).

The inlet values of the turbulent quantities were set to $k_{inl}=1.5(Tu*V_0)^2$ and $\omega_{int}=(k_{inl})^{1/2}/l_{t,inlet}$, were V_0 denotes the mean velocity in the symmetry plane. In addition to the specified k and ω profiles, the vortex method of Fluent was used to generate fluctuations at the jet exit. At the outflow boundaries, pressure outlet boundary conditions were applied with a zero normal gradient condition for the modelled scalars. At the walls, no-slip conditions were used with the turbulent quantities computed according to k=0, $\omega=10(6\nu)/(\beta_0(\Delta y)^2)$, where Δy is the first point away from the wall. Periodic boundary conditions have been applied in the spanwise z direction.

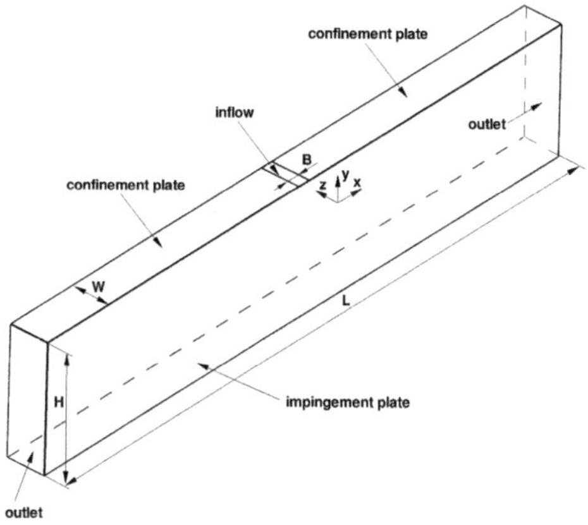

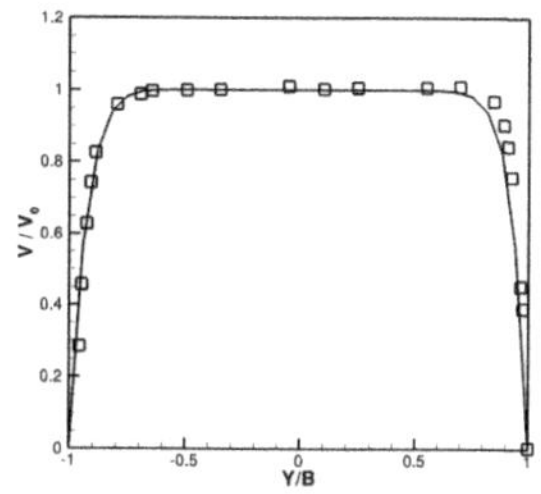

Fig. 1 (left) scheme of the computational domain (right) profile of the mean v-velocity component imposed at the jet exit

In the present hybrid RANS/LES simulations the number of grid points has been reduced in the x-direction, compared to the LES simulations of Beaubert and Viazzo (2003). The grid points have been clustered towards the walls (in order to fulfil the condition $y^+<3$) and in the shear layer of the jet, since these are the regions of high velocity gradients. A uniform grid spacing was used in the spanwise z direction.

The computations have been preformed with the Fluent code ver. 6.3.26. The bounded central differencing scheme was applied for approximation of the convective terms in the momentum equations and the second order upwind scheme was applied for approximation of the convective terms in the energy, k- and ω-equations.

5 Results

Fig. 2 (a) shows the profiles of mean and (b) fluctuating y-velocity obtained with the hybrid RANS/LES formulations (M1-model –solid line, M2-model-dashed line, M3-model – dashed dotted line) for the first case, the experimental data of Maurel and Solliec (2001), (symbols), the LES results of Beaubert and Viazzo (2003), (solid line together with the symbols) and 2D RANS results obtained with the k ω model (dashed-double dotted line). For the hybrid methods, the data have been averaged in time and in the spanwise z direction.

With RANS, the length of the jet core is strongly overpredicted with respect to the measured value and the value computed by LES and hybrid RANS/LES. As shown in Fig. 2 (b) the fluctuating velocity component predicted by RANS, $(2k/3)^{1/2}$, slightly decays with increasing distance from the jet exit while in reality the fluctuations rise already at (H+y)/B=2. The poor performance of the RANS model is due to too weak production of turbulent kinetic energy in the shear layer of the jet. This makes the predictions by 2D RANS for large distance between jet exit and impingement plate very erroneous. The results obtained with the hybrid RANS/LES methods are close to each other and they show good agreement with experimental data and LES.

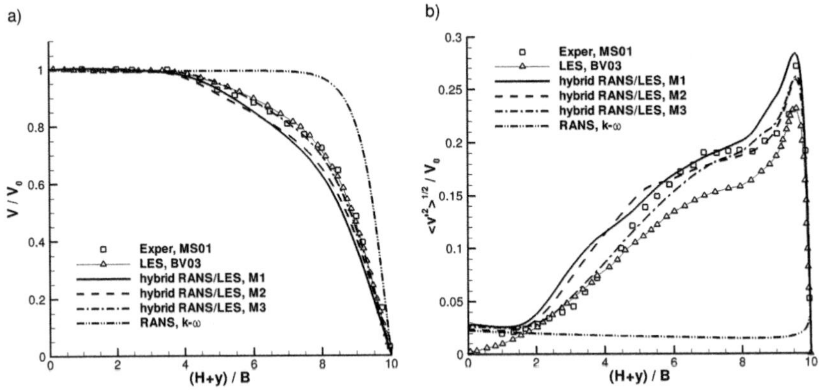

Fig. 2 Mean (a) and fluctuating (b) velocity profiles along a symmetry plane for H/B=10, Re=13500. For hybrid RANS/LES, resolved fluctuating velocity profiles are shown

Fig. 3 shows the profiles of the skin friction (scaled with $Re_b=b_pV_0/\nu$, where b_p is the half-width of the impingement pressure profile) along the impingement plate. The hybrid RANS/LES results are again compared to experimental data (Tu and Wood, 1996), to LES results (Beaubert and Viazzo, 2003) and to the 2D RANS results.

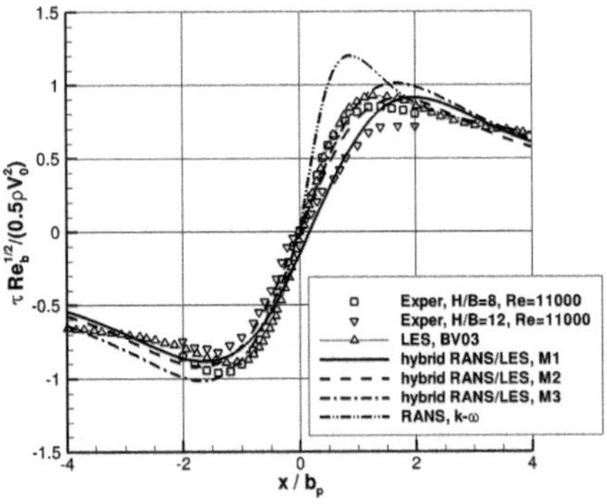

Fig. 3 Distribution of $\tau \mathrm{Re}_b^{1/2}/\left(0.5\rho V_0^2\right)$ along the impingement plate for H/B=10, Re=13500

Note the differences in the nozzle-plate distances and Reynolds number between the experimental data of Tu and Wood (1996): H/B=8 and 12, Re=11000 and the present results: H/B=10, Re=13500.

The peak values of $\tau \mathrm{Re}_b^{1/2}/\left(0.5\rho V_0^2\right)$ predicted by RANS are higher than the peak values obtained with the hybrid RANS/LES and LES methods. As discussed above, this is due to overprediction of the length of the core, so underestimation of the jet expansion. The RANS solution agrees very well with LES at larger distances from the symmetry plane (for $x/b_p>2$). The results obtained with the hybrid RANS/LES formulations show good agreement with the experimental data and the LES results.

Fig. 4 shows profiles of computed mean x-velocity components and the comparison with the experimental data of Asforth-Frost et al. (1997) and Zhe and Modi (2001) along a line perpendicular to the impingement plate at two distances from the symmetry plane for simulation of the plane impinging jet at H/B=9.2 and Re=20000. First of all, quite large differences are visible between the experimental data of Asforth-Frost et al. (1997) and Zhe and Modi (2001). As shown in Fig. 4 (a) at distance x/B=1, the mean velocity profile obtained with the RANS k-ω model is strongly overpredicted. This is caused by the too slow spreading of the mean velocity profiles in the developing wall jet region into the free stream as a result of too weak turbulence mixing in the shear layer of the jet. It means that the k-ω model is not able to reproduce the near-wall characteristics of the turbulent boundary layer due to overprediction of the length of the jet core. With the hybrid RANS/LES models much better correspondence is obtained with the experimental data. As explained before, this is the consequence of the LES representation of the shear layers of the jet which results in more reliable prediction of the mean velocity profiles along the impingement plate.

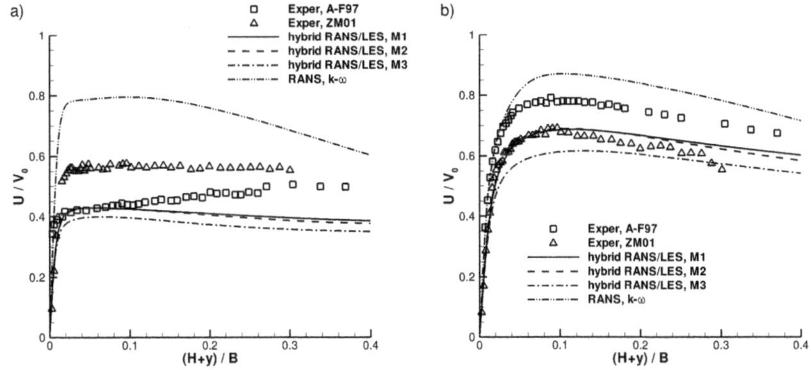

Fig. 4 Variation of the mean x-velocity component along a line perpendicular to the impingement plate at two distances from the symmetry plane (a) *x/B*=1 and (b) *x/B*=5 for *H/B*=9.2, *Re*=20000

The mean velocity profile obtained with the hybrid M3 model is slightly lower than predicted by the other models. This is already an indication that this model reacts somewhat differently. In order to identify the cause of the difference, contour plots of modelled turbulent kinetic energy (k) are shown in Fig. 5 in the x-y plane located at z/B=π/2. The instantaneous fields of modelled k obtained with the M1 model (Fig. 5, a) and the M2 model (Fig. 5, b) are quite similar. The instantaneous field of k obtained with the M3 model strongly differs from the others. Much higher levels of modelled turbulent kinetic energy are obtained with the M3 model. This is due to the different way of modelling of the destruction term in the k-equation.

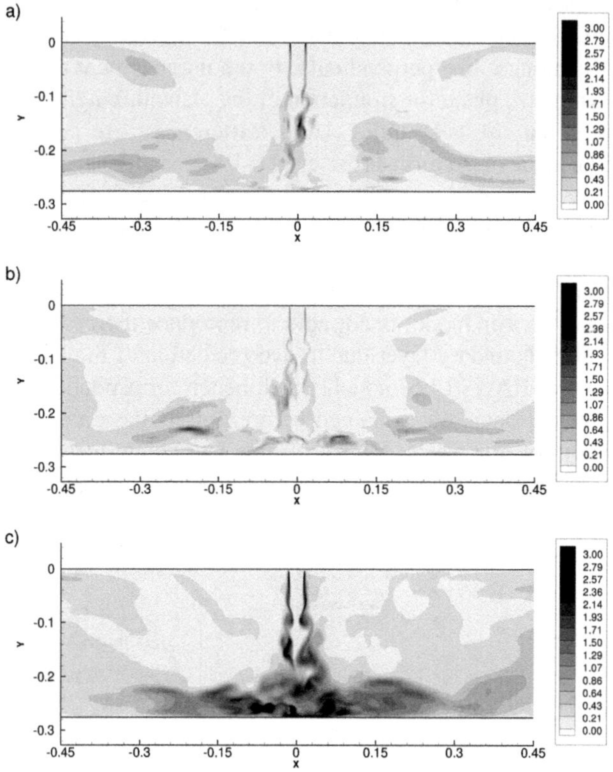

Fig. 5 Instantaneous field of modelled k obtained by (a) M1 model (b) M2 model and (c) M3 model for H/B=9.2, Re=20000

Fig. 6 (a) shows the skin friction coefficient c_f along the impingement plate obtained with the RANS k-ω model and the hybrid RANS/LES formulations. RANS predicts the skin friction coefficient far too high in the impingement region and also in the developing wall jet region. As mentioned above, this is caused by underprediction of the turbulence mixing in the shear layer of the jet which results in too slow spreading of the mean velocity profiles into the free stream as the flow turns into a wall jet. With the hybrid RANS/LES methods much better correspondence between computations and experiment is obtained. The peak value of c_f at x/B=2 is slightly overpredicted with all hybrid RANS/LES formulations but they are able to reproduce quite well the asymptotic behaviour for x/B>8. Fig. 6 (b) shows the distribution of the Nusselt number along the impingement plate and the comparison with the experimental data of Ashforth-Frost et al. (1997). RANS is not able to correctly reproduce the monotonic decay of the Nusselt number profile with increasing distance from the symmetry plane, even if the stagnation point Nusselt number is correctly predicted. The M3 hybrid RANS/LES model gives a too high level of Nusselt number. The M1 and M2 models give good correspondence between simulations and experiments. The overprediction of the heat transfer rate at x/B<6 by the M3 model is caused by the large subgrid turbulent viscosity in the near-wall region of the developing wall jet.

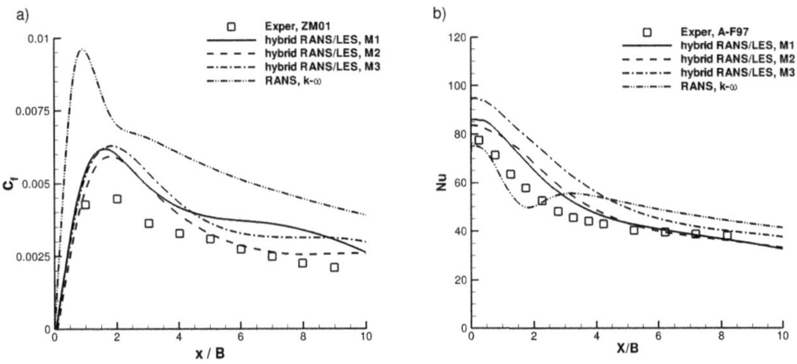

Fig. 6 (a) Distributions of the skin friction coefficient and (b) of the Nusselt number along the impingement plate for *H/B*=9.2, *Re*=20000

6 Conclusions

The performance of hybrid RANS/LES models based on the k-ω model of Wilcox (2008) for simulation of plane jets impinging onto a flat plate, with large distance between jet exit and impingement plate, has been analysed. Different ways of substitution of the turbulent length scale by the local grid size have been tested. With the M1 model, the local grid size Δ is introduced in the destruction term of the k-equation while for the M2 model it is introduced in both the destruction term of the k-equation and in the definition of ν_t in LES mode. In the M3 model, the k-equation is unaltered and the local grid size is only used to damp ν_t in LES mode.

There is not much difference in the prediction by the models of mean velocity profiles, profiles of fluctuating velocity components and skin friction distribution on the plate. In particular, the profiles in the jet approaching the plate are very close to each other. The M3 model has a tendency to overpredict the mixing of the developing wall jet with the surrounding fluid in the region downstream of the impact zone. This is caused by its much higher level of subgrid turbulent kinetic energy in the impact region, compared to the other models. This higher level also has as consequence an overprediction of the heat transfer in the impact region.

The results demonstrate that it is essential to take into account the relation between cut-off length (grid size), specific dissipation rate and subgrid turbulent kinetic energy in the destruction term of the equation for subgrid turbulent kinetic energy. Leaving this term unchanged with respect to the RANS model clearly leads to overprediction of the subgrid turbulent kinetic energy. The differences between the M1 and M2 models are minor. This shows that using the relation between grid size, specific dissipation rate and subgrid turbulent kinetic energy a second time in the definition of the subgrid viscosity is not essential.

The results show that application of a hybrid RANS/LES model is extremely effective for simulation of plane impinging jets at high nozzle-plate distances. This is due to the ability of hybrid RANS/LES models to correctly predict the turbulent mixing in the shear layers of the jet.

Acknowledgement. The authors acknowledge the support from the research project 'Novel Multiscale Approach to Transport Phenomena in Electrochemical Processes' (IWT, contract: MuTEch SBO 040092).

References

Ashforth-Frost, S., Jambunathan, K., Whitney, C.F.: Velocity and turbulence characteristics of a semiconfined orthogonally impinging slot jet. Experimental Thermal and Fluid Science 14, 60–67 (1997)

Batten, P., Goldberg, U., Chakravarthy, S.: Interfacing statistical turbulence closures with large-eddy simulation. AIAA Journal 42(3), 485–492 (2004)

Beaubert, F., Viazzo, S.: Large eddy simulations of plane turbulent impinging jets at moderate Reynolds numbers. Int. J. Heat Fluid Flow 24, 512–519 (2003)

Jaramillo, J.E., Perez-Segarra, C.D., Rodriguez, I., Oliva, A.: Numerical study of plane and round impinging jets using RANS models. Numer. Heat Transfer, Part B 54, 213–237 (2008)

Kok, J.C., Dol, H., Oskam, H., van der Ven, H.: Extra-large eddy simulation of massively separated flows, AIAA Paper 2004-0264 (2004)

Maurel, S., Solliec, C.: A turbulent plane jet impinging nearby and far from a flat plate. Exper. in Fluids 31, 687–696 (2001)

Strelets, M.: Detached eddy simulation of massively separated flows, AIAA Paper 2001-0879 (2001)

Tsubokura, M., Kobayashi, T., Taniguchi, N., Jones, W.P.: A numerical study on the eddy structures of impinging jets exited at the inlet. Int. J. Heat Fluid Flow 24, 500–511 (2003)

Tu, C.V., Wood, D.H.: Wall pressure and shear stress measurements beneath an impinging jet. Experimental Thermal and Fluid Sci. 13, 364–373 (1996)

Wilcox, D.C.: Formulation of the k-ω turbulence model revisited. AIAA Journal 46(11), 2823–2837 (2008)

Yan, J., Mocket, C., Thiele, F.: Investigation of alternative length scale substitutions in detached-eddy simulation. Flow, Turbulence and Combustion 74, 85–102 (2005)

Zhe, J., Modi, V.: Near wall measurements for a turbulent impinging slot jet. Trans. of the ASME, J. Fluid Eng. 123, 112–120 (2001)

Large Eddy Simulations: Wall Functions with Forcing Given by Backscatter from a Scale-Similarity Model

L. Davidson

Abstract. Developing accurate wall functions for LES is a great challenge. This is indeed also the case also for RANS. It can be argued that it should be easier to develop accurate wall functions for LES than for RANS. In a well-resolved LES using wall functions, the task of the wall function is to give an instantaneous wall shear stress; this does probably not need to be *exactly* correct. Compare with inlet fluctuating boundary conditions: there are many (infinite) ways to impose instantaneous fluctuations that will yield accurate mean flow and Reynolds stresses. The present work presents LES with wall functions. Forcing taken from back scatter from a scale-similarity model is used in the wall-adjacent cells. The object is to increase and enrich the resolved turbulence in the wall region. The new approach is applied for fully developed channel flow and 3D hill flow. For channel flow, it is indeed found that the results are improved using forcing. However, for 3D hill flow, it is found that the effect of forcing is negligible.

1 Introduction

The main hindrance for doing accurate Large Eddy Simulations (LES) of wall-bounded flows is the requirement of high grid resolution in the near-wall region. The grid needs to be fine not only in the wall-normal direction but also in the wall-parallel directions. Furthermore, the size of the grid cells scale with Reynolds number meaning that the problem becomes worse the higher the Reynolds number. One way to circumvent the requirement of fine grid in the near-wall region is to use wall functions. The use of wall functions allows us to reduce the number of cells. Needless to say, we put large – even unrealistic – requirements on wall functions: we model the entire inner wall region consisting of the viscous region, the buffer region and the inner logarithmic region with one single computational cell. In addition to saving computational cells, wall function grids entail another attractive feature, namely the fact that the aspect ratio of the wall-adjacent cells is much smaller than in hybrid LES-RANS and wall-modelled DES. For a structured grid, the ratio of

Lars Davidson
Division of Fluid Dynamics, Department of Applied Mechanics,
Chalmers University of Technology, SE-412 96 Göteborg, Sweden
lada@chalmers.se

S.-H. Peng et al. (Eds.): Progress in Hybrid RANS-LES Modelling, NNFM 111, pp. 271–281.
springerlink.com

the streamwise cell side to the wall-normal one can in hybrid LES-RANS and wall-modelled DES easily reach several hundred. Furthermore, high aspect ratios often deteriorate the convergence rate of the iterative solution procedure.

Figure 1 presents LES with wall functions of channel flow at four different Reynolds numbers. As can be seen, the velocity is overpredicted. The reason is that the total (i.e. modelled plus resolved) shear stress is too low; or, rather, that too a large velocity gradient is required to obtain the $y-1$ profile of the total shear stress dictated by the time-averaged streamwise momentum equation. To reduce the U^+ velocity, either the modelled or the resolved shear stress has to be increased. One way to generate additional resolved turbulence is to use forcing. This has previously been used both in LES and hybrid LES-RANS Piomelli et al (2003); Batten et al (2004); Temmerman et al (2005); Davidson and Dahlström (2005); Davidson and Billson (2006); Keating and Piomelli (2008).

The present work is the first study in which forcing is applied together with wall functions. The object is to increase the resolved turbulence in the wall region. This will be done by adding backscatter (forcing) from a scale-similarity model in the wall-adjacent cells.

A dissipative scale-similarity subgrid model was recently proposed in which only the dissipative part of the subgrid stresses was added to the momentum equations Davidson (2010). This was achieved by adding the gradient of the scale-similarity stress only when its sign agreed with that of the corresponding viscous term. This idea was later used the other way around as forcing in hybrid LES-RANS Davidson (2009): only the part of a subgrid stress term that corresponds to back scatter was added to the momentum equations. The forcing triggers resolved turbulence.

In the present work, the latter approach is used in connection with wall functions based on the instantaneous log-law. Forcing is applied in the cells adjacent to the walls to stimulate the generation of resolved turbulence. The new approach is evaluated for fully developed channel flow at $Re_\tau = 4000$, 8000, 16000 and 32000, and for 3D hill flow.

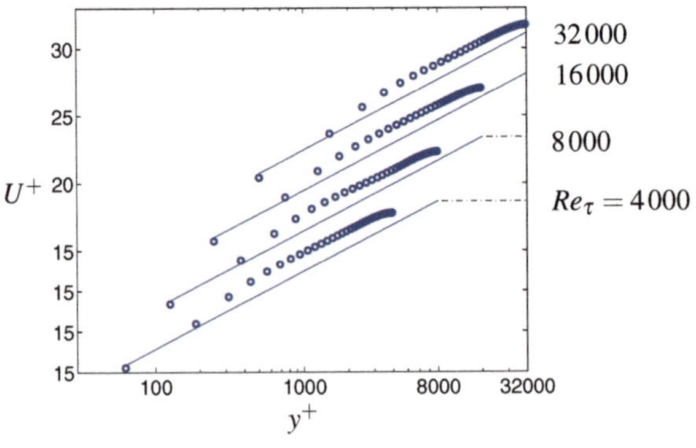

Fig. 1 LES using wall functions. Channel flow. ○: LES; ——: $U^+ = (\ln y^+)/0.4 + 5.2$

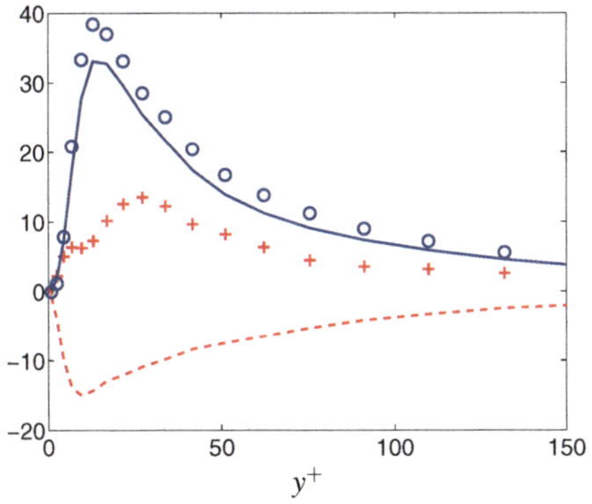

Fig. 2 Dissipation terms from DNS data taken from Davidson and Billson (2006). 96^3 mesh data filtered onto a 48^3 mesh. $Re_\tau = 500$. ——: ε^+_{SGS}; - - -: ε^-_{SGS}; ○: ε^D_{SGS}; +: ε_{SGS}

2 Scale-Similarity Model

2.1 Dissipative Model

The filtered Navier-Stokes read ($\bar{.}$ denotes filtering)

$$\frac{\partial \bar{u}_i}{\partial t} + \frac{\partial}{\partial x_j}(\bar{u}_i \bar{u}_j) + \frac{1}{\rho}\frac{\partial \bar{p}}{\partial x_i} = \nu \frac{\partial^2 \bar{u}_i}{\partial x_k \partial x_k} - \frac{\partial \tau_{ik}}{\partial x_k} \tag{1}$$

where τ_{ik} denotes the SGS stress tensor, which is modelled with the scale-similarity model Speziale (1985)

$$\tau_{ik} = \overline{\bar{u}_i \bar{u}_k} - \bar{\bar{u}}_i \bar{\bar{u}}_k \tag{2}$$

This model is not sufficiently dissipative. Let us take a closer look at the equation for the resolved, turbulent kinetic energy, $K = \langle u'_i u'_i \rangle /2$, which reads ($\langle . \rangle$ denotes averaging in time)

$$\begin{aligned}
&\frac{\partial K}{\partial t} + \frac{\partial}{\partial x_j}(\bar{u}_j K) + \langle u'_k u'_i \rangle \frac{\partial \langle \bar{u}_i \rangle}{\partial x_k} + \frac{\partial \langle p' u'_i \rangle}{\partial x_i} + \frac{1}{2}\frac{\partial \langle u'_k u'_i u'_i \rangle}{\partial x_k} = \\
&\nu \left\langle \frac{\partial^2 u'_i}{\partial x_k \partial x_k} u'_i \right\rangle - \left\langle \left(\frac{\partial \tau_{ik}}{\partial x_k} - \left\langle \frac{\partial \tau_{ik}}{\partial x_k} \right\rangle \right) u'_i \right\rangle = \underbrace{\nu \left\langle \frac{\partial^2 u'_i}{\partial x_k \partial x_k} u'_i \right\rangle}_{\varepsilon^{non}} - \left\langle \frac{\partial \tau_{ik}}{\partial x_k} u'_i \right\rangle = \\
&\nu \frac{\partial^2 K}{\partial x_k \partial x_k} - \underbrace{\nu \left\langle \frac{\partial u'_i}{\partial x_k} \frac{\partial u'_i}{\partial x_k} \right\rangle}_{\varepsilon} - \underbrace{\left\langle \frac{\partial \tau_{ik}}{\partial x_k} u'_i \right\rangle}_{\varepsilon_{SGS}}
\end{aligned} \tag{3}$$

The first term on the right side of the second line is the non-isotropic (i.e. the true) viscous dissipation, ε^{non}; this is predominately negative. The first term on the last line is the isotropic viscous diffusion term and the second term, ε, is the viscous dissipation term, which is always positive. The last term, ε_{SGS}, is a source term arising from the SGS stress tensor, which can be positive or negative. When it is positive, forward scattering takes place (i.e. it acts as a dissipation term); when it is negative, back scattering occurs.

Figure 2 presents SGS dissipation, ε_{SGS} in Eq. 3, computed from filtered DNS data. The forward scatter, ε^{+}_{SGS}, and back scatter, ε^{-}_{SGS}, are defined as the sum of all instants when ε_{SGS} is positive and negative, respectively. ε^{D}_{SGS} is computed with τ^{D}_{ik}, see Eq. 7. As can be seen, the scale-similarity model is slightly dissipative (i.e. $\varepsilon_{SGS} > 0$), but the magnitudes of the forward and the back scatter dissipation are both much larger than ε_{SGS}.

One way to make the SGS stress tensor strictly dissipative is to set the back scatter to zero, i.e. $\max(\varepsilon_{SGS}, 0)$. This can be achieved by setting $\partial\tau_{ik}/\partial x_k = 0$ when its sign is different from that of u'_i (see the last term in Eq. 3). This would work if we were solving for K. Usually we do not, and the equations that we do solve (the filtered Navier-Stokes equations) are not directly affected by the dissipation term, ε_{SGS}.

Instead we have to modify the SGS stress tensor as it appears in the filtered Navier-Stokes equations, Eq. 1. The second derivative on the right side of Eq. 1 is usually called a *diffusion* term because it acts like a diffusion transport term. However, when analyzing the stability properties of discretized equations to an imposed disturbance, u', using Neumann analysis (see, for example, Chapter 8 in Hirsch (1988)), this term is referred to as a *dissipation* term. In stability analysis the concern is to dampen *numerical* oscillations; in connection with SGS models, the aim is to dampen *turbulent resolved* fluctuations. It is shown in Neumann analysis that the viscous diffusion term in the Navier-Stokes equations is always dissipative, i.e. it dampens numerical oscillations. It is interesting to note the difference in tradition between researchers in numerical analysis and researchers in SGS modelling. The former always consider dissipation by investigating the Navier-Stokes equations. When they devise new discretization schemes to add dissipation (see, for example, Jameson's method Jameson (1982); Esfahaniana and Akbarzade (2008)), they add an additional flux that corresponds to a (numerical) diffusion term. On the other hand, when researchers in SGS modelling want to add dissipation, they consider the equation for turbulent kinetic energy, K. Increased dissipation in their thinking means a larger dissipation term in the K equation. The new scale-similarity model Davidson (2010, 2009) was developed by using the approach of both researchers in numerical analysis as well researchers in SGS modelling. The latter approach is applied in choosing the model for the SGS stress tensor, i.e. by the using the scale-similarity model. The sign of the gradient of the SGS stress tensor in the filtered Navier-Stokes equations is chosen by making use of numerical analysis: it is determined so that it behaves as a forcing (or dissipative) term by making sure that is has the opposite (or same) sign as the viscous term in the filtered Navier-Stokes equations.

The viscous dissipation in the K equation stems from the viscous diffusion term in the filtered Navier-Stokes equation. The latter term is always dissipative. This is best seen by looking at Eq. 3. The viscous diffusion term in the filtered Navier-Stokes equation for u'_i appears in the first term on the second line in Eq. 3. To ensure that $\varepsilon_{SGS} > 0$, we set $-\partial \tau_{ik}/\partial x_k$ to zero when its sign is different from that of the viscous diffusion term (cf. the two terms on the right side of the second line in Eq. 3). This is achieved by defining a sign function

$$M_{ik}^D = \text{sign}\left(-\frac{\partial \tau_{ik}}{\partial x_k}\frac{\partial^2 u_i'}{\partial x_k \partial x_k}\right), \text{no summation on } i,k \tag{4}$$

where $M_{ik}^D = \pm 1$. The problem is that we do not know u_i' $(= \bar{u}_i - \langle \bar{u}_i \rangle)$ because $\langle \bar{u}_i \rangle$ is not known until the simulations have been carried out. Fortunately, the sign of the second derivative of the resolved velocity fluctuation, u_i', is in most cases the same as that of the resolved velocity, $\bar{u}_i$. The correlation of the signs of the two second derivatives is larger than 95% in channel flow for $y^+ > 40$ Davidson (2009). Hence Eq. 4 can be replaced by

$$M_{ik}^D = \text{sign}\left(-\frac{\partial \tau_{ik}}{\partial x_k}\frac{\partial^2 \bar{u}_i}{\partial x_k \partial x_k}\right), \text{no summation on } i,k \tag{5}$$

Each component of the divergence of SGS stress tensor in Eq. 1 is then simply multiplied by

$$\tilde{M}_{ik}^D = \max(M_{ik}^D, 0) \tag{6}$$

so that

$$\frac{\partial \tau_{ik}^D}{\partial x_k} = \tilde{M}_{ik}^D \frac{\partial \tau_{ik}}{\partial x_k}, \text{no summation on } i,k \tag{7}$$

By using the limiter, $\tilde{M}_{ik}^D$, we omit the back scatter caused by the SGS stresses. Figure 2 includes ε_{SGS}^D which is computed with τ_{ik}^D.

2.2 Back Scatter Model

A dissipative scale-similarity model was presented above. Via the sign function, only the dissipative part of the subgrid stress term, $-\partial \tau_{ik}/\partial x_k$, was included in the momentum equations. The subgrid stress term can also be used as a forcing term. In this case, only the back scatter part is included. This is easily accomplished by replacing Eqs. 5, 6 and 7 by

$$\begin{aligned} M_{ik}^B &= -\text{sign}\left(-\frac{\partial \tau_{ik}}{\partial x_k}\frac{\partial^2 \bar{u}_i}{\partial x_k \partial x_k}\right), \text{no summation on } i,k \\ \tilde{M}_{ik}^B &= \max(M_{ik}^B, 0) \\ \frac{\partial \tau_{ik}^B}{\partial x_k} &= \tilde{M}_{ik}^B \frac{\partial \tau_{ik}}{\partial x_k}, \text{no summation on } i,k \end{aligned} \tag{8}$$

where superscript B indicates back scatter. In this way, the subgrid tensor term is included whenever the sign of $-\partial \tau_{ik}/\partial x_k$ is *opposite* to that of the viscous diffusion. This means that the subgrid stress term, $-\partial \tau_{ik}^B/\partial x_k$, in the momentum equations acts as a *counter-gradient* diffusion term.

Note that the components of $\partial\tau_{ik}^B/\partial x_k$ must be added as a *volume* source (i.e. we cannot integrate the term using the Gauss divergence theorem) because it may be zero and non-zero, respectively, in two adjacent cells. As a consequence, the back-scatter term is not conservative.

3 Equations

3.1 The Turbulence Model

A one-equation model is employed, which reads

$$\frac{\partial k_{sgs}}{\partial t}+\frac{\partial}{\partial x_j}(\bar{u}_j k_{sgs})=\frac{\partial}{\partial x_j}\left[(\nu+\nu_{sgs})\frac{\partial k_{sgs}}{\partial x_j}\right]+P_{k_{sgs}}-C_\varepsilon\frac{k_{sgs}^{3/2}}{\Delta} \tag{9}$$
$$P_{k_{sgs}}=-2\nu_{sgs}\bar{s}_{ij}\bar{s}_{ij},\nu_{sgs}=C_k k^{1/2}\Delta,\Delta=(\Delta V)^{1/3},C_k=0.07,C_\varepsilon=1.05$$

3.2 Wall Functions

The wall shear stress is implemented by prescribing the turbulent viscosity at the wall. This is a convenient way of implementing the wall shear stress in general curvi-linear grids. The wall shear stress is defined as

$$\tau_w=\mu\frac{\partial\bar{u}}{\partial y}=\mu_{t,w}\frac{\bar{u}_P}{y_P} \tag{10}$$

where y_P and $\bar{u}_P$ denotes the distance from the wall to the wall-adjacent cell center and the wall-parallel velocity, respectively. Using the definition of the friction velocity, $\tau_w=\rho u_\tau^2$, we obtain

$$\mu_{t,w}\frac{\bar{u}_P}{y_P}=\rho u_\tau^2\rightarrow\mu_{t,w}=\frac{u_\tau}{\bar{u}_P}\rho u_\tau y_P \tag{11}$$

Substituting $u_\tau/\bar{u}_P$ with the log-law, $\bar{u}_P/u_\tau=\ln(Ey^+)/\kappa$, we can finally write

$$\mu_{t,w}=\frac{\kappa\rho u_\tau y_P}{\ln(Ey^+)} \tag{12}$$

We also need a boundary condition for k_{sgs}. In the log-law region, Eq. 9 reduces to the production and the dissipation term, i.e.

$$\rho C_\varepsilon\frac{k_{sgs}^{3/2}}{\Delta}=\rho P_{k_{sgs}}=\mu_{sgs}\left(\frac{\partial\bar{u}}{\partial y}\right)^2 \tag{13}$$

Multiplying both sides by μ_{sgs} gives

$$\left(\mu_{sgs}\frac{\partial\bar{u}}{\partial y}\right)^2=\rho^2 C_\varepsilon\frac{k_{sgs}^{3/2}}{\Delta}C_k k_{sgs}^{1/2}\Delta \tag{14}$$

Table 1 Time averaged and RMS wall shear stress. F denotes forcing

Re_τ	4000	4000F	8000	8000F	16000	16000F	32000	32000F
$\langle\tau_w\rangle$	1	1.16	1	1.18	1	1.18	1	1.21
$\tau'_{w,rms}$	0.23	0.49	0.25	0.51	0.19	0.5	0.19	0.54

The left side is equal to the shear stress up to the power of two, which in the log-law region can be approximated as $\tau_w^2 = \rho^2 u_\tau^4$, which gives

$$k_{sgs,P} = (C_k C_\varepsilon)^{-0.5} u_\tau^2 \tag{15}$$

The wall-adjacent cells should be located in the logarithmic region, i.e. with y^+ values between approximately 30 and 200; the upper limit increases with increasing Reynolds number.

3.3 The Momentum Equations

The incompressible momentum equation with an added SGS viscosity reads

$$\frac{\partial \bar{u}_i}{\partial t} + \frac{\partial}{\partial x_j}(\bar{u}_i \bar{u}_j) = -\frac{1}{\rho}\frac{\partial \bar{p}}{\partial x_i} + \frac{\partial}{\partial x_j}\left((\nu + \nu_{sgs})\frac{\partial \bar{u}_i}{\partial x_j}\right) - \frac{\partial \tau_{ij}^B}{\partial x_j} \tag{16}$$

where the last term is the forcing term, which is non-zero only in the wall-adjacent cells; $\partial \tau_{ij}^B / \partial x_j$ is obtained from Eqs. 2 and 8. The forcing term includes the wall-normal derivate, $\partial \tau_{ij}^B / \partial x_2$. Hence we need to prescribe a numerical wall boundary condition for τ_{ij}^B. The second derivative of τ_{ij}^B is set to zero at the wall. For τ_{12}^B, for example, this is accomplished by setting the first derivative at the wall equal to that at the wall-adjacent node. For an equidistant grid this gives

$$\tau_{12}^B\big|_{j=1} = 1.5\tau_{12}^B\big|_{j=2} - \tau_{12}^B\big|_{j=3} \tag{17}$$

where $j = 1$ and $j = 2$ denote the wall node and the wall-adjacent node, respectively.

4 Results

4.1 Channel Flow

Fully developed channel flow was computed for four different Reynolds numbers, $Re_\tau = u_\tau \delta / \nu$ (δ denotes the half width of the channel), namely $Re_\tau = 4000, 8000,$ 16000 and 32000. The computational extent is $x_{max} = 3.2\delta$, $y_{max} = 2\delta$, and $z_{max} = 1.6\delta$. Constant grid spacing is used in all coordinate directions. The results do not change when the extent and the number of nodes are doubled in x and z directions. Periodic boundary conditions are applied in the streamwise and spanwise directions.

Figure 3a presents the velocity profiles. As can be seen, U^+ decreases when forcing is used, mainly because the wall shear stress increases. Table 1 presents the wall shear stresses and their fluctuations, $\tau'_{w,rms} = \langle \tau_w^2 - \langle \tau_w \rangle^2 \rangle^{1/2}$. It shows that the wall shear stress increases by approximately 20% when forcing is applied and that the resolved fluctuations of the wall shear stress increase by 100% or more.

The resolved turbulent kinetic energy presented in Fig. 3b exhibits a slight decrease when forcing is applied. Forcing usually generates additional resolved turbulence. However, since the forcing is applied only in the wall-adjacent cells, it increases only the wall shear stress and generates additional fluctuations in the wall shear stress.

4.2 3D Hill Flow

The flow around the 3D hill was also computed using wall functions. The Reynolds number is 130000 based on the hill height. The coarse mesh has $160 \times 80 \times 130$ cells

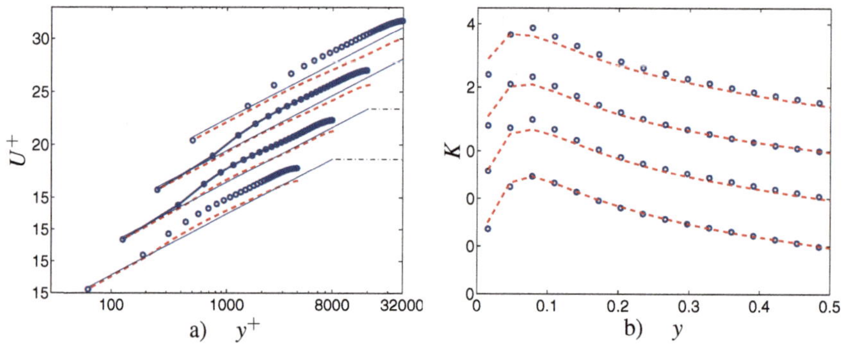

Fig. 3 Channel flow. Mean velocity and resolved turbulent kinetic energy, $K = 0.5\langle u'_i u'_i \rangle$; —— : $U^+ = (\ln y^+)/0.4 + 5.2$; - - - : forcing; ○: no forcing. From bottom to top in each figure: $Re_\tau = 4000$, 8000, 16000, 32000

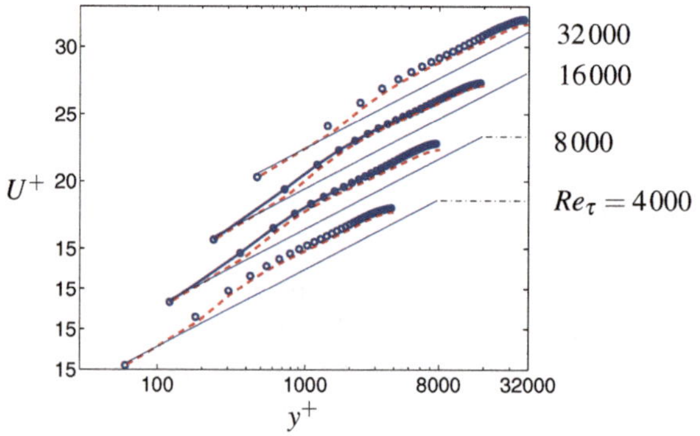

Fig. 4 Channel flow. Mean velocity. —— : $U^+ = (\ln y^+)/0.4 + 5.2$; - - - : dissipative scale-similarity model, i.e. $\partial \tau^D_{ij}/\partial x_j$; ○: standard scale-similarity model, i.e. $\tilde{M}^D_{ik} \equiv 1$

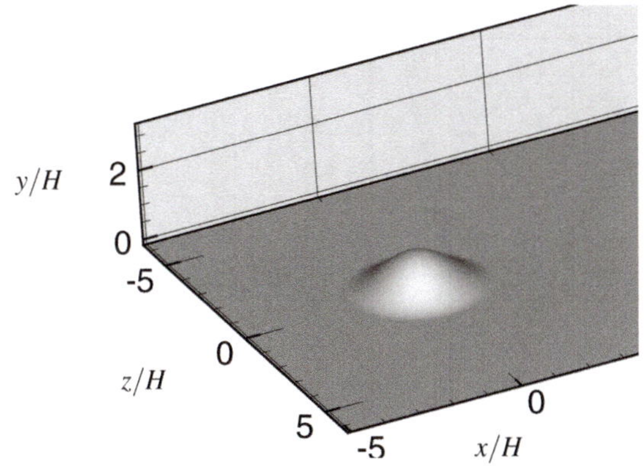

Fig. 5 3D hill. Computational domain. Hill crest at $x = z = 0$. Extension of hill: $-2 \leq x/H \leq 2$; $-2 \leq z/H \leq 2$. Inlet and outlet located at $x/H = -4.1$ and $x/H = 15.4$, respectively

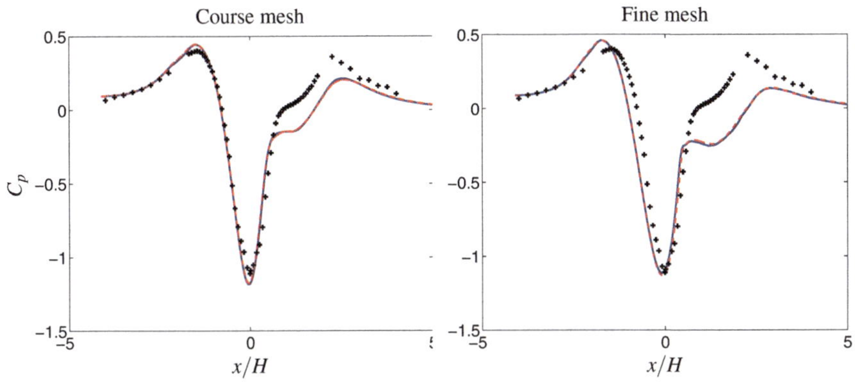

Fig. 6 3D hill flow. Pressure coefficient. —— : no forcing; - - -: forcing; +: experiments Simpson et al (2002)

(streamwise, x; wall-normal, y; spanwise, z). The fine mesh has $210 \times 80 \times 210$ cells (it is refined in the region $0 < x/H < 2$, $-2 < z/H < 2$). The wall-adjacent cells are located so that $30 < y^+ < 60$. A weak stretching of approximately 3% is used in the wall-normal direction, which gives $(\Delta x/\Delta y)_{max} < 6$ and $(\Delta x/\Delta y)_{max} < 4$ in the hill region ($|x| < 4$, $|z| < 2$) for both meshes. Forcing cannot be applied over the entire low-wall region because this causes divergence. Instead, forcing is applied only outside the hill region, i.e. $|x| < 2$ and $|z| < 2$. Synthetic fluctuations are used at the inlet for creating inlet turbulence Davidson (2007).

The pressure coefficient for both meshes is presented in Fig. 6. As can be seen, the effect of forcing is negligible for both meshes. The agreement with experiments is poor; it should be mentioned that good agreement with experiments using wall functions has been shown in the literature Tessicini et al (2007). These authors used

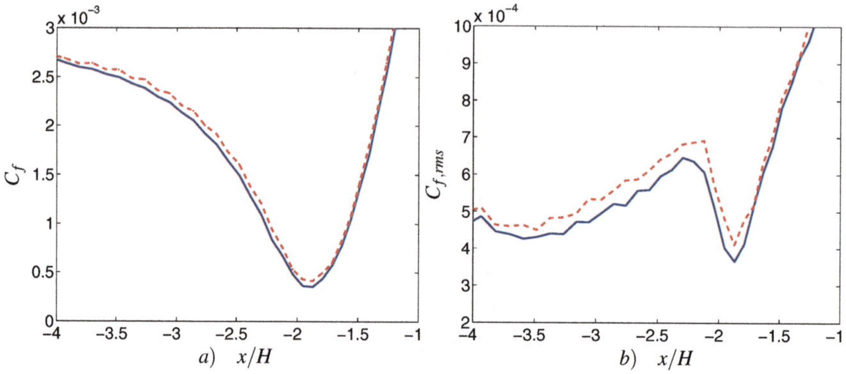

Fig. 7 3D hill flow. Wall friction coefficient (a) and its root-mean-square (b). Coarse mesh. Zoom in front of the hill. —— : no forcing; - - - : forcing

a different SGS model (a dynamic Smagorinsky) but the mesh ($192 \times 96 \times 192$ cells) was fairly similar to the present fine mesh.

Figure 7 presents the wall friction coefficient and its RMS in front of the hill. As expected, forcing does give a small increase in C_f and a slightly larger $C_{f,rms}$.

5 Concluding Comments

LES with wall functions have been presented. A new approach improving wall functions has been proposed: forcing is applied at the cells adjacent to the walls. The method was validated for fully developed channel flow and 3D hill flow. For channel flow, forcing increases the resolved fluctuations (RMS) of the wall shear stress by more than a factor of two, and the resolved wall shear stresses are increased by some 20%.

However, for the 3D hill flow, the effect of forcing is unfortunately negligible. There is a small increase in the wall friction coefficient and its resolved fluctuations, but there is no effect on the mean flow. One reason may be that the separation process to a large degree is governed by geometry. This flow is indeed an extremely challenging flow Garca-Villalba et al (2009) and maybe it should not be expected that wall-functions should be able to handle this flow. It is expected that the new wall functions work better in flows where the separation process is governed by the delicate balance between momentum diffusion and pressure gradient. However, the author is not aware of any such test cases at high Reynolds number.

References

Batten, P., Goldberg, U., Chakravarthy, S.: Interfacing statistical turbulence closures with large-eddy simulation. AIAA Journal 42(3), 485–492 (2004)

Davidson, L.: Using isotropic synthetic fluctuations as inlet boundary conditions for unsteady simulations. Advances and Applications in Fluid Mechanics 1(1), 1–35 (2007)

Davidson, L.: Hybrid LES-RANS: back scatter from a scale-similarity model used as forcing. Phil. Trans. of the Royal Society A 367(1899), 2905–2915 (2009)

Davidson, L.: A dissipative scale-similarity model. In: Armenio, V., Geurts, B., Fröhlich, J. (eds.) DLES7: Direct and Large-Eddy Simulations 7. ERCOFTAC series, vol. 13, pp. 261–266 (2010)

Davidson, L., Billson, M.: Hybrid LES/RANS using synthesized turbulence for forcing at the interface. International Journal of Heat and Fluid Flow 27(6), 1028–1042 (2006)

Davidson, L., Dahlström, S.: Hybrid LES-RANS: An approach to make LES applicable at high Reynolds number. International Journal of Computational Fluid Dynamics 19(6), 415–427 (2005)

Esfahaniana, V., Akbarzade, P.: The Jameson's numerical method for solving the incompressible viscous and inviscid flows by means of artificial compressibility and preconditioning method. Applied Mathematics and Computation 206(2), 651–661 (2008)

García-Villalba, M., Rodi, W., Leschziner, M.: Large-eddy simulation of separated flow over a three-dimensional axisymmetric hill. Journal of Fluid Mechanics 627, 55–96 (2009)

Hirsch, C.: Numerical Computation of Internal and External Flows: Fundamentals of Numerical Discretization, vol. 1. John Wiley & Sons, Chichester (1988)

Jameson, A.: Steady-state solution of the Euler equations for transonic flow. Advances in Scientific Computing, pp. 37–69 (1982)

Keating, A., Piomelli, U.: A dynamic stochastic forcing method as a wall-layer model for large-eddy simulation. Journal of Turbulence 7(12), 1–25 (2008)

Piomelli, U., Balaras, E., Pasinato, H., Squire, K., Spalart, P.: The inner-outer layer interface in large-eddy simulations with wall-layer models. International Journal of Heat and Fluid Flow 24, 538–550 (2003)

Simpson, R., Long, C.H., Byun, G.: Study of vortical separation from an axisymmetric hill. International Journal of Heat and Fluid Flow 23(5), 582–591 (2002)

Speziale, C.: Galilean invariance of subgrid-scale stress models in the large-eddy simulation of turbulence. Journal of Fluid Mechanics 156, 55–62 (1985)

Temmerman, L., Hadiabdi, M., Leschziner, M., Hanjalic, K.: A hybrid two-layer URANS-LES approach for large eddy simulation at high Reynolds numbers. International Journal of Heat and Fluid Flow 26(2), 173–190 (2005)

Tessicini, F., Li, N., Leschziner, M.: Large-eddy simulation of three-dimensional flow around a hill-shaped obstruction with a zonal near-wall approximation. International Journal of Heat and Fluid Flow 28(5), 894–908 (2007)

LES of the Non-isothermal Flow in Rotating Cavity

E. Tuliszka-Sznitko, A. Zieliński, and W. Majchrowski

Abstract. The Large Eddy Simulation with subgrid modelling based on the spectral Chebyshev-Fourier approximation was performed to investigate the 3D turbulent and transitional non-isothermal flows within a rotor/stator cavity. In the LES we used a version of the dynamic Smagorinsky eddy viscosity model proposed by Meneveau et al. (1996), in which the Smagorinsky coefficient at a given position x depends on the history of the flow along the fluid particle pathline. Computations were performed for the cavity of the aspect ratio L=5.0 and curvature parameters Rm=1.8, 3.0 and 5.0. Distributions of the heat flux components, the Reynolds stress tensor components, the turbulent Prandtl number and other structural parameters obtained in the present simulation compare favorably with the experimental data published in the literature.

1 Introduction

The instability structures of the flow in the rotor/stator and rotor/rotor cavity have been investigated since the sixties of the last century, mostly with reference to applications in turbomachinery. Additionally, the flow between rotating disks is one of the simplest 3D flows, highly suitable for investigating the effect of mean flow parameters on the turbulence and transitional structures. The flow in rotor/stator cavity was investigated experimentally and numerically among others by Gauthier et al. (2002), Moisy et al. (2004), Serre and Pulicani (2001), Serre et al. (2004), Lygren and Anderson (2004), Séverac and Serre (2007), Tuliszka-Sznitko et al. (2002, 2008, 2009a, b). Lygren and Anderson (2004) performed the LES of the flow in the open rotor/stator cavity using three different models, and compared the results with those obtained by DNS. Séverac and Serre (2007) performed numerical computations for the enclosed cavity for the Reynolds number up to $Re = \Omega R_1^2/\nu = 1000000$ ($L = (R_1 - R_0)/h = 5.0$, $Rm = (R_1 + R_0)/(R_1 - R_0) = 1.8$) using the Spectral Vanishing Viscosity (SVV) method, and compared the results with their LDV experimental data.

The non-isothermal flow conditions were also considered in some investigations (Tuliszka-Sznitko et al. 2002, 2008, 2009a, Randriamampianina et al. 1987), which showed that the thermal effects and the rotation-induced buoyancy influence the stability characteristics and the critical conditions. Tuliszka-Sznitko et al. (2009a) performed the LES of the non-isothermal flow in the rotor/stator cavity with heated

E. Tuliszka-Sznitko · A. Zieliński · W. Majchrowski
Poznan University of Technology, Institute of Thermal Engineering, Poland
ewa.tuliszka-sznitko@put.poznan.pl

S.-H. Peng et al. (Eds.): Progress in Hybrid RANS-LES Modelling, NNFM 111, pp. 283–292.
springerlink.com

stator and outer cylinder, delivering distributions of the local Nusselt numbers along the stator and rotor for different configurations and Reynolds numbers. Pellé and Harmand (2007) performed measurements over the rotor (in the rotor/stator configuration), using a technique based on infrared thermography. A very detailed experimental investigation of the turbulent flow around a single heated rotating disk was performed by Elkins and Eaton (2000).

In the present paper, we investigate the flow in the rotor/stator cavity of the aspect ratio L=5.0 and curvature parameters Rm=1.8, 3.0 and 5.0, using the LES based on a version of the dynamic Smagorinsky eddy viscosity model proposed by Meneveau (1996), in which the averaging is performed over the fluid particle pathline. The objective of our investigations was to compute the turbulent heat fluxes and the Reynolds stress tensor components, the turbulent Prandtl number and other structural parameters. These data can be useful for heat transfer modeling and can help to understand modifications in the turbulence structures introduced by the mean flow skewing.

2 Mathematical and Geometrical Models

We investigate the non-isothermal flows in the cavity between stationary and rotating disks of the inner and outer radius R_0 and R_1, respectively. The interdisks spacing is denoted by 2h. The rotor rotates at uniform angular velocity $\Omega = \Omega e_z$, e_z being the unit vector on the axis. The outer cylinder is attached to the stator and the inner one is attached to the rotor, (Fig.1).

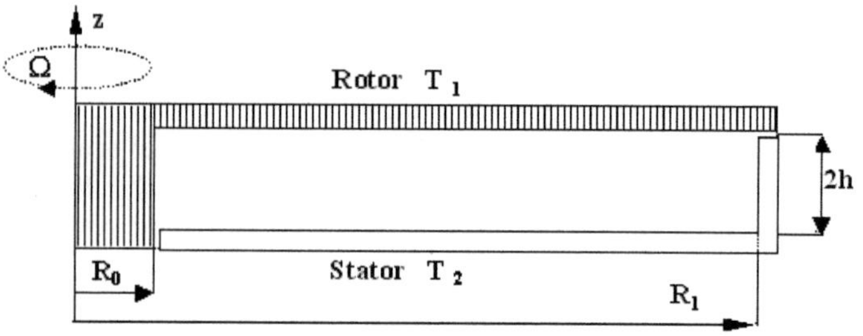

Fig. 1 Schematic picture of the rotating cavity

The flow is described by the Navier-Stokes continuity and energy equations, written in a cylindrical coordinate system (R, φ, Z) with respect to a rotating frame of reference. To take into account the buoyancy effects induced by the involved body forces i.e. the Coriolis force and the circumferential force (the gravity force is neglected, for its magnitude is small compared to the centrifugal force), the Boussinesq approximation is invoked. The time, length and velocity are normalized as follows: $(\Omega)^{-1}$, h and ΩR_1. The dimensionless temperature is defined in the following manner: $\Theta = (T - T_1)/(T_2 - T_1)$, where T_1 is the temperature of the rotor and the inner cylinder, and T_2 is the temperature of the stator and the outer cylinder. The dimensionless axial and radial coordinates are $z = Z/h, z \in [-1,1]$, $r = (2R - (R_1 + R_0))/(R_1 - R_0)$; $r \in [-1,1]$. The dimensionless components of the velocity vector in radial, azimuthal and axial directions are denoted by u, v, w and dimensionless pressure is denoted by p. The flow is governed by the following dimensionless parameters: aspect ratio, $L = (R_1 - R_0)/h$, curvature parameter, $Rm = (R_1 + R_0)/(R_1 - R_0)$, Reynolds number, $Re = \Omega R_1^2/\nu$ and thermal Rossby number, $B = \beta(T_2 - T_1)$, where $\beta = -1/\rho_r(\partial\rho/\partial T)_p$, ρ is density. The no-slip boundary conditions are used with respect to all rigid walls, u=w=0. For the azimuthal velocity component, the boundary conditions are as follows: $v = 0$ on the rotating disk and the inner cylinder, and v=-(Rm+r)/(Rm+1) on the stator and the outer cylinder. The periodicity condition is used in azimuthal direction. The thermal boundary conditions are as follows: $\Theta = 1$ for z=-1.0 and for $-1.0 \le r \le 1.0$, $\Theta = 0$ for z=1.0 and for $-1.0 \le r \le 1.0$, $\Theta = 1$ for r=1.0 and for $-1.0 \le z \le 1.0$, $\Theta = 0$ for r=-1.0 and for $-1.0 \le z \le 1.0$.

In the LES we use a version of the dynamic Smagorinsky eddy viscosity model proposed by Meneveau et al. (1996), in which the required averaging is performed over the fluid particles pathlines, instead of averaging over the direction of statistical homogeneity. The Smagorinsky coefficient is determined by minimizing the modeling error over the pathlines of the fluid particles. The numerical algorithm used for the LES of the non-isothermal flow in the annular cavity, proposed in the papers Tuliszka-Sznitko et al., (2008, 2009a, b), is an extended version of the DNS algorithm developed by Serre and Pulicani (2001) and Randriamampianina et al. (1989). The numerical solution is based on a pseudo-spectral collocation Chebyshev-Fourier-Galerkin approximation. In the time approximation we use a second-order semi-implicit scheme, which combines an implicit treatment of the diffusive term and an explicit Adams-Bashforth extrapolation for the non-linear convective terms. In the non-homogeneous radial and axial directions, Chebyshev polynomials are used with the Gauss-Lobatto distributions to ensure high accuracy of the solution inside the very narrow boundary layers at the disks.

3 Results

Computations have been performed for annular cavities of the aspect ratio L=5.0 and curvature parameters Rm=1.8, 3.0 and 5.0 with the heated stator and the outer cylinder. The considered flows are of Batchelor type, it means that flows consist of two separate boundary layers and an inviscid core. Positive thermal Rossby number B>0 means that the buoyancy-driven secondary flow enforces the basic rotation-driven flow. Figs.2a and b present the axial profiles of the radial and azimuthal velocity components obtained in the middle section of cavities (L=5.0, Rm=5.0, B=0.1) for different Reynolds numbers. From Fig.2 we can see that the flow is pumped radially outward along the rotor and recirculates along the stator. Figs.3a, b and c show the iso-lines of the dimensionless temperature obtained in the azimuthal (stator boundary layer) and meridian sections of cavities

(L=5.0, Rm=1.8, B=0.1, Re=300000), (L=5, Rm=3.0, B=0.1, Re=250000) and (L=5, Rm=5.0, B=0.1, Re=300000), respectively. From Fig.3 we can see that turbulence is concentrated in the stator boundary layer.

The areas of maximum turbulence are also visible in Figs.4a and b, where the axial profiles of the azimuthal and radial Reynolds stress tensor components obtained for (L=5.0, Rm=5.0, B=0.1) and for different Reynolds numbers are analyzed. We can see again that the stator is more turbulent than the rotor. The comparison of the Reynolds stress tensor components profiles obtained for L=5.0, Rm=1.8, Re=100000, B=0.0 (Tuliszka-Sznitko et al. 2009b) with the experimental data and SVV data obtained by Séverac and Serre (2007) for the same parameters shows good agreement.

Correlation coefficients and structural parameters are the most desired parameters for modeling purposes because these parameters are less sensitive to geometrical parameters than the turbulent stresses and heat fluxes. Fig.5 shows the ratio of the wall parallel normal stresses to the vertical normal stress $(\overline{u'^2}+\overline{v'^2})/\overline{w'^2}$ in terms of the axial coordinate normalized by the boundary layer thickness. The $(\overline{u'^2}+\overline{v'^2})/\overline{w'^2}$ parameter is a measure of the coherence of the turbulent structures. In the stator boundary layer this parameter reaches a peak (not visible in Fig.5) near the disk and then decreases rapidly to the value of about 2 near the edge of the boundary layer, showing that the vertical motion is weak close to the disk. In Fig.5 our results are compared with the 2DTBL results obtained by Klebanoff (1954) and 3DTBL results obtained by Bradshaw and Pontikos (1985), Anderson and Eaton (1989), Littell and Eaton (1991) and Elkins and Eaton (2000). In the papers by Klebanoff (1954), Bradshaw and Pontikos (1985) and Anderson and Eaton (1989) the value of the parameter $(\overline{u'^2}+\overline{v'^2})/\overline{w'^2}$ starts between 5 and 8 near the wall and then decreases to about 2. The distributions of the parameter $(\overline{u'^2}+\overline{v'^2})/\overline{w'^2}$ obtained by Elkins and Eaton (2000) and Littell and Eaton (1991) for the heated rotating disk boundary layer differ from the results obtained in our investigations and from other results presented in Fig.5: in their distributions, the values of the parameter start from about 5 and increase to the values between 8 and 10 near the edge of the boundary layer. The results obtained by Elkins and Eaton (2000) and Littell and Eaton (1991) show that the vertical motion is strong in the area near the rotating disk but weakens in the outer part of the boundary layer.

Similarly to other 3DTBLs, the a_1 parameter (defined as the ratio of shear stress vector magnitude to twice the turbulent kinetic energy) obtained in our investigations is reduced significantly below the limit of 0.15, typical for 2DTBL, showing that the boundary layers in rotating cavities are less effective in creating shear stresses from the turbulent motion.

Fig.6 shows distributions of correlation coefficient $R_{\varphi\Theta}$, defined as:

$$R_{\varphi\Theta} = (\overline{v'\Theta'}^2 + \overline{u'\Theta'}^2)^{0.5} / (\overline{\Theta'^2}(\overline{v'^2} + \overline{u'^2}))^{0.5} \tag{1}$$

In the 2DTBLs $R_{\varphi\Theta}$ is constant with the values of approximately 0.7-0.8. For the heated rotating disk, Elkins and Eaton (2000) obtained $R_{\varphi\Theta} \approx 0.5$ in the inner part of the boundary layer and $R_{\varphi\Theta}$ decreased with increasing z. In our investigations we obtained the maximum value of parameter $R_{\varphi\Theta}$ near the stator. This value then decreases to about 5 near the edge of the boundary layer.

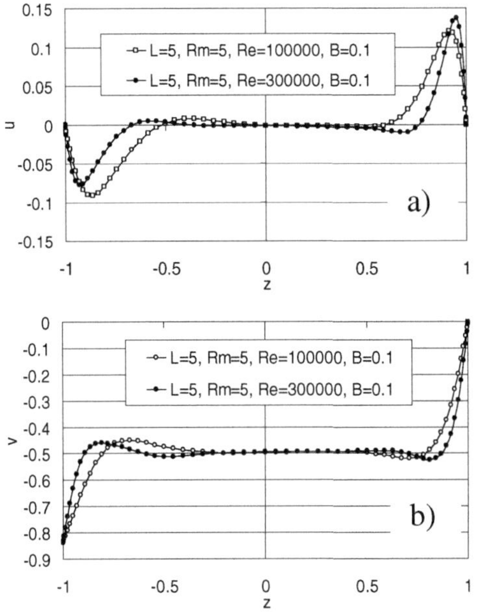

Fig. 2 The axial profiles of the radial (a) and azimuthal (b) velocity components obtained in the middle section of cavity (L=5, Rm=5.0, B=0.1) for different Reynolds numbers

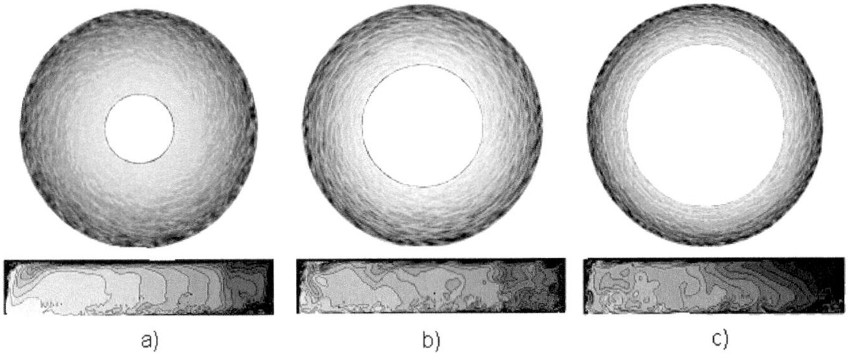

Fig. 3 The iso-lines of the temperature in azimuthal (stator boundary layer) and meridian sections of cavities: a) L=5.0, Rm=1.8, B=0.1, Re=300000, b) L=5.0, Rm=3.0, B=0.1, Re=250000, c) L=5.0, Rm=5.0, B=0.1, Re=300000

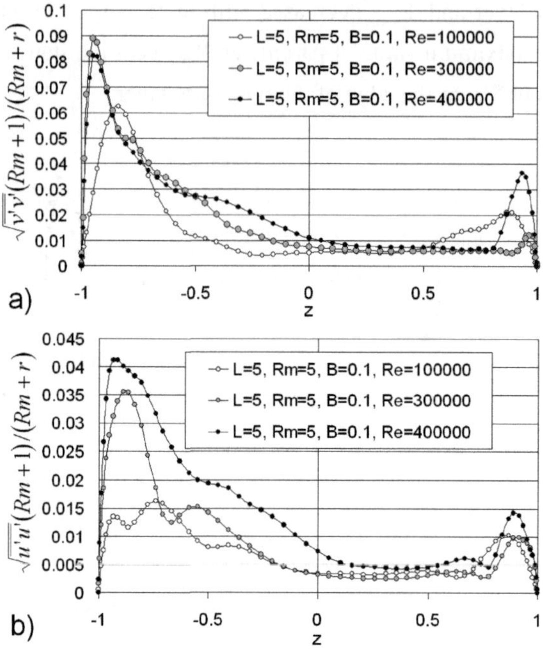

Fig. 4 The axial profiles of the azimuthal (a) and radial (b) Reynolds stress tensor components obtained for different Re. The middle section of the cavity (L=5.0, Rm=5.0)

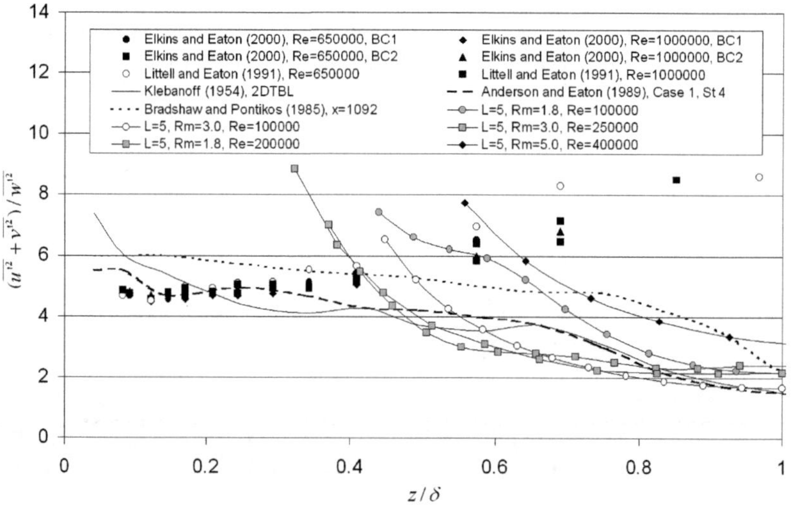

Fig. 5 The structural parameter $(u'^2 + v'^2)/w'^2$ in terms of z/δ obtained in the middle section of different configurations (stator boundary layer). Comparison with experimental results taken from the literature

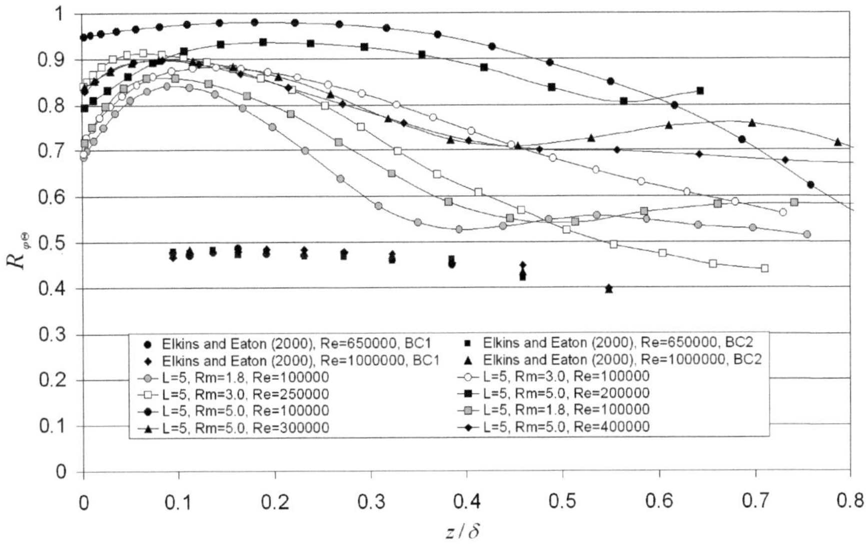

Fig. 6 The coefficient $R_{\varphi\Theta}$ in terms of z/δ obtained in the middle section of different configurations (stator boundary layer). Comparison with experimental results taken from the literature

The turbulent Prandtl number is defined as the ratio of the eddy diffusivity for momentum to the eddy diffusivity for heat:

$$\mathrm{Pr_t} = (\varepsilon_M / \varepsilon_H) = (-\overline{w'v'} / \partial\overline{v} / \partial z) / (\overline{w'\Theta'} / \partial\overline{\Theta} / \partial z) \tag{2}$$

The above definition is not a strict definition for strongly 3DTBLs, however, we used it to compare our results with data published by Elkins and Eaton (2000) for the flow along a single heated rotating disk. In many 2DTBLs the turbulent Prandtl number equals 1 in the area near the wall and decreases to 0.8 with increasing z. Elkins and Eaton (2000) showed that for a single rotating disk $\mathrm{Pr_t}$ equals 1 near the disk and then decreases to the value of about 0.6 at half the width of the boundary layer δ. In Fig.7 we present the comparison of axial distributions of $\mathrm{Pr_t}$ obtained in the present investigations in the middle sections of the cavities (stator boundary layer) with the results obtained by Elkins and Eaton (2000) and Littell and Eaton (1991) for the flow around a single heated rotating disk and with the distributions obtained in some 2DTBL investigations (Wroblewski and Eibeck, 1990, Subramanian and Antonia, 1981, Gibson and Verriopoulos, 1984). We can see that the results obtained in the present paper for the stator boundary layer are in good agreement with those obtained for the heated rotating disk and for 2DTBLs.

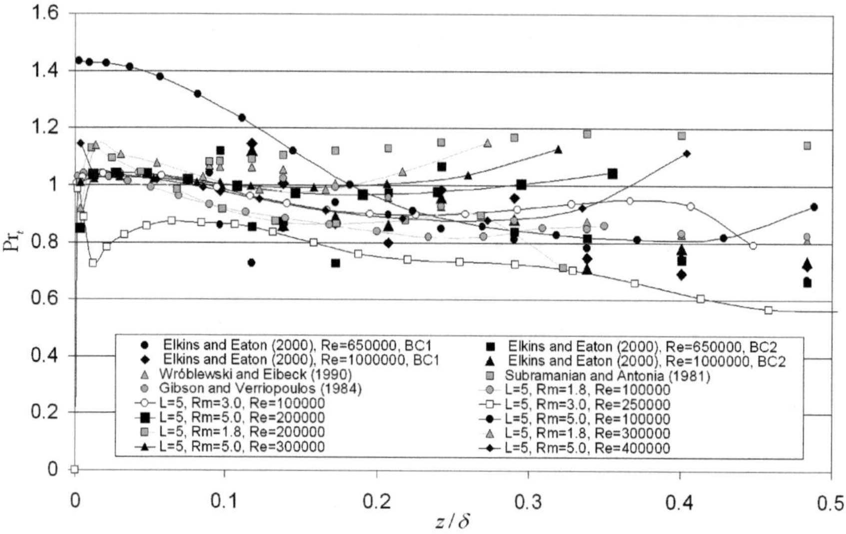

Fig. 7 The turbulent Prandtl Pr_t number in terms of z/δ obtained in the middle section of different configurations (stator boundary layer). Comparison with experimental results taken from the literature

4 Conclusions

In the paper we presented the Large Eddy Simulation of the non-isothermal transitional and turbulent flow in sealed cavities of the aspect ratio L=5.0 and curvature parameters Rm=1.8, 3.0 and 5.0. In the LES we used a version of the dynamic Smagorinsky eddy viscosity model proposed by Meneveau (1996), in which the Smagorinsky coefficients are averaged along the fluid particles pathlines. This algorithm (Tuliszka-Sznitko et al. 2008, 2009a, b) turned out to be very effective and allowed us to perform computations for high Reynolds numbers.

Computations have been performed for the thermal Rossby number from the range B=0.0-0.2 (with the heated stator and outer cylinder) and for Reynolds numbers up to Re=400000. The investigated flows belong to the Batchelor family (the flows are divided into two boundary layers separated by a central rotating inviscid core). We have found that the fluid turbulence concentrates in the stator boundary layer and its intensity increases towards the outer cylinder. We focused on the analysis of the correlation coefficients and structural parameters which are discussed in the light of experimental results obtained by Elkins and Eaton (2000) for a single rotating disk heated by uniform flux. The value of the a_1 parameter is between 0.9-0.13, which is below the typical value obtained for 2DTBLs (a_1 =0.15). The reduced value is attributed to the fact that three-dimensionality weakens the turbulent structure near the wall. The obtained distributions of the structural parameter $(\overline{u'^2} + \overline{v'^2}) / \overline{w'^2}$ in the stator boundary layer agree well with the results obtained by Klebanoff (1954), Bradshaw and Pontikos (1985) and Anderson and Eaton (1989) in the outer boundary layer but are in contrast to the distributions obtained by Elkins and Eaton (2000) and Littell and Eaton (1991) for the heated single rotating disk. The distributions of the turbulent Prandtl number in the stator boundary layer obtained in the present paper are in good agreement with all cited results. The correlation coefficient $R_{\varphi\Theta}$ reaches a maximum in the areas near the wall and then gradually decreases.

Acknowledgement. The computations were carried out in the Poznan Supercomputing and Networking Center, which is gratefully acknowledged. The authors thank prof. Patrick Bontoux from CNRS-Universites d'Aix-Marseille and prof. Akira Murata from the Tokyo University for fruitful discussions. The work is supported by COST/78/2006.

References

Anderson, S.D., Eaton, J.K.: An experimental investigation of pressure driven three dimensional boundary layers. J. Fluid Mech., 202, 263–294–362 (1989)

Bradshaw, P., Pontikos, N.S.: Measurements in the turbulent boundary layer on an 'infinite" swept wing. J. Fluid Mech. 159, 105–130 (1985)

Elkins, C.J., Eaton, J.K.: Turbulent heat and momentum transport on a rotating disk. J. Fluid Mech. 402, 225–253 (2000)

Gauthier, G., Gondret, P., Moisy, F., Rabaud, M.: Instabilities in the flow between co- and counter-rotating disks. J. Fluid Mech. 473, 1–21 (2002)

Germano, M., Piomelli, U., Moin, P., Cabot, W.: A dynamic sub-grid eddy viscosity model. Phys. Fluids, A 3(7), 1760–1765 (1991)

Gibson, M.M., Verriopoulos, C.A.: Turbulent boundary layer on a middle curved surface: Part 2: temperature field measurements. Exps. Fluids 2, 73–80 (1984)

Klebanoff, P.S.: Characteristics of turbulence in a boundary layer with zero pressure gradient, NACA TN 3178 (superseded by NACA Rep. 1247) (1954)

Lilly, D.: A proposed modification of the Germano subgrid-scale closure method. Phys. Fluids, A 4(3), 633–635 (1992)

Littell, H.S., Eaton, J.K.: An experimental investigation of the three-dimensional boundary layer on a rotating disk, Stanford Univ. of Mechanical Engineering. Thermosciences Div. Rep. MD-6 (1991)

Lygren, M., Andersson, H.I.: Large eddy simulations of the turbulent flow between a rotating and a stationary disk. ZAMP 55, 268 (2004)

Meneveau, C., Lund, T.S., Cabot, W.H.: A Lagrangian dynamic subgrig-scale model of turbulence. J. Fluid Mech. 319, 353–385 (1996)

Moisy, F., Doaré, O., Pasutto, T., Daube, O., Rabaud, M.: Experimental and numerical study of the shear layer instability between two counter-rotating disks. J. Fluid Mech. 507, 175–202 (2004)

Pellé, J., Harmand, S.: Heat transfer measurements in an opened rotor–stator system air-gap. Exp. Therm. Fluid Sci. 31, 165–180 (2007)

Randriamampianina, A., Bontoux, P., Roux, B.: Ecoulements induits par la force gravifique dans une cavité cylindrique en rotation. Int. J. Heat Mass Transfer 30(7), 1275–1292 (1987)

Serre, E., Pulicani, J.P.: A three-dimensional pseudospectral method for rotating flows in a cylinder. Computers & Fluids 30, 491 (2001)

Serre, E., Tuliszka-Sznitko, E., Bontoux, P.: Coupled numerical and theoretical study of the transition flow between a rotating and stationary disk. Phys. Fluids 16(3), 688–707 (2004)

Séverac, E., Serre, E.: A spectral viscosity LES for the simulation of turbulent flows within rotating cavities. J. Comp. Phys. 226(2), 1234–1255 (2007)

Subramanian, C.S., Antonia, R.A.: Effect of Reynolds number on a slightly heated turbulent boundary layer. Int. J. Heat Mass Transfer 24, 1833–1846 (1981)

Tuliszka-Sznitko, E., Serre, E., Bontoux, P.: On the nature of the boundary layers instabilities in a flow between a rotating and a stationary disc. C.R. Acad. Sci. Paris II B – Mech. 330(2), 91–99 (2002)

Tuliszka-Sznitko, E., Zielinski, A.: DNS/LES of transitional flow in rotating cavity. Int. J. Transport Phenomena 10(3), 223–234 (2008)

Tuliszka-Sznitko, E., Zielinski, A., Majchrowski, W.: LES and DNS of the non-isothermal transitional flow in rotating cavity. Int. J. Heat and Fluid Flow 30(3), 534–548 (2009a)

Tuliszka-Sznitko, E., Zielinski, A., Majchrowski, W.: Large Eddy Simulation of transitional flows in rotor/stator cavity. Archives Mech. 61(2), 93–118 (2009b)

Wroblewski, D.E., Eibeck, P.A.: An experimental investigation of turbulent heat transport in a boundary layer with an embedded streamwise vortex. PhD Thesis, University of California at Berkeley, Mechanical Engineering Department (1990)

Turbulence Modelling for Supersonic Separated Flows

A. Shams and P. Comte

Abstract. Shock-induced flow separation takes place in rocket nozzles when operate under over-expanded conditions. Numerical prediction of these supersonic separated flows has been a challenge for the existing RANS models and their respective LES/RANS techniques. In the present research work an extensive and comparative study of these turbulence modelling methodologies has been undertaken to investigate free shock separation flow configuration in thrust optimized contour (TOC) nozzle.

1 An Overview

Accurate prediction of turbulent flows with adverse pressure gradient and separation has been an area of interest over decades. Significant disagreement with the experimental data has been observed, especially in the presence of strong shock-waves and their interaction with the turbulence (boundary layers) Knight et al (2002). This shock-wave / turbulent boundary layer interaction is complicated by the simultaneous presence of flow separation and lead to high pressure and heat loads. These play an important role in the design and operation of aerospace vehicles (high-speed) and propulsion systems. Supersonic separated flow in rocket nozzles is one of the examples of such complex flow configurations and its correct prediction has become a challenge for existing turbulence models.

An adverse pressure gradient of sufficient strength can cause the boundary layer to separate. This flow separation is undesirable due to it unsteady and non-symmetric nature, which is responsible for the dangerous side-loads. Two-types of flow separation, free shock separation (FSS) and restricted shock separation (RSS), occur in rocket nozzle Nave L. H (1973). These flow separations have been the subject of several theoretical, experimental and numerical studies since the last few decades. A number of turbulence models by different researcher have been used to understand the physics of these flow regimes Chen et al (1994)-Shams and Comte (6-9 July, 2009). Such type of shock-induced separated flows contains different flow discontinuities and turbulence model based on linear approximation across the shock produces some non-physical behaviour. This non-physical behavior of the

Afaque Shams · Pierre Comte
LEA, Université de Poitiers, ENSMA, CNRS UMR-6609, CEAT,
43 route de l'Aérodrome, F-86036 Poitiers, France
afaque.shams@lea.univ-poitiers.fr,
Pierre.Comte@lea.univ-poitiers.fr

S.-H. Peng et al. (Eds.): Progress in Hybrid RANS-LES Modelling, NNFM 111, pp. 293–304.
springerlink.com

turbulence models sometimes completely changes the flow physics. Several modifications have been proposed to improve the prediction capabilities of these models e.g. realizability constraint, compressibility correction, length scale modification, rapid compression correction Coakley and Huang (1992)-Liou et al (2000). Few of them are used and discussed in this research work.

In the present study we have focused our attention to the numerical prediction of free shock separated flows in a thrust optimized contour nozzle. This nozzle is experimentally investigated at the Laboratoire d'Etudes Aérodynamiques (LEA), Poitiers, France, and is denoted as LEATOC nozzle. There is strong experimental evidence that in LEATOC nozzle the appearance of FSS regime is limited for the NPR range ≤ 24 Girard et al (2001) & Nguyen (2003). A variety of RANS models and hybrid (RANS / LES) techniques are used for the flow predicted in LEATOC nozzle of axi-symmetric configuration, and the details of the issues and problems encountered by using these techniques are documented in this draft. The paper is organized as: First the flow physics of free shock separation is given followed by the detail of turbulence models and hybrid techniques used for its prediction. Then numerical methods and finally results section in which following issues / problems are discussed: (i) behaviour of RANS models across the shock, (ii) grid sensitivity in the shock region, (ii) boundary layer profile, (iv) evolution of mean wall pressure and (v) wall shear stresses.

2 Free Shock Separated (FSS) Flow in Rocket Nozzle

Depending upon the nozzle contour two types of flow separation takes place in rocket nozzles namely, Free Shock Separation (FSS) and Restricted Shock Separation (RSS) Nave L. H (1973). Numerical investigation of these flow configurations undergoes a number of complexities. In the present study we have focused on the numerical and turbulence modelling issues and problems related to FSS.

In free shock separation (see Fig. 2 & 6), the over-expanded flow fully separates from the wall. The resulting stream wise wall pressure evolution is mainly governed by the physics of the shock-wave/boundary layer interaction occurring in any supersonic flow separation. The separation and the subsequent formation of a recirculation zone give rise to an oblique shock wave near the wall. This separation shock interacts with the Mach disk, and a reflected shock appears from the triple point. Due to this oblique shock, the wall pressure suddenly rises and reaches a plateau. Downstream of this plateau, the wall pressure slowly increases but remains less than the ambient pressure. This loss in wall pressure occurs due to the existence of the small recirculation zone near the nozzle lip.

3 Turbulence Modelling

Free shock separation exhibit relatively strong shock-wave / turbulent boundary layer interaction followed by massive separation. A number of numerical studies by different researchers have been performed in the past for such complex flow regimes Chen et al (1994)-Shams and Comte (6-9 July, 2009) and few of them has been accepted. Among them two-equation $k-\omega$ models have shown some improved results as compare due to their known ability to predict pressure gradient flows.

Two different turbulence modelling methodologies have been used in this research work, (i) Reynolds Averaged Navier-Stokes (RANS) and (ii) Hybrid RANS-LES.

3.1 RANS Approach

Five different types of RANS turbulence models are used for the numerical prediction of shock-induced free shock separated flow. All these models are based on the eddy-viscosity assumptions and are given as follows: for details see their corresponding references: (i) Spalart Allmaras (SA) Model Spalart and Allmaras (1992), (ii) k-ε model with Chien correction Chien (1982), (iii) Wilcox k-ω model Wilcox (1988), (iv) SST k-ω model Menter (1994) and (v) k-ω with realizability correction Moore and Moore (1999).

3.2 Hybrid RANS-LES Approach

Since hybrid RANS-LES technique is well suited for such massively separated flow configurations. Hence among hybrid methods three models are used for the present research work, which are (i) Delayed detached eddy simulation (DDES) based on k-ω model (ii) realizable delayed detached eddy simulation (RDDES) by Shams et al. Shams A (10-12 June, 2009) and (iii) limited numerical scales (LNS) approach by Batten et al. Batten et al (2002).

4 Numerical Methods

4.1 Space Discretization

The aforementioned turbulence modelling methodologies have been implemented in the TGNS3D code developed at CEAT/LEA (Poitiers, France). This code solves the Navier-Stokes equations on multiblock structured grids. The Navier-Stokes equations are descretized in space using a second-order cell centered finite volume method. Roe's flux difference splitting scheme is employed to obtain the flux at the cell interface. The monotone upstream-centered schemes for conservation laws (MUSCL) approach has been used to extend the spatial accuracy to second order with the combination of limiters, which are minmod and Van Albada, to prevent the numerical instabilities in shock regions.

4.2 Time Discretization

Explicit schemes are not efficient enough for such type of high accelerating flow simulations, hence implicit schemes are required. A second order two step implicit scheme has been implemented for time discretization. Diagonally dominant alternating direction implicit (DDADI) approach has been used for the inversion of a large sparse matrix system for the implicit scheme.

4.3 Results and Discussion

Numerical calculations are performed on a thrust optimized contour (TOC) nozzle. This nozzle is experimentally investigated at the Laboratoire d'Etudes Aérodynamiques (LEA), Poitiers / France, and is denoted as LEATOC nozzle Girard et al (2001). In series of experiments performed by Nguyen Nguyen (2003), it has been found that during start-up process free shock separated regimes appears only for the NPR range ≤ 24. When NPR reaches that limit this free separated jet reattach back to the nozzle wall and the resulting flow regime is known as Restricted Shock Separation (RSS). Hence for the present study numerical investigation of this very LEATOC nozzle is performed for NPR's < 24. During this numerical prediction of Free Shock Separation (FSS) flow regimes we have encountered several issues related to the aforementioned turbulence modelling methodologies and are discussed in detail in the following sections.

4.4 Behavior of RANS Model across the Shocks

In eddy-viscosity models, the Reynolds stresses are linearly related to the mean strain rates via eddy viscosity. These model works well in certain cases where the turbulent time scale is of the same order in magnitude of the mean strain. However, the present flow configuration exhibits strong shock-wave / boundary layer interaction (i.e. highly non-equilibrium flow) and the time scale of the mean distortion is significantly smaller than that of the turbulence. As a result, the equilibrium concept of eddy viscosity breaks down and the models yield unrealistically high values of production of turbulent kinetic energy (TKE). Figure 1, gives the solution obtained for SA and k-ε model with Chien correction at NPR=15.5 and 20 respectively. There is an experimental evidence that during the start-up at this NPR in LEATOC nozzle only FSS flow regime appears Nguyen (2003), but due to the high value of unrealistic turbulence predict by these models across the shock cause the free annular supersonic jet to reattach back to the nozzle wall. Several modifications such as realizability constraint, compressibility correction and production limiters have been proposed to overcome this problem. SST k-ω model limits this production with respect to the dissipation term i.e. $P_k = min(P_k, 20.D_k)$ Menter (1994). A similar kind of limiter is proposed by Wallin Wallin (1999): $P_k = min(P_k^{model}, P_k^{lim})$ where $P_k^{lim} = \rho K \sqrt{P_k^{model}/\mu_t}$. This limiter is not active for $P_k^{model} < \varepsilon / C_\mu$. Recently Cyril et al. Cyril and Abderrahmane (2004) limited the ratio of μ/μ_t to 90000, for such type of separated flows in thrust ideal contour (TIC) nozzle. These types of modifications enable the model to successfully reproduce FSS flow configuration at NPR=15.5 as depicted from Fig. 2. Comparison of iso-contours of turbulent kinetic energy in Fig. 3 & 4 shows the level of decrease in turbulent eddy-viscosity and kinetic energy value by using these limiters. One can notice that along the nozzle axis downstream of the shock a very high level of turbulence produced by linear eddy-viscosity model. This turbulence produces radial momentum away from the nozzle axis sufficient enough for the flow reattachment.

Fig. 1 Numerical solution (iso-Mach contours) obtained by S-A and $k-\varepsilon$ models at NPR=15 (left) & 20 (right) respectively

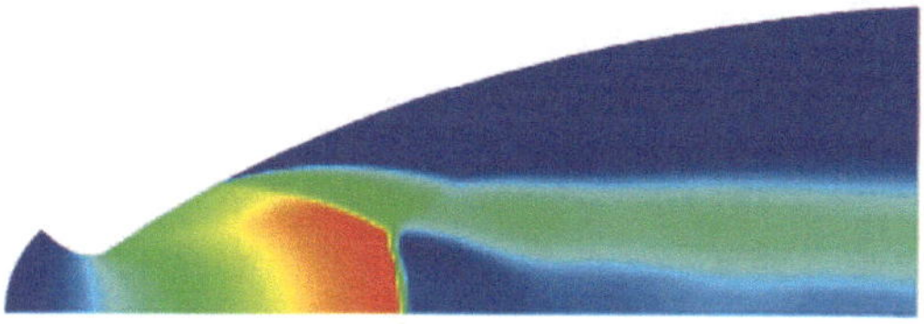

Fig. 2 Iso-Mach contour of FSS flow regime at NPR=15.5: $k-\omega$ with production limiter

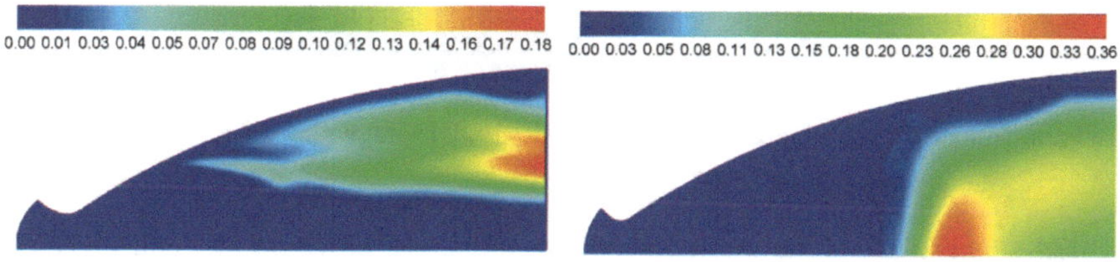

Fig. 3 Iso-contours of turbulent viscosity (μ_t) for RANS model with (left) and without (right) production limiter

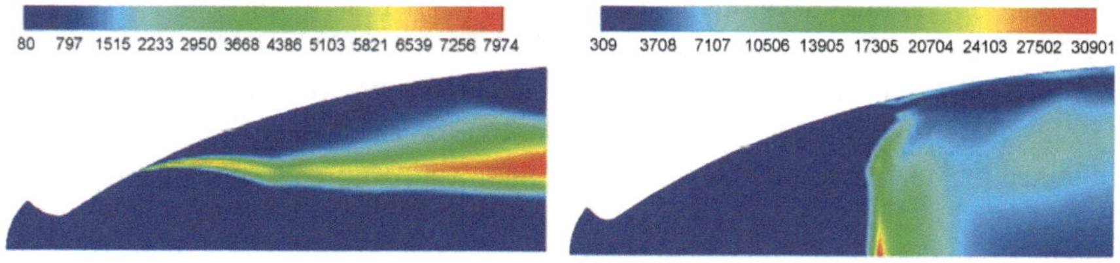

Fig. 4 Iso-contours of turbulent kinetic energy (TKE) (μ_t) for RANS model with (left) and without (right) production limiter

4.5 Grid Sensitivity in the Shock Region

The second problem was identified by the turbulence models with respect to the grid refinement across the shock. These supersonic separated flows exhibit several discontinuities and require fairly refined grid to have a considerably well prediction of flow features. It is known that across the shock eddy-viscosity model gives higher amplification of TKE, this is because of the fact that the production term varies $\propto 1/\delta^2$: where δ is the shock thickness. In addition pressure-dilatation term assumes very large negative value. The balance of these two terms gives high value of TKE downstream of the shock. Whereas the dissipation term have very negligible contribution in the shock region.

Fig. 5 Mesh (500*200) inside the nozzle (left) and respective obtained solution (right) at NPR=15.5

Fig. 6 Numerical solution on fine mesh at NPR=15.5: $k-\omega$ with realizability correction

Let suppose Δx is the mesh size where the shock wave occurs, and inversely varies with the production term. This suggests that more your refine the grid more non-physical behaviour of eddy-viscosity model you get across the shock. This constraint also prevents to get the grid convergence in these scenarios. Such an example can be seen in Fig. 5 with very refined mesh of 500×200: Δx inside the nozzle is of the order of boundary layer thickness. Figure 5 (right), shows the solution obtained by the Wilcox $k-\omega$ model and predict RSS flow regime at NPR=15.5 which is not physical. Whereas for a model with realizability correction Moore and Moore (1999), the production term does not depend on the mesh size. Numerical solution obtained by Wilcox $k-\omega$ model with this realizability correction on this fine mesh is presented in Fig. 6. One can notice a successfully reproduced FSS flow regime and by using this correction model is no more dependent on grid size in the flow discontinuity region. Hence grid convergence is also possible. To understand the effect of three-dimensional numerical solution in this range of NPR, a 3D numerical simulation is carried out by using RDDES. Figure 7, presents the successfully produced free shock separation at NPR=15.5. This also confirms the fact that at this NPR numerically obtained flow regime, during the start-up process in LEATOC nozzle, other than FSS is non-physical.

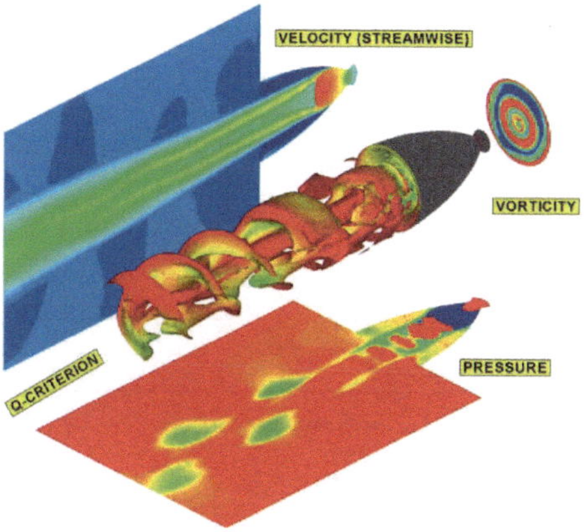

Fig. 7 Three dimensional view of free shock separation (FSS): RDDES, 4.3 million grid points

In Fig. 8, evolution of turbulent kinetic energy and eddy-viscosity profiles along the shock are plotted for the discussed RANS models. Eddy viscosity model, k-ε with Chien correction, yield a very high level of turbulence downstream of shock. This unrealistic level of turbulence has been well controlled by using production limiters, but even though becomes sensitive for the case of mesh refinement and produce non-physical results. On the other hand as we have seen that model with realizability correction does not depend on this grid sensitivity problem and no such unrealistic production of turbulence has been observed.

4.6 Boundary Layer Profile

Figure. 9 present comparison of velocity profile in the attached boundary layer for these turbulence models. $K-\omega$ models (with and without correction) and their extended DES versions have shown the same profile for boundary layer. Whereas LNS and $k-\varepsilon$ models do not able to recover the velocity profile and shows a sudden increase in velocity as in the case of high acceleration. Due to this very reason LNS hybrid technique (which is based on the $k-\varepsilon$ model near the wall region) under-predicts the separation point location (see Fig. 11).

4.7 Wall Shear Stresses

As a result of numerous studies on Hybrid RANS-LES techniques in the recent past, several shortcomings have been identified. One of them is modelled stress depletion which has also been encountered in this study. Due to ambiguous grid distribution

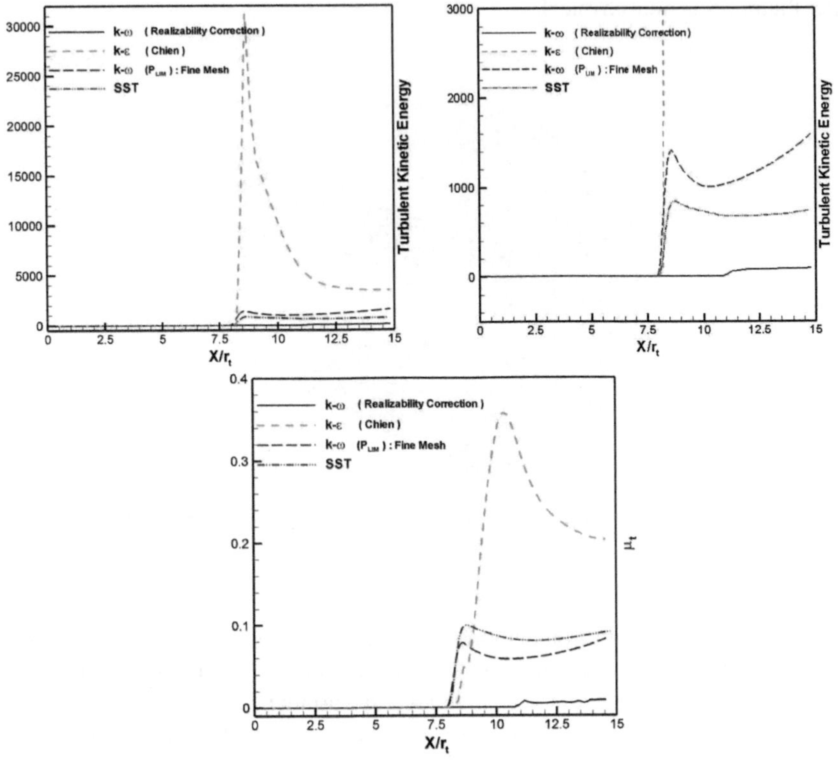

Fig. 8 Evolution of TKE (Top) and μ_t (Bottom) across the shock

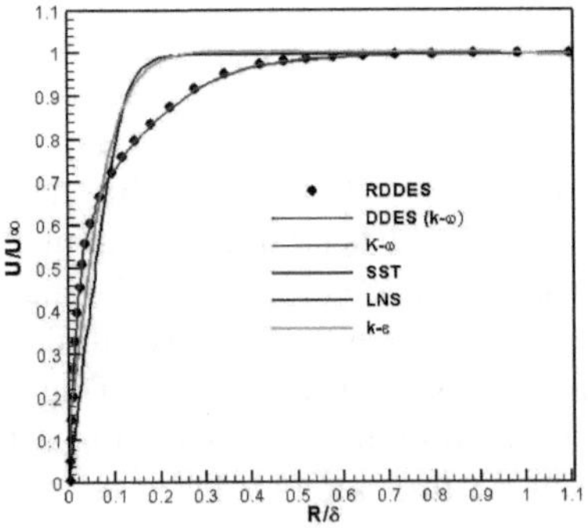

Fig. 9 Evolution of velocity profile inside boundary

near the wall LES mode is switched-on inside the boundary layer and defined fine grid for RANS model is not sufficient enough to resolve stresses for LES mode

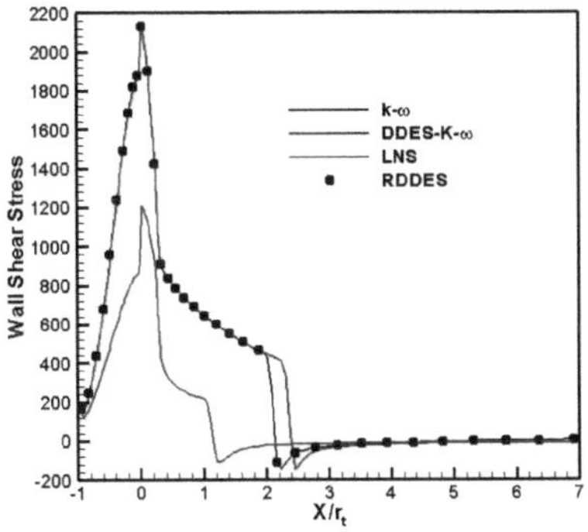

Fig. 10 Shear stress distribution along the nozzle wall for free shock separated flow at NPR=15.5

and shows depletion in the resolved wall stresses. Figure 10, shows the evolution of wall shear stresses obtained for the hybrid techniques used here and comparison has been made with the RANS model. LNS hybrid technique which does not have any defined shield to protect the immature switching into LES mode inside the boundary layer undergoes this very MSD problem.

Whereas, in the case of DDES we do not see this problem and shows the same level of wall shear stresses as by $k-\omega$ model. Although a small difference is the shear stress level between RDDES and DDES $k-\omega$ is due to the difference in the prediction of separation line.

5 Mean Wall Pressure

Numerically obtained mean pressure along the nozzle wall is compared with the experimental data available at LEA, Poitiers / France Nguyen (2003). DDES based on $k-\omega$ model over-predicts the separation point location due to the high level of turbulence produced across the shock. RDDES, realizable $k-\omega$ and SST have well predicted the separation point but slight under-prediction of plateau pressure have shown by SST model (see Fig. 11). LNS model have shown a huge under-prediction of separation point location and consequently high level of plateau pressure downstream of it. On may conclude that this onset separation shown by LNS is due to the combined effect of MSD problem and high accelerating behaviour in the near wall layer.

Data on free shock separated flow configuration has been obtained from a number of sources Dumnov (1998)-Keanini and Brown (2007) and are plotted in Fig. 12. This represents the flow in a variety of nozzle geometries, both in full scale/sub-scale models and hot/cold fire conditions. It is worthwhile to mention here that working fluid for most of the experiments is air except the Bloomer's hot fire measurements with $k = 1.2$: rocket fuel and liquid oxygen. In addition, obtained numerical

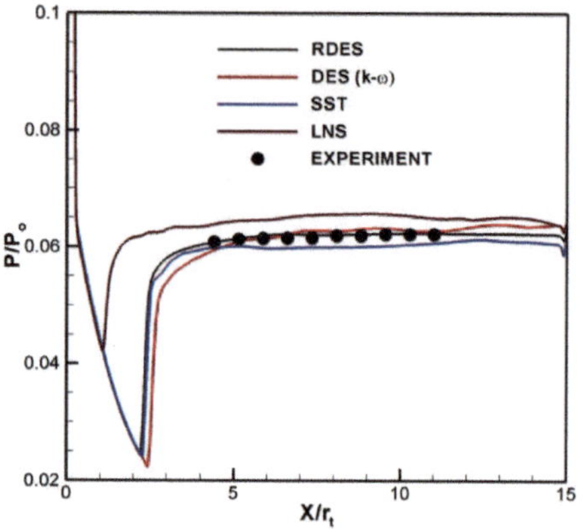

Fig. 11 Evolution of mean wall pressure

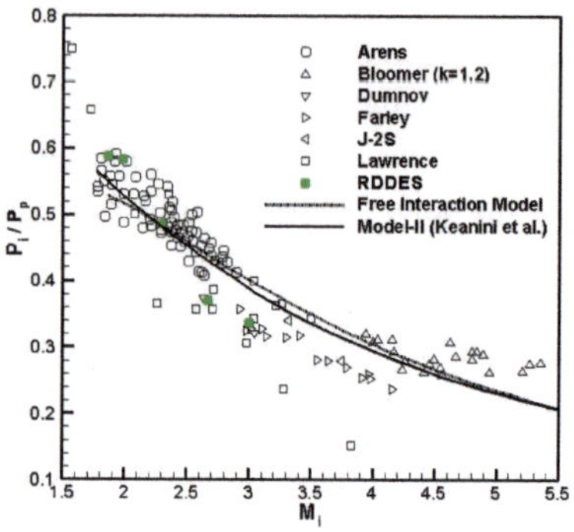

Fig. 12 Comparison of RDDES with separation measurements in rocket nozzles and analytical models

results of RDDES for various NPR in the range of free shock separation are compared with the free interaction model and recently proposed model by Keanini et al. Keanini and Brown (2007). The separation pressure ratio in figure indicate that RDDES provides a fairly good prediction in comparison with the variety of experimentally obtained data in a wide range of incipient Mach number (M_i).

6 Conclusions

Numerical investigation of free shock separated flow was undertaken on a thrust optimized contoured (LEATOC) nozzle. A variety of RANS models and their extended hybrid (RANS/LES) approaches were used to predict this complex flow regime. A comparative study of these models has been performed and has shown that standard turbulence models based on Boussinesq assumption yield unrealistically high value of turbulence kinetic energy across the strong shock. This sometimes may change the flow physics. However, production limiter across the flow discontinuities seems to give respectively better results. In these models production of TKE varies inversely with the shock thickness and also restricts us to achieve grid convergence in these scenarios. No such problem has been encountered when models are used with modification based on realizability constraint. Hence obtained results in comparison with the standard models are in excellent agreement with the experimental and grid convergence is achievable. MSD problem has been observed in the case of LNS Limiter numerical scales due to the early switching of LES mode inside the boundary layer. As a results wall shear stress and separation point are highly underpredicted as compared to the RDDES and DDES $k-\omega$, which uses a shielding function to avoid this problem.

Acknowledgments. The Higher Education Commission (HEC), Government of Pakistan, is acknowledged for the research grant. The first author would like to thank T. Alziary de Roquefort for many discussions related to turbulence modelling and numerical methods for supersonic nozzle flows.

References

Batten, P., Goldberg, U., Chakarvarthy, S.: Lns - an approach towards embedded les. AIAA Paper 2002-0427 (2002)

Chen, C.L., Chakravarthy, S.R., Hung, C.M.: Numerical investigation of separated nozzle flows. AIAA J. 32, 1836–1843 (1994)

Chien, Y.: Predictions of channel and boundary layer flows with a low-reynolds number turbulence model. AIAA Journal 20, 33–38 (1982)

Coakley, T.J., Huang, P.G.: Turbulence modelling for high speed flows. AIAA Paper 92-0436 (1992)

Cyril, P., Abderrahmane, N.: Flow separation in truncated ideal contour nozzle. Journal of Turbulence 5(1) (2004)

Dumnov, G.E.: Unsteady side-loads acting on the nozzle with developed separation zone. AIAA Paper 98-3619 (1998)

Girard, S., Deniau, H., Nguyen, A.T.: Alziary de roquefort t, etude de l'écoulement dans une tuyére propulsive à contour parabolique en régime surdétendu. In: 37eme Colloque d'Aérodynamique Appliquée de l'AAAF: Aérodynamique et Propulsion des Véhicules à grande vitesse, Aracachon, France (2001)

Keanini, R.G., Brown, A.M.: Scale analysis and experimental observations of shock-induced turbulent boundary layer separation in nozzles. European Journal of Mechanics B/Fluids 26, 494–510 (2007)

Knight, D., Yan, H., Panaras, A., Zheltovodov, A.: Rto wg: Cfd validation for shock wave turbulent boundary layer interactions. AIAA Paper 2002-0437 (2002)

Liou, W.W., Huang, G., Shih, T.H.: Turbulence model assesment for shock wave/turbulent boundary layer interactionin transonic and supersonic flows. Computers and Fluids 29, 275–299 (2000)

Menter, F.R.: Two-equation eddy-viscosity turbulence models for engineering applications. AIAA Journal 32, 1598–1605 (1994)

Moore, J.G., Moore, J.: Realizability in two equation models. AIAA Paper 99-3779 (1999)

Nave, L.H., Coffey, G.A.: Sea-level side-load phenomenon in high area ratio rocket engines. AIAA Paper 73-1284 (1973)

Nguyen, A.T.: Découlement instationaire et charges latérales dans les tuyéres propulsives. Ph.D Thesis, Université de Poitiers (2003)

Shams, A., Comte, P.: Formation of cap-shock pattern. In: 3rd European Conference for Aero-Space Sciences (EUCASS), Versailles-Paris, France, July 6-9 (2009)

Shams, A.C.P.: Realizable detached eddy simulation (rdes) for strong shock-wave/boundary layer interaction. In: Third Symposium on Hybrid RANS-LES Methods, Gdansk, Poland, June 10-12 (2009)

Spalart, P.R., Allmaras, S.R.: A one-equation turbulence model for aerodynamic flows. AIAA Paper 92-0439 (1992)

Wallin, S.: An efficient explicit algebraic reynolds stress k-ω model (earsm) for aeronautical applications. Report, The Aeronautical Research Institute of Sweden (FFA), Bromma (1999)

Wilcox, D.C.: Reassessment of the scale deterring equation for advanced turbulence models. AIAA Journal 26, 1299–1310 (1988)

Evaluation of Hybrid RANS-LES Models in Ventilated Enclosures

F. Brännström and D. Müller

1 Introduction

During the last few years the interest in more detailed knowledge in ventilated enclosure flows and their numerical implementation raised due to the need of thermal comfort and its relation to energy consumption. There is a need for transient simulations, which predict the frequencies and amplitudes of large scale structures and the turbulent intensities in combination with local flow direction, velocity and temperature. The existence of low-frequency transient flow structures is well known in HVAC (see Zhang et al. Zhang et al (2000)) and has to be taken into account.

The aim of this work is to examine assets and drawbacks of different hybrid LES-RANS methods for airflow inside ventilated enclosures. Apart from the SST-SAS model (SAS) of Menter Menter et al (2003) the filtered SST model (SSTF) of Gyllenram et al. Gyllenram and Nilsson (2007) and a hybrid model based on $y*$ (yStar) is evaluated. The main focus will be on the differences of flow development and the distribution of hybrid RANS-LES regions.

2 Numerical Method

All calculations are done using OpenFOAM[1]. The code employs the finite volume discretization method. For the SAS and SSTF calculations a second order accurate hybrid scheme is employed for the momentum and turbulence equation, which introduces some upwinding to the scheme. The effect of different schemes in combination with the yStar model on the flow domain was investigated. For this evaluation three different schemes were chosen. One simulation was done with central differencing scheme for the whole domain. The result was compared to a scheme which introduces a small amount of upwinding to the central differencing scheme to stabilize the calculation. In addition a locally blended scheme with second order upwind

F. Brännström · D. Müller
RWTH Aachen
E.ON Energy Research Center, Jägerstr. 17/19, D - 52066 Aachen, Germany
f.braennstroem@gmx.de

[1] www.openfoam.com

S.-H. Peng et al. (Eds.): Progress in Hybrid RANS-LES Modelling, NNFM 111, pp. 305–314.
springerlink.com

in RANS region and central differencing scheme in the LES region was applied. The transition for this scheme was defined by the interface position of the RANS-LES region. The LES simulation was done with a central differencing scheme for comparison. For all simulations an implicit Crank-Nicholson time differencing scheme and SIMPLE for pressure-velocity coupling were used. The code handles the collocated grid arrangement using Rhie-Chow interpolation.

3 Evaluated Hybrid Turbulence Models

All three hybrid RANS-LES models are based on the default $k-\omega$-SST model available in OpenFOAM.

3.1 SST-SAS Model

The idea of SAS is based on Menter et al. (2003). They introduced the van Kármán length scale $L_{vK} = \kappa S/|U''|$ in Rotta's (1972) turbulence scale equation model by relaxing overly restricted assumptions. The model adds an additional production term to the ω-equation, which reacts sensitively to unsteady fluctuations. This production term increases ω in regions, where the flow switches to unsteady behavior, resulting in a reduced turbulent kinetic energy and turbulent viscosity. This promotes the momentum equation to allow unsteady flow situations. The length scale based on velocity gradients is much smaller than time-averaged velocities for turbulence resolving equations.

3.2 Filtered SST-k-ω Model

The filtering approach, applied to the SST model, resolves flow scales in time and space depending on the time step and mesh size. It is based on the filtering approach of Willems Willems (1996) for $k-\varepsilon$ model. Gyllenram Gyllenram (2008) generalized the filter for the SST model. The upper limit of the turbulent length scale is set by the filter width Δ_f, which is defined as

$$\Delta_f = \gamma \max\left\{ |\underline{U}|\, \delta t, \Delta^{1/3} \right\}. \tag{1}$$

The coefficient γ is used to include a limited number of cells to actually resolve turbulent structures; in this work $\gamma = 3$. In addition the time step δt sets a limit to the resolved length scale by using $|\underline{U}|\, \delta t$ as the shortest distance over which a particle can move each iteration. This approach incorporates a CFL-number based selection of the filter width. The turbulent viscosity, as calculated in the SST model, is scaled by α

$$\nu_T = \alpha \frac{a_1 k}{\max(a_1 \omega, SF_2)} \quad \text{with} \quad \alpha = \min\left[(\Delta_f/L_t)^{4/3}, 1.0\right] \tag{2}$$

using the length-scale $L_t = \sqrt{k}/\beta^* \omega$.

3.3 Zonal Approach Based on y*

In the terminology of Fröhlich et al. Fröhlich and von Terzi (2008) this zonal approach is a combination of a weighted sum of RANS and LES for the transition region and a layering approach of RANS and LES used by Breuer et al. Breuer et al (2008) and Knieser Knieser (2008). The actual interface position f depends on y^* and is smoothed by a hyperbolic tangent function:

$$y^* = \frac{y\sqrt{k}}{\nu}, \quad f = \tanh\left(A \cdot \left(y^* - y^*_{interface}\right) + 1\right)/2 \tag{3}$$

with $A = 0.04$ and $y^*_{interface} = 100$ for this evaluation. For the LES region a simple Smagorinsky model with a cube-root-rule filter was used. The base RANS simulation coexists to the LES simulation in the complete flow domain. Both models are combined by a weighted sum:

$$\nu_t = (1 - f) \cdot \nu_t^{RANS} + f \cdot \nu_t^{LES}. \tag{4}$$

No special treatment was done for the two interface conditions RANS/LES and LES/RANS. The calculated interface position is not averaged in time with the current implementation.

4 Description of Test Cases and Numerical Setup

4.1 Nielsen Test-Case: 2D Wall Jet in Enclosure

The 2D wall jet test case is frequently used as a validation case in the HVAC field (e.g. Davidson et al. Davidson and Nielsen (1998)). Though, earlier calculations did not focus on the transient flow structure with respect to thermal comfort. The Reynolds-number based on the inlet height and bulk inlet velocity $U_{in} = 0.45\,\mathrm{m/s}$ is equal to 5000. Fig. 2 describes the experimental setup. The simulation for this evaluation was done with a hexahedral cartesian mesh with 2.4 Mio cells. Preliminary calculations showed, that for this flow RANS methods can be successful for the prediction of mean velocities. The minimum physical simulation time was for all setups around 250 sec. Based on the inlet velocity this will give a simulation time of about 12 characteristic time units.

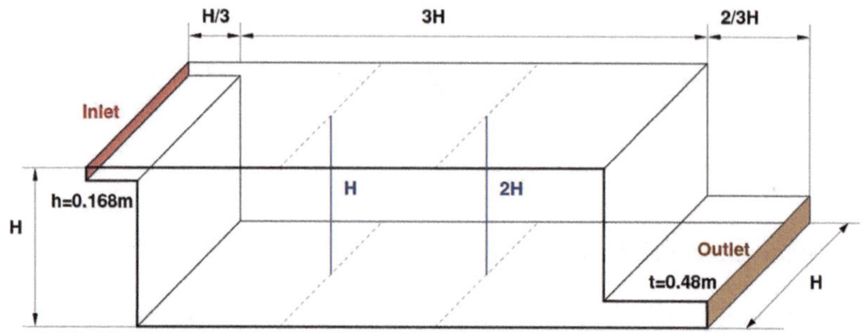

Fig. 1 Description of experimental setup of the Nielsen test-case

4.2 Hurnik Test Case – Ventilated Enclosures with Round Jet as Inlet

As a second test case a room with a round inlet at the front-wall was chosen Hurnik and Blaszczok (2002), the Reynolds-number is around 19800. The experimental setup Hurnik and Blaszczok (2002) provides an evenly distributed velocity inlet, which allows the use of a block profile with a short inlet section for the simulations. Data of the mean x-velocity at two sample lines are available. The position is $x = 10\,\text{d}$ and $x = 25\,\text{d}$ with $d = 0.0385\,\text{m}$ as the diameter of the inlet. The physical simulation time was in the range of 10 to 80 sec, which corresponds to a characteristic time unit of 26 to 210 based on the inlet velocity. Similar to the first test case, standard RANS methods are able to predict the mean velocity at the two sample lines.

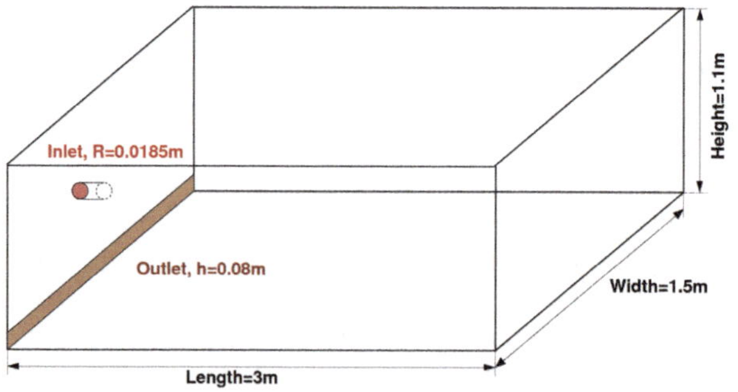

Fig. 2 Description of experimental setup of the Hurnik test-case

5 Results

5.1 Test Case of Nielsen – 2D Wall Jet in Enclosure

As the distribution of eddy viscosity has direct influence on the flow field in the room, the distribution of RANS- and LES-content will be evaluated. Fig. 3 shows the distribution of eddy viscosity for the SAS and SSTF simulation.

In Fig. 3 the different regions of influence of the SAS and SSTF are compared to the base URANS-SST simulation. The SAS model reacts to the flow unsteadyness by enlarging the dissipation via the SAS production term P_{SAS}, which gets active in regions of the shear layer of the 2D wall jet and a bit in the main vortex in the right end of the room. The SAS model has a minor effect on the eddy viscosity compared to the SST-URANS simulation; though the model was not intended for this kind of low-Re flow. The overall influence on the eddy viscosity is much larger for the SSTF model, which is active on the whole flow domain in this case. This is more suitable for thermal comfort predictions, where one needs to take into account the secondary flow region as well. The reduction of eddy viscosity is directly transferred to the flow development as observed by the faster transition to turbulent behavior with the SSTF model in Fig. 4. The probes show the behavior of the flow in the middle of the room at x-position $x = 6\,\text{m}$ at the height of 0.6 m. The SAS simulation stays almost constant for the simulation time of around 200 sec. In comparison, the yStar model as a zonal approach inhibits a different behavior. As seen in Fig. 5, there is a large LES region in the flow domain with just a small band of RANS behavior close to the wall, which makes the model attractive for thermal comfort predictions. The black contour lines indicate the actual position of the switch to complete LES behavior. The model captures the complete comfort zone in LES mode by defining the transition region at $y* = 100$. This value can be further adjusted to move the interface

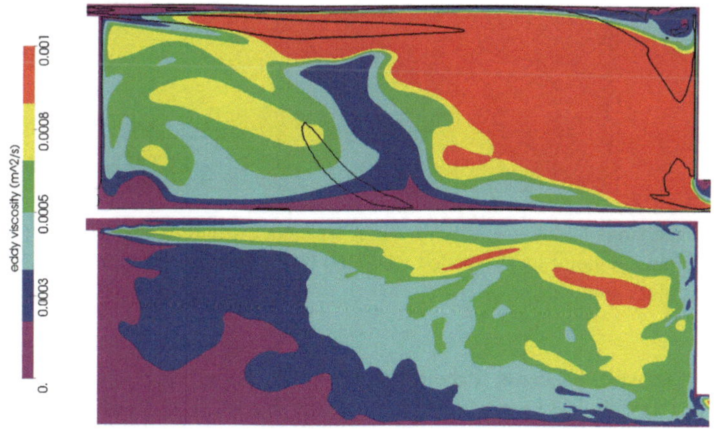

Fig. 3 Comparison of eddy viscosity (top: SAS with black contour lines indicating large production of specific turbulent dissipation due to L_{vK}; bottom: SSTF)

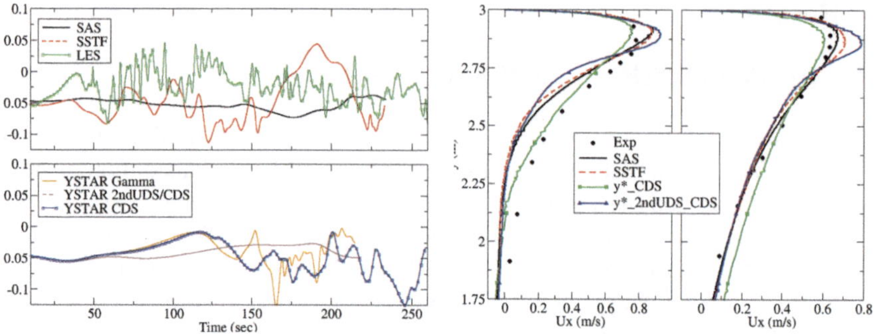

Fig. 4 Left: flow development (x-velocity); right: mean x-velocity at x-position H, $2H$ as described in Fig. 2

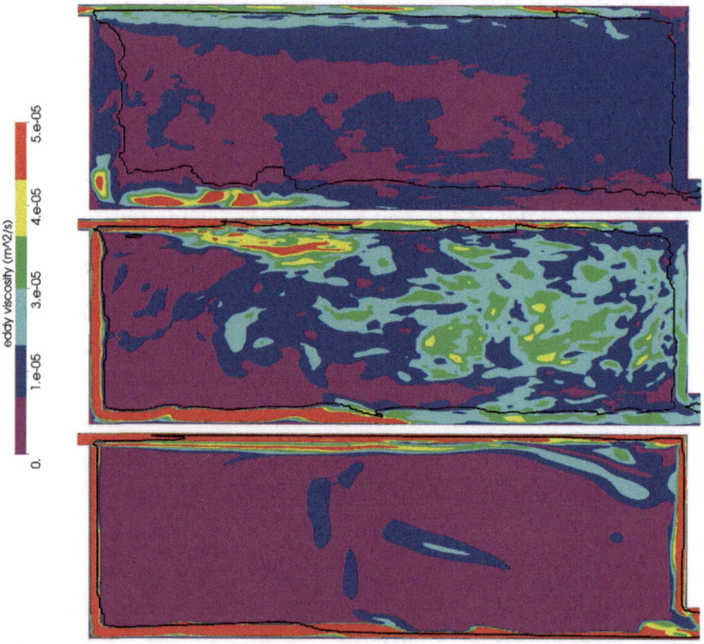

Fig. 5 Comparison of eddy viscosity for the yStar model with different schemes (top: CDS, middle: Gamma, bottom: local blended CDS/2nd UDS; interface position of RANS-LES is included as black contour line

closer to the wall. It seems to be sensitive to the underlying scheme. In general, the use of CDS should be preferred for regions in LES mode. For this case the influence of the numerical scheme is highlighted in Fig. 5. The comparison is done for CDS, Gamma and a local blended scheme using second order upwind in RANS mode and CDS in LES mode. The switch of the schemes for each mode corresponds to the local value of $y*$, which declares the RANS and LES zones. Comparing the turbulent viscosity levels of the yStar calculations with the other SAS and SSTF simulations, it can be observed that the level of subgrid viscosity is around two orders of magnitude lower for the yStar model. The scales in the figures are adjusted for better comparison of the different yStar simulations. For the yStar simulation the Gamma scheme introduces more numerical diffusion due to the upwind content. A complete different distribution of eddy viscosity is given by the simulation with the locally blended model. There is a strong jump in eddy viscosity close to the wall, which corresponds closely to the RANS-LES interface for this simulation. This high eddy viscosity seems to stabilize the flow field, which lowers the subgrid viscosity in the LES region tremendously. The additional diffusion of the Gamma scheme is reflected on the flow development as seen in Fig. 4. Similar to the local blended scheme this results in a slower development of turbulent structures. As we can see in the same figure, the calculated mean x-velocity is in good agreement to the experimental results for all models. At position H, the yStar model in combination with a central differencing scheme gives the best result, whereas the SAS model fits slightly better to the experiment at position $2H$.

5.2 Test Case of Hurnik – Ventilated Enclosures with Round Jet as Inlet

The investigation of Hurnik's test case aims at the comfort regions of the flow, i.e. the secondary flow regions outside the jet. For the hybrid calculations, no mesh independence study with a refinement of $\sqrt{2}$ was done. Rather the mesh sensitivity was studied with three different structured hexahedral grids in the range from 1.6 Mio , 2.7 Mio and 3.4 Mio cells. The finest mesh with 3.4 Mio cells is based on the coarse mesh with an additional refinement region around the jet region.

In Fig. 6 the effect of the SAS model for the coarse and fine mesh on the turbulent viscosity is compared to the base SST-URANS (1.6 Mio cells) simulation. The SAS model provides additional turbulent dissipation mainly in the jet region and reduces the eddy viscosity in this region. It reacts to the finer mesh resolution with a further reduction of turbulent viscosity, though the mesh influence is minor. A view on the interface region and the turbulent viscosity for the yStar simulations is outlined in Fig. 7. The model runs inside the black contour lines in complete LES mode. As expected and as in the first test case, the yStar model defines a large LES region, which corresponds well to the comfort zone, with a thin layer of RANS near the wall for each mesh. For this simulation the interface position was set to a value of $y* = 60$. The mesh dependence is shown via the filter size of the underlying Smagorinsky subgrid-model. Especially the additional mesh refinement for the fine mesh results in less turbulent viscosity and a non-smooth behavior at the mesh refinement interface. In this sense, the mesh dependence of the SAS model is small. Though the calculated eddy viscosity for the yStar model is again two order of magnitudes longer. As seen in Fig. 8 the overall simulation time is neither enough for a well resolved flow field nor for sufficient averaging. Nevertheless, the diagram gives an insight into the development and illustrates the steady behavior of the SAS model in the first 40 sec for the finest mesh with 3.4 Mio cells. The probe is located outside

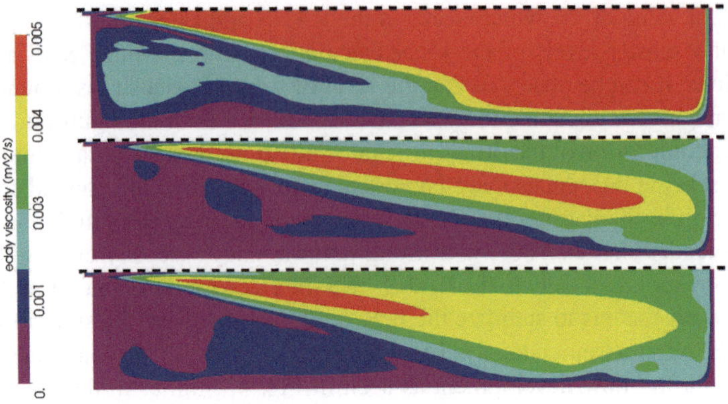

Fig. 6 Comparison of eddy viscosity at a half-centerline-plane normal to the width of the room; top: SST (1.6 Mio), middle: SAS (1.6 Mio); bottom: SAS (3.4 Mio)

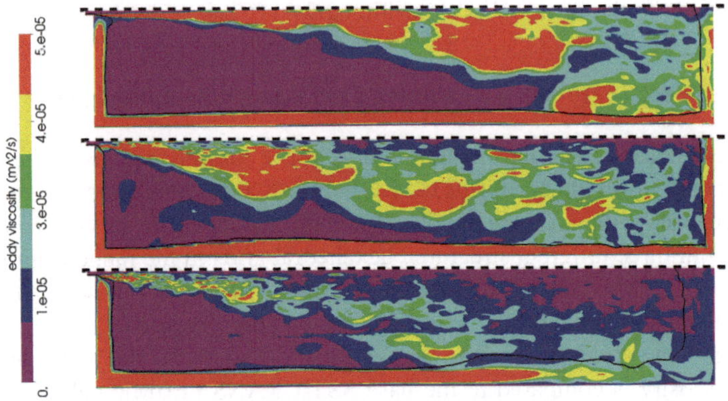

Fig. 7 Comparison of eddy viscosity and RANS-LES interface (black contour line); top: yStar (1.6 Mio), middle: yStar (1.6 Mio); bottom: ystar (3.4 Mio)

of the jet in x-direction at $x = 1.5\,\mathrm{m}$ and with a distance of 0.2 m to the side walls at a height of 0.55 m. Compared to this, the yStar simulation needed around 6 sec for the fine mesh and around 15 sec for the coarser meshes to develop transient flow behavior. This behavior goes along with the stronger reduction of turbulent viscosity in the room. Even though the SAS model reacts to transient flow structures in the jet region, it is not able to predict the transient secondary flow regions.

The SAS simulations with the coarser 1.6 Mio meshes just ran around 10 sec physical time each. No fluctuations were registered in this time period. With respect to the short averaging time, a comparison of the mean velocity with experimental data is not suitable. Base simulations with a URANS-SST model did not predict any fluctuations for a physical simulation time of around 200 sec and is in this aspect not suitable for thermal comfort predictions.

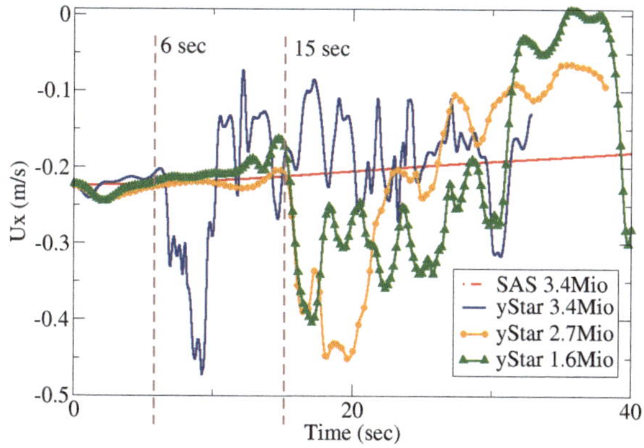

Fig. 8 Flow development based on initial SST-URANS results of the yStar model (mesh 2.7 Mio and 3.4Mio) and the SAS model with 3.4 cells

6 Conclusion – Future Work

The use of three different hybrid RANS-LES turbulence models was evaluated for two different generic ventilated enclosures with respect to thermal comfort needs. For both test cases the prediction of mean velocities at defined positions can be handled with RANS models. Though transient methods are needed for large scale unsteadyness inhibited in ventilation airflow to predict local velocity, turbulent intensities and local flow structure. This evaluation focused on the distribution of RANS-LES regions for the SAS, SSTF and yStar models. Further the influences of the convection scheme and underlying mesh were compared with respect to flow development and eddy viscosity.

For both test cases, it is observed, that the use of y^* as the interface condition for RANS-LES transition reduces the modelled viscosity in a large portion of the room by an order of magnitude compared to the initial URANS-SST simulation. The use of SSTF or SAS turbulence model results only in a small reduction of eddy viscosity. The convection scheme has a direct influence on the eddy viscosity and it is known, that a central differencing scheme is needed for LES regions. The use of the global blended scheme Gamma raises the eddy viscosity level, whereas a locally blended scheme with second order upwind for the RANS region and central differencing scheme for the LES part stabilizes the flow with a reduced eddy viscosity. The secondary flow outside the jet region does not predict any unsteadyness in the simulation time. Overall these simulations did not enhance the prediction of mean velocity, but rather showed some discrepancies, which could be partly blamed on flows, that were not developed completely, and some meshing issues for the second test case.

Hybrid methods seem to be a very interesting concept in the context of thermal comfort predictions and can be advantageous due to less restrictions for mesh expansion and resolution to pure LES for flows in ventilated compartments like defrosting and wall jets with high velocities close to the wall. The use of a model with the interface position based on y^* seems to be a good base for thermal comfort predictions. It provides a large LES region for the domain and allows the development of unsteadyness in the secondary flow region in this study.

References

Breuer, M., Jaffrezic, B., Arora, K.: Hybrid LES-RANS technique based on a one-equation near-wall model. Theor. Comput. Fluid Dyn. 22, 157–187 (2008)

Davidson, L., Nielsen, P.: Large eddy simulations of the flow in a three-dimensional ventilated room. In: Proceedings of Roomvent, Stockholm (1998)

Fröhlich, J., von Terzi, D.: Hybrid LES/RANS Methods. Progress of Aerospace Sciences 44, 349–377 (2008)

Gyllenram, W.: Analytical and Numerical Studies of Internal Swirling Flows. PhD thesis, Chalmers University of Technology, Gothenburg (2008)

Gyllenram, W., Nilsson, H.: Design and validation of a scale-adaptive filtering technique for LRN turbulence modeling of unsteady flow. Journal of Engineering Mathematics 130 (2007)

Hurnik, M., Blaszczok, M.: Comparison of CFD predictions and LDA measurements of airflow in a ventilated room. In: Proceedings of Roomvent, Copenhagen, Denmark (2002)

Knieser, B.: Ein hybrides LES/RANS-Verfahren für konjugierte Impuls-, Wärme- und Stoffübertragung mit Relevanz zu Brennkammerkonfigurationen. PhD thesis, Technische Universität Darmstadt (2008)

Menter, F.R., Kuntz, M., Bender, R.: A Scale-Adaptive Simulation Model for Turbulent Flow Predictions. AIAA Paper 2003–0767 (2003)

Willems, W.: Numerische Simulation turbulenter Scherströmungen mit einem Zwei-Skalen Turbulenzmodell. PhD thesis, RWTH Aachen (1996)

Zhang, G., Morsing, S., Bjerg, B., Svidt, K., Strøm, J.S.: Test room for validation of airflow patterns estimated by computational fluid dynamics. Journal of Agric. Engineering Res. 76, 141–148 (2000)

Experimental and Numerical Investigation of Transonic Bump Flow

J.-C. Huang, G. Barakos, E. Benard, G. Barbut, Y. Hoarau, M. Braza, and P. Derian

Abstract. A combined experimental and numerical investigation of transonic shock wave/boundary layer interaction over a wall-mounted bump has been undertaken looking at the details of the flow and the capability of CFD in predicting such complex, yet fundamental, aerodynamic interactions. The transonic wind tunnel of the Queens University of Belfast (QUB) has been employed for experiments that were combined with CFD investigations carried out by the University of Liverpool (ULIV) and the Institut de Méchanique des Fluides de Toulouse (IMFT). Zonal LES methods based on wall modelling, and computations using the Organised Eddy Simulation method were used for the numerical solutions. The shock unsteadiness was not resolved by any of the employed statistical turbulence closures. On the contrary, attempts to use turbulence simulation resulted in unsteady flow solutions. Based on the obtained results, a flow mechanism has been suggested which was found to be in agreement with previous experimental observations. This work was carried out as part of the UFAST F6 EC research project in unsteady shock/boundary-layer interaction.

1 Introduction

The initial objective of the experiment was to simulate low-frequency shock oscillations resulting from unsteady shock/boundary layer interaction (Levy, 1982) and control these using synthetic jets. The target configuration was a circular arc aerofoil due to its documented shock unsteadiness at transonic flow

J.-C. Huang
School of Aeronautical Engineering, Queen's University of Belfast, Belfast BT9 5AH, United Kingdom
ghuang@engr.uky.edu

G. Barakos
Department of Engineering, University of Liverpool, Liverpool, L69 3GH, United Kingdom

E. Benard
Department of Aerospace Engineering, University of Glasgow, Glasgow, G20 6QD, United Kingdom

G. Barbut · Y. Hoarau · M. Braza · P. Derian
Institut de Méchanique des Fluides de Toulouse, Allée du Prof. Camille Soula, 31400 Toulouse, France

S.-H. Peng et al. (Eds.): Progress in Hybrid RANS-LES Modelling, NNFM 111, pp. 315–328.
springerlink.com

conditions. However, after initial considerations, the size of the wind tunnel section of QUB was found to be small for a full aerofoil experiment and a compromise had to me made. It was therefore decided to produce a half-model of the aerofoil and mount it on the lower tunnel wall like a bump. Experience with such flows was also available since wall-mounted bumps were already used for experimentation in shock/boundary layer interaction (Delery et al, 1986; Johnson et al, 1982; Delery, 1985). In their work, Delery and Marvin (1986) note that Shock/Boundary-Layer Interaction (SBLI) can considerably increase the size of the dissipative flow region which would affect the whole flow field and generally lead to losses is terms of lift, drag increase, and buffet. All these phenomena are of profound relevance to the aerospace industry.

Several works have so far provided good insight into steady-state SBLI and resulted in established databases for validation of CFD methods and development of better turbulence models for complex transonic flows. Amongst others, the works of Bachalo and Johnson (1982) and Delery (1985, 1986) are the most popular examples of shock/boundary layer interaction. One of the strengths of these works is that they provided detailed flow measurements near the interaction region including surface pressure and boundary layer data. These data have been used as a benchmark against which many CFD codes and turbulence models are assessed (Barakos et al, 1998). Less information is, however, published for unsteady SBLI and there is little evidence of unsteadiness for transonic flows over wall-mounted bumps.

2 Summary of the QUB Experiments

The QUB tunnel is shown in Figures 1 (a,b) with the wall-mounted bump. Two initial configurations were proposed with solid and perforated upper walls. The perforated wall was made out of two slots covered with porous material. The cross section of the test area was 101.6 mm by 101.6 mm at the inlet, and had upper and lower plenum chambers each 101.6 mm wide and 24.7 mm depth (approximately). The test section had a length of 979 mm and linked the tunnel nozzle with the diffuser. Figure 1 shows that the test section consists of two parallel sidewalls (solid with two port-holes for the schlieren/PIV system), a top wall which could be contoured or replaced by a perforated roof, and the bottom wall with a bump. The top and the bottom walls of the test section were not parallel in order to compensate for the blockage effect due to boundary layer growth.

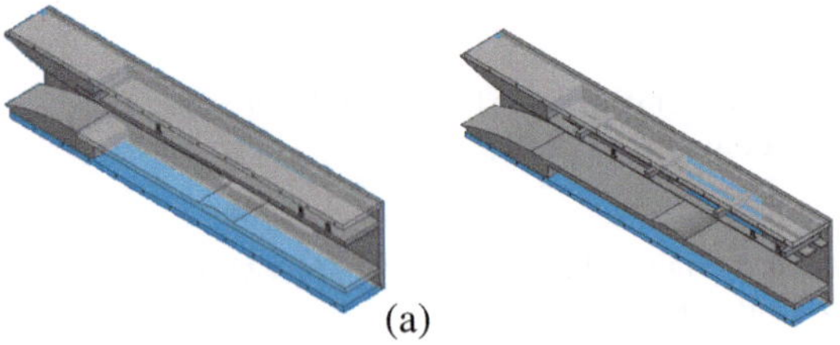

(a) (b)

Fig. 1 (a) solid and (b) porous upper walls for the QUB wind tunnel. (c, d) Bump model with dimensions in mm

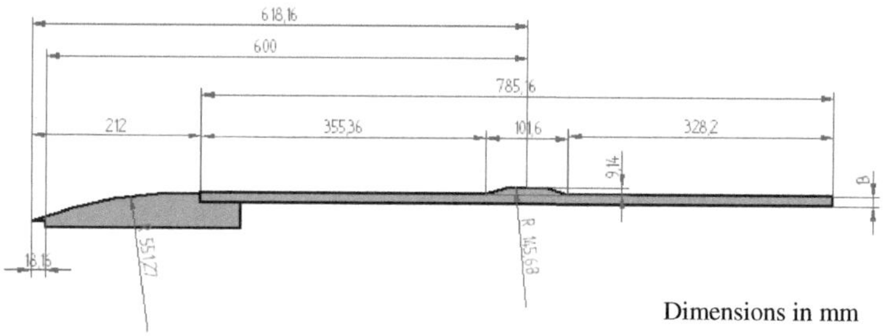

(c)

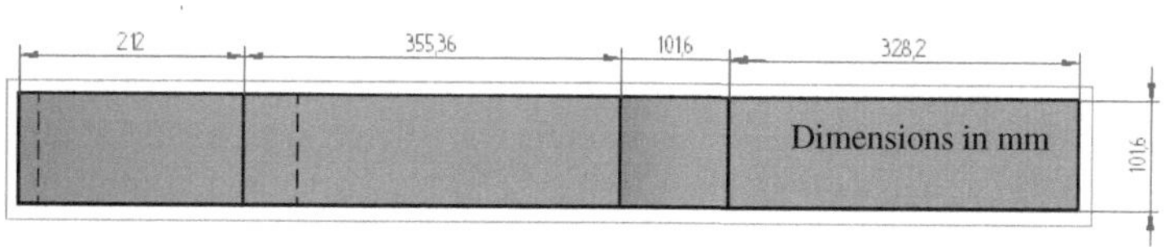

(d)

Fig. 1 (*continued*)

The lower wall of the tunnel was occupied by the bump which is shown in Figures 1 (c,d). The model was a bi-convex semi profile of 9% thickness, 101.6 mm of length, and its maximum thickness was 9.144 mm. The model was spanning the test section. The radius of the bump was 145.68 mm and the centre of the bump was located at 600 mm in front the entrance of the test section. Therefore, the leading edge of the bump was located at X=549.2 mm. The stagnation conditions for the experiment were near ambient pressure and temperature: $p_o \approx 0.99 \pm 0.04 \times 10^5\,Pa$ and $T_o \approx 290 \pm 5\,K$. The relative humidity was kept below 18% and the free stream turbulence conditions for the tunnel can be considered as isotropic, with a level of 0.35% velocity fluctuations. The active part of a test run was in the range of 8 to 10 seconds. In the case of the porous wall configuration, the shock location on the model was measured to be at approximately 65% chord length, corresponding to a free-stream Mach number M_∞= 0.78 (at station X=382 mm) and peak Mach number M_{peak}=1.31 for 30% of humidity. For the contoured wall case, the shock location on the model was measured to be at 65% of the chord length, peak Mach number M_{peak}=1.37 and free-stream Mach number M_∞= 0.785 (18% of humidity). The reattachment point was located between X=670 and 690 mm (or 120 to 140% of the chord). Preliminary results gave an upstream momentum based Reynolds' number (calculated at X=382 mm) approximately equal to $R_\theta \approx 11000$.

Due to problems with tunnel humidity, the first sets of data were obtained with the porous tunnel walls and comparisons against CFD computations using free-stream conditions were not encouraging. This prompted a detailed study by QUB of the effect of humidity on the obtained results. This study is reported in reference by Amecke (1986). A level of humidity of less than 18% was required for direct comparisons against CFD obtained using dry air conditions.

3 Contoured Tunnel Wall

Preliminary results obtained using the URANS approach indicated that the flow experiments of QUB will probably result in steady-state flow configurations due to the half-aerofoil model to be employed and the difficulties to obtain high Mach number due to interactions taking place on the upper wall of the tunnel. Therefore, a requirement for high peak Mach number in the tunnel (in the range of 1.2 to 1.40 was put forward. This Mach number would allow for flow separation and could lead to shock oscillation, unsteadiness and buffet. To compromise between the requirement for a transonic interaction with a well-identified supersonic pocket, the need to reach high peak Mach number, and the need for solid walls that could be easily modelled within the URANS framework, a decision was made to contour the upper wall of the tunnel. For the solid wall configuration ULIV computed a contoured wall shape which could minimise the influence of the upper wall and result in a good transonic interaction. It has to be clarified, however, that the issues of mean-flow unsteadiness for this test case still remained. The one-step method of Amecke (1986) has been used to this end along with 2D CFD results. The outcome of this effort can be seen in Figure 2. Several possible curves have been considered and a compromise had to be made between the obtained peak Mach number and the required wall deflection.

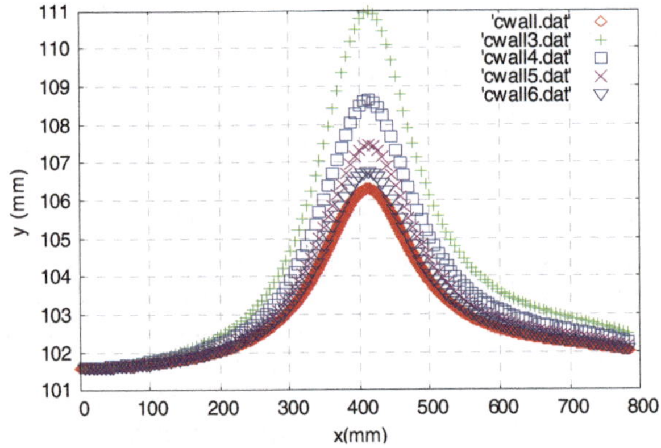

Fig. 2 Contoured wall results for the QUB test case, curve 6 has been used for computations

The obtained experimental results for Curve 6 of Figure 2 are shown in Figure 3. The results suggest that a well-defined supersonic flow region is established and a decision was therefore made for Curve 6 to be used.

Once a first round of test experiments was complete, a comparison between the CFD predictions and the data was made for the peak Mach number. This set of results is presented in Table 1 and as can be seen the CFD predictions were accurate resulting in a peak Mach number of 1.365. The shock was located at about 77.5% of the bump and the Mach number near the top wall remained at about 0.88. This way, substantial flow separation has been obtained and at the same time, the interaction remained transonic. The final configuration ascertained the existence of a subsonic flow around the shock and the provision of a clean path for subsonic flow to move around the supersonic pocked without the formation of a normal shock between the upper and lower walls of the tunnel. In Figure 3, the experiments for the contoured and porous upper walls are compared (Huang et al, 2007). The isentropic Mach number along the bump is very similar for the two cases and the Schlieren images suggest that the contoured wall is indeed effective for the selected set of flow conditions.

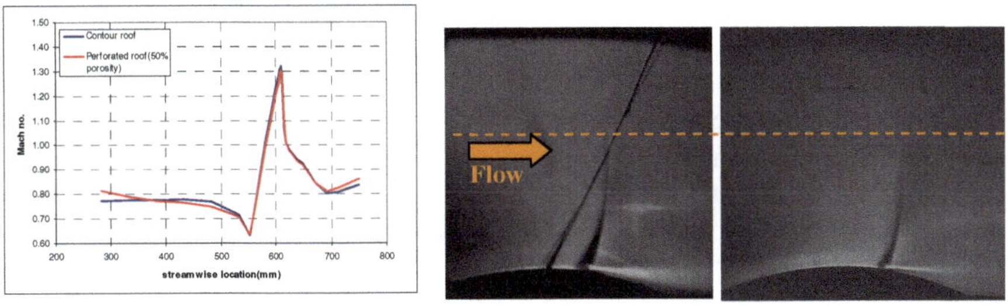

30% Relative Humidity Perforated roof Contoured roof

Fig. 3 Comparison between isentropic Mach number on the bump for the perforated and contoured upper wall cases (Huang et al, 2007)

Table 1 Summary of obtained results for the 2D contoured wall cases. Curve 6 was finally used

	M_∞	M_{max}	%Shock Location	M_{max} at top wall	Height of supersonic region (mm).
Curve 5	0.77	1.32	73.5	0.83	43.6
Curve 5	0.8	1.43	82.0	0.93	80.0
Curve 6	0.77	1.365	77.4	0.877	53.7
Curve 6	0.8	1.475	86.0	1.1	Supersonic region reached the top wall
Experiment	0.785	1.37	64	-	-

4 URANS Results

IMFT and ULIV employed similar tools for the CFD analysis of this test case. In particular, IMFT used the NSMB flow solver (Martinat et al, 2008) which allows for multi-block structured grids to be used. The solver uses MPI for parallel execution. Time marching was done via the dual-time stepping method and a third order Runge-Kutta scheme. The spatial discretisation employed an implicit Roe scheme with the Van Leer limiter. Central differences were used for the discretisation of the viscous fluxes. NSMB has a variety of turbulence models including the option of performing Organised Eddy Simulations (OES).

ULIV used the Parallel Multi-Block (PMB) solver which also follows the multi-block structured approach. The solver uses an implicit time marching method, and the resulting linear system of equations is solved using a pre-conditioned Generalised Conjugate Gradient (GCG) method. For unsteady simulations, implicit dual-time stepping was used based on Jameson's (1993) pseudo-time integration approach. From the beginning, the solver was designed with parallel execution in mind and for this reason a divide-and-conquer approach was used to allow for multi-block grids to be computed on distributed-memory machines and especially low-cost Beowulf clusters of personal computers.

A first set of meshes was prepared by ULIV. All meshes included side walls and provided local refinement near the shock/boundary layer interaction region. The first cell above the lower wall was at y^+ of about 0.9 and the stream wise and span wise spacing in the near-bump region were of the order of Δx^+ and Δy^+ of 10 for the finest of the meshes. A simple multi-block topology was used since the geometry was smooth and the number of the employed blocks was increased to allow for parallel execution on about 64 processors with load differences of less than 5% between the CPUs. The mesh size was approaching 3.5 million cells.

For the obtained URANS results a variety of turbulence models was employed. Although Liverpool used mainly the standard k-ω model (Wilcox, 2006), results were also obtained using cubic Reynolds stress expansion (Wallin et al, 2000) as well as the new k-ω formulation of Wilcox (2006). Results are shown in Figure 4 where 3D comparison of the isentropic Mach number along the bump is made for the new k-ω formulation of Wilcox. The same figure shows the recirculation flow region and its size. Overall, the agreement for the shock position and the downstream pressure is good and all models managed to predict the mean separated flow with fair accuracy. Table 2 presents a comparison for the separation and re-attachment locations.

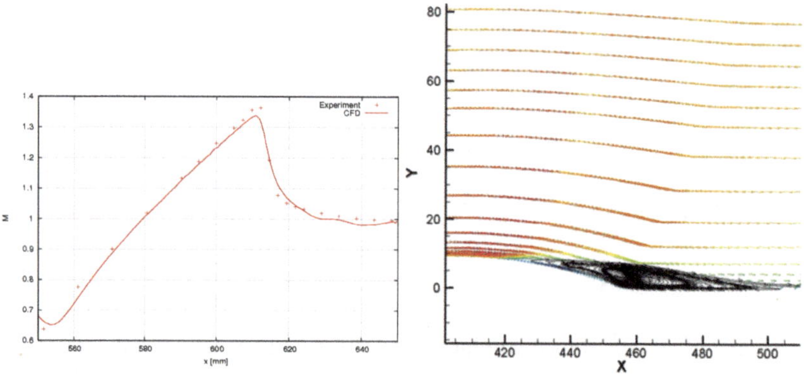

Fig. 4 Left: Isentropic Mach number along the mid-plane of the lower wall. Results are shown for the new k-ω model of Wilcox (2006). Right: Substantial flow re-circulation can be seen near the trailing edge of the bump

On the other hand, none of the URANS predictions provided any shock unsteadiness and regardless of the employed time step and grid size, the models failed to show the shock excursions reported by QUB. This was a disappointing result since the overall flow configuration was well-predicted. Figure 5 shows experimental results documenting the shock unsteadiness and identifying a shock motion frequency around 350Hz. The unsteadiness was evident both in the pressure transducer signals and the obtained Schlieren images. Apart from this result, the work with the URANS models provided good insight in the flow configuration near the vertical tunnel walls and highlighted the need for three-dimensional computations and adequate resolution of the corner flow formed before the shock and on each side of the interaction region. URANS computations (Spalart, 1994; Abe et al, 2003) were also used by IMFT, and the separated flow region, as well, as the sonic surfaces are shown in Figure 6. It appears that the models were at least able to predict the development of the separation between two corner flow regions adjacent to the side walls of the tunnel. As was the case for results obtained by Liverpool, a tight sonic region is present above the bump which extends up to the side walls of the tunnel and its shape is influenced by the corner vortices.

Table 2 Comparison between experiments and CFD for the separation and re-attachment location along the mind-plane of the bump (a distance of 1 corresponds to the trailing edge of the bump, all distances measured from the bump leading edge)

	Separation point	**Re-attachment point**
Experiment Method 1	0.64	1.40
Experiment Method 2	0.63	1.32
URANS (k-ω)	0.65	1.48
Wall-function LES	0.62	1.49 (1.42 min 1.51 max)

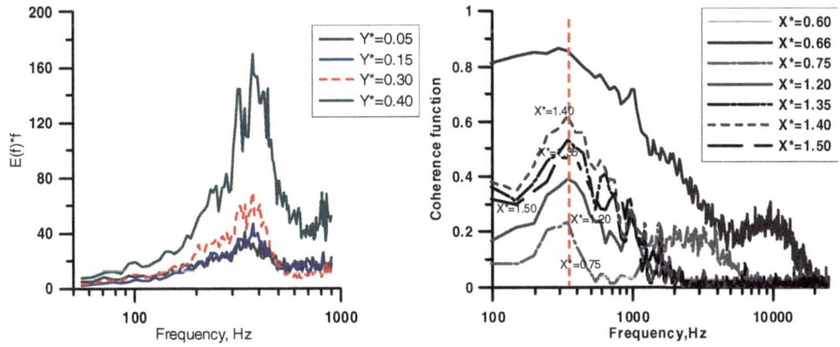

Fig. 5 Left: Shock wave motion spectra extracted from Schlieren images. The dominant frequencies are in the range of 325 to 400 Hz. Right: Coherence function between shock wave and re-circulation region. The peak value is at 350Hz

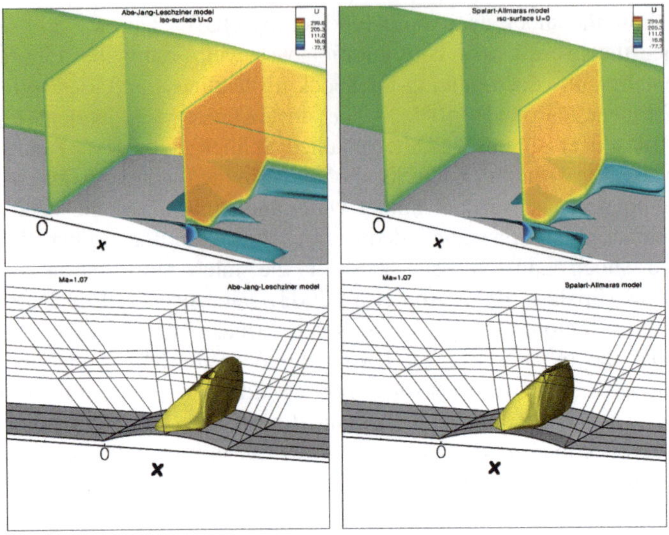

Fig. 6 Separated flow region and sonic surfaces predicted with the Abe-Jang-Leschziner [14] model (left) and the Spalart-Allmaras [13] model (right)

The span-wise development of the separated flow is clearly visible in Figure 6 and results from the Abe et al. model [14] are similar to the predictions of the popular Spalart-Allmaras (1994) one-equation model. As was the case for the ULIV results, a central separated flow region is present and large corner vortices appear along the test section. These tend to dominate the flow as the bump is approached and near their peak they occupy more space than the central reversed flow region.

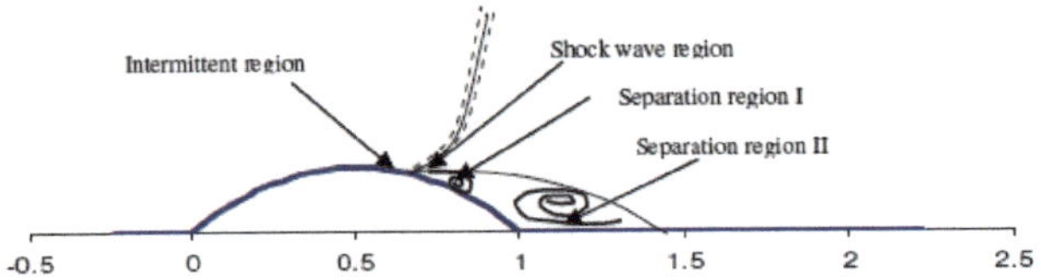

Fig. 7 Schematic of the interaction between shock and separated flow regions suggested by Pearcey et al. (1959)

5 Turbulence Simulation Efforts

The results obtained using the URANS models presented in the previous section were encouraging and show that most of the flow features were well-resolved. This set of features lacked, however, unsteadiness and for this reason, turbulence simulation was considered as a way to better study this flow. The information available from the experiments of QUB suggested a relatively high frequency of shock oscillation at between 325 and 400 Hz. This created a problem for URANS and LES simulations since not only fine meshes were required to resolve the walls of the tunnel but also very fine time steps to allow for the resolution of this frequency along with long computations so that a substantial length of the time signal can be collected.

A literature survey (Pearcey, 1959; Liu, 1987; Bron, 2003) for unsteady shock/boundary layer interaction, revealed a that the type of flow configuration obtained in the wind tunnel of QUB was studied in the past by Pearcey (1959) and his co-workers. The flow unsteadiness was driven by an interaction between two separated flow regions behind the shock as the schematic of Figure 7 shows. To reveal the details of the flow near the interaction region, a more detailed study was required. Since little has been published about LES for transonic flow, the computations for this case were based on works from the literature where steady shock-boundary layer interactions were simulated using LES (Wollblad, et al, 2006).

The first decision to be made was to isolate the interaction region near the lower wall, using a zonal approach. A mesh of 432x135x120 cells in the steam-wise, normal and span-wise directions was used which was load-balanced on 96 CPUs. For this work, the compressible version of the Smagorinsky model was used as formulated by Erlebacher et al. (1992) with a modified eddy viscosity as suggested by Nicoud and Ducros (1999). An upstream boundary was imposed and white noise at the level of 0.2% of the kinetic energy was added. This was just enough to generate some forcing in the boundary layer before the main interaction between the boundary layer and the shock. The zone boundary followed the approach described by Temmerman et al (2000). An attempt was also made to re-scale DNS data from the ERCOFTAC database, but this was abandoned due to the need to scale the DNS for time and space. The downstream condition followed the approach described by Wollblad et al (2006). From the LES results, the time histories of certain probes were kept. This was not possible for the full domain, due to disk space limitations, and for this reason, 100 POD modes were generated and used for the various flow re-constructions. A time step of $\Delta t=10^{-5}$ seconds was used and about 0.3 seconds of the flow were computed. The last part of this signal (about 0.2 seconds) was kept for comparisons. A fourth order central scheme was used for the spatial discretization.

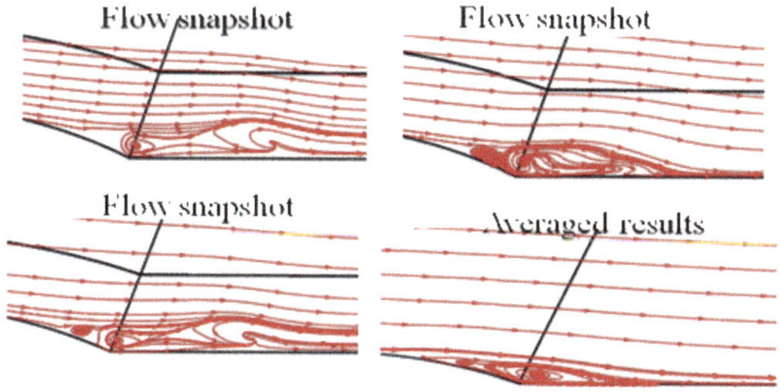

Fig. 8 Flow snap-shots and averaged results for the region around the trailing edge of the bump near the mid-plane of the tunnel

Figure 8 presents a set of flow snapshots visualised using streamlines. As can be seen, the instantaneous results reveal a complex flow with several vortices present. On the average though the flow appears to be organised in a well-identified recirculation area downstream the shock. The presence of separation near the foot of the shock is in qualitative agreement with the observations of Pearsey (1959). Figure 9 presents the predicted isentropic Mach number and the results are in good agreement with the experiments. Table 2 shows the predicted separation and re-attachment points.

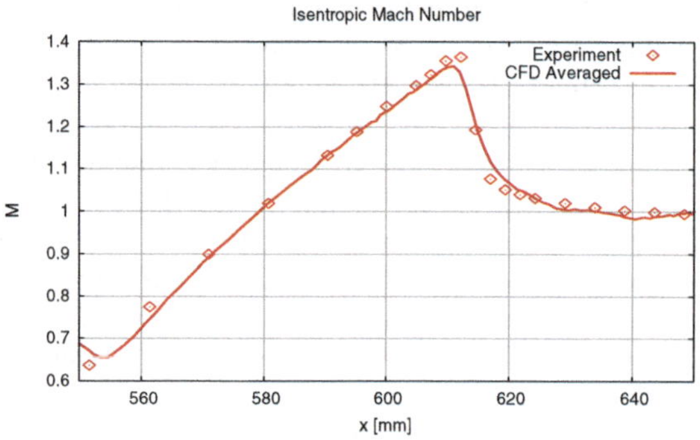

Fig. 9 Comparison between CFD and experiment for the averaged isentropic Mach number in the centre of the bump

Using the unsteady flow snapshots re-constructed from POD Figure 10 presents iso-surfaces of negative stream-wise velocity. The size of the central separated flow region changes with time and the corner flow appears to be somehow less extended in comparison to the URANS results. It is not, however, clear from the figure whether two or more separated flow regions exist downstream the shock. Figure 11 presents comparisons between the experiments and CFD for the pressure spectra at three locations along the mid-plane of the tunnel. There is clearly some similarity in the results especially near the 350 Hz region.

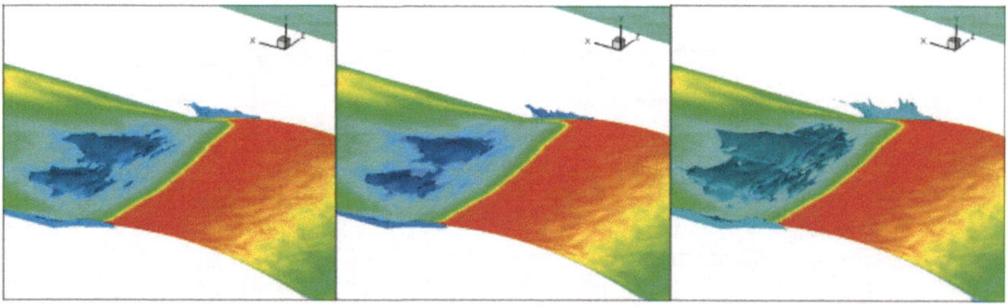

Fig. 10 Instantaneous flow visualisation for the QUB best case: The colour on the surface represents pressure and the iso-surfaces negative stream-wise velocity at the level of 0.01 of the free-stream

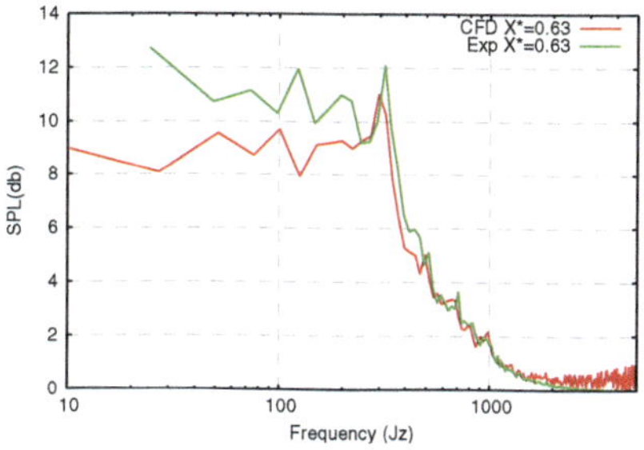

(a) Upstream

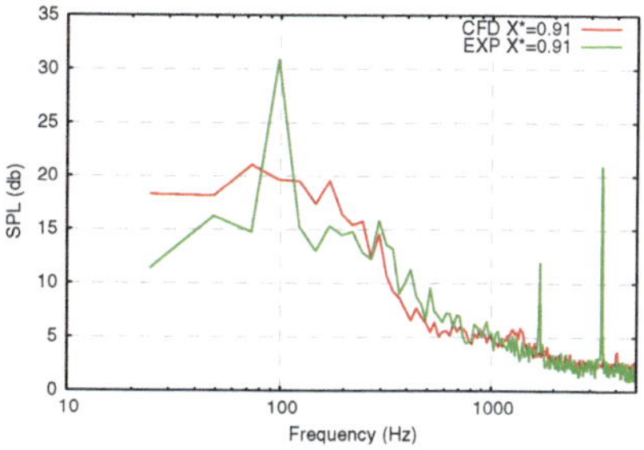

(b) Near the trailing edge

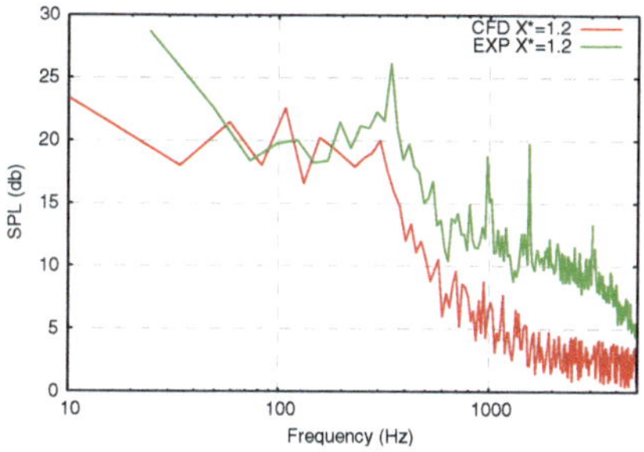

(c) Near the re-attachment region

Fig. 11 Comparisons between CFD and experiments for the unsteady pressure load at three stations along the bump

The OES technique (Braza et al, 2006; Bourguet et al, 2008) was pioneered by IMFT and is a good alternative to URANS for this test case. Figure 12 presents flow visualization obtained using the Q criterion for the mid-plane of the tunnel. Shedding of vortices behind the shock can be identified. On the same figure the wall-function LES results of ULIV were re-constructed after filtering with a band-pass filter between 200 and 500 Hz. One can see that flow structures near the trailing edge of the bump as well as further upstream near the foot of the shock, survived the filtering. As shown, a small separated flow region appears just at the foot of the shock. This structure appears to travel downstream towards the trailing edge and eventually merges with the larger trailing edge separation. From these comparisons, it is evident that the employed wall-function-based LES and the OES methods gave results with similar physics and perhaps captured some of the flow mechanism for this unsteady SWBLI.

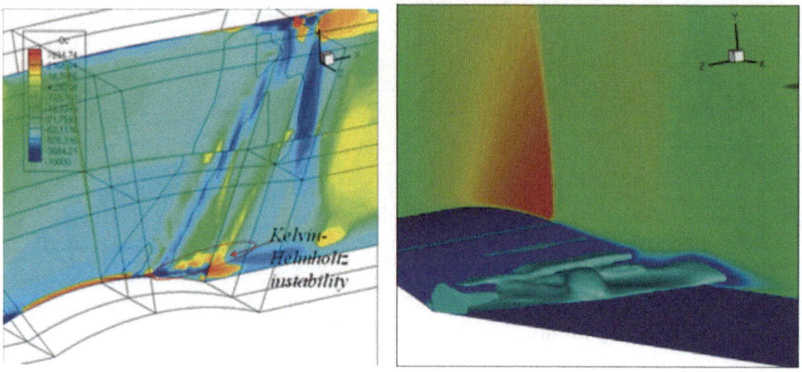

Fig. 12 Filtered flow solution near the interaction region suggesting the presence of a complex flow region similar to Pearcey's observations (1959). Right flow recirculation regions filtered between 300 and 400 Hz (zonal LES). Left: OES results showing Q criterion and the presence of shedding from the shock region

6 Conclusions and Future Steps

A fundamental SWBLI flow has been investigated and a substantial amount of results are available in the database of UFAST. One of the key issues in this investigation is that the low-speed shock oscillation was not achieved and for this reason, the CFD task to simulate the detailed shock motion at high frequencies was challenging. URANS provided adequate results for most cases, when the separation length, shock strength and surface pressure were compared with the experiments. It was not possible, at least with the available URANS models to resolve the high-frequency shock oscillations. The presence of the corner flow, and the difficulty of URANS in dealing with it, was also highlighted. On the other hand, URANS proved to be a good tool for computing the shape of the upper wall of the tunnel, and this is a good example of how CFD and tunnel experimentation can be used in a synergetic way.

LES for this test case was also challenging since transonic LES is currently in its infancy with few publications available in the literature. The configuration also included side walls with corner flow regions, and required a special technique for CFD computations. For this purpose, a zonal method based on wall-modelled LES was adopted and the obtained results shed some light in the fundamentals of the interaction. In contrast to previous cases studied, the interaction for the QUB test case appears to be driven by the complex unsteady flow pattern in the recirculation region behind the shock. Along the same lines, the OES method was used and again flow unsteadiness was predicted.

The work for this test case should certainly be extended both at the CFD and experimental fronts. By reducing the Reynolds number one could perhaps obtain similar physics and make a full LES possible. This would certainly be at the centre of any future investigations. The expansion of the comparisons against experiments should also be pursued. At the moment, only surface pressure and a few time series of pressure transducers are available. Pressure surveys normal to the wall along with entry and exit conditions could be useful for CFD purposes. In addition, some better form of measuring the flow unsteadiness could provide valuable data for CFD comparisons. This could be in the form of time-resolved PIV or LDA.

Acknowledgements. This work was supported by the 6th European Research Effort under the UFAST research project on unsteady shock-boundary layer interaction and its control.

References

Abe, K., Yang, J., Leschziner, M.A.: An Investigation of Wall Anisotropy Expressions and Length Scale Equations for Non-linear Eddy Viscosity Models. J. Heat, Fluid Flow 24(2), 181–198 (2003)

Amecke, J.: Direct Calculation of Wall Interferences and Wall Adaption for Two-dimensional Flow in Wind Tunnels with Closed Walls, NASA Technical Memorandum TM 88523 (1986)

Barakos, G., Drikakis, D.: Assessment of Various Low-Reynolds Number Turbulence Models in Shock-Boundary Layer Interaction. Comp. Meth. Appl. Mech. Engng. 160(1-2), 155–174 (1998)

Bourguet, R., Braza, M., Harran, G., El Akoury, R.: Anisotropic Organised Eddy Simulation for the Prediction of Non-equilibrium Turbulent Flows Around Bodies. J. Fluids and Structures 24(8), 1240–1251 (2008)

Braza, M., Perrin, R., Hoarau, Y.: Turbulence Properties in the Cylinder Wake at High Reynolds Number. J. Fluids and Structures 22(6-7), 757–771 (2006)

Bron, O.: Numerical and Experimental Study of the Shock-Boundary Layer Interaction in Transonic Unsteady Flow. Ph.D. thesis, Royal Institute of Technology, Sweden (2003)

Delery, J., Marvin, J.G.: Shock-Wave Boundary Layer Interactions, North Atlantic Treaty Organization: Advisory Group for Aerospace Research and Development, AGARDograph No. 280 (1986)

Delery, J.: Shock Wave/Turbulent Boundary Layer Interaction and its Control. Progress in Aerospace Sciences 22, 209–228 (1985)

Erlebacher, G., et al.: Toward the Large-Eddy Simulation of Compressible Turbulent Flows. J. Fluid Mechanics 238, 155–185 (1992)

Huang, J.C., Gault, R.I., Benard, E., Raghunathan, A.: Effect of Humidity on Transonic Bump Flow. Journal of Aircraft 45(6), 2092–2100 (2008)

Huang, J.C., Benard, E.: QUB Report on WP-2 Basic Experiments Data-bank Input, UFAST Deliverable D2.1.7 (2007)

Jameson, A.: Computational Algorithms for Aerodynamic Analysis and Design. Applied Numerical Mathematics 13(5), 383–422 (1993)

Johnson, D.A., Horstman, C.C., Bachalo, W.D.: Comparison between Experiment and Prediction for a Transonic Turbulent Separated Flow. AIAA J. 20(6), 737–744 (1982)

Levy, L.L.: Experimental and Computational Steady and Unsteady Transonic Flows about a Thick Airfoil. AIAA J. 16(6), 564–572 (1982)

Liu, X., Squire, L.C.: An Investigation of Shock/Boundary Layer Interactions on Curved Surfaces at Transonic Speeds. J. Fluid Mech. 187, 467–486 (1987)

Martinat, G., Braza, M., Hoarau, Y., Harran, G.: Turbulence Modelling of the Flow Past a Pitching NACA0012 Airfoil at 105 and 106 Reynolds Numbers. Journal of Fluids and Structures 24(8), 1294–1303 (2008)

Nicoud, F., Ducros, F.: Sub-grid Scale Stress Modelling Based on the Square of the Velocity Gradient Tensor. Flow Turbulence and Combustion 62(3), 183–200 (1999)

Pearcey, H.H.: Some Effects of Shock-Induced Separation of Turbulent Boundary Layers in Transonic Flow Past Aerofoils, Aeronautical Research Council Reports and Memoranda No. 3180 (1959)

Spalart, P.R., Allmaras, S.R.: A One-Equation Turbulence Model for Aerodynamic Flows. La Recherche Aerospatiale 1, 5–21 (1994)

Temmerman, L., Leschziner, M.A., Hanjalic, K.: A-priori Studies of a Near-Wall RANS Model Within a Hybrid LES/RANS Scheme. In: Rodi, W., Fueyo, N. (eds.) Engineering Turbulence Modelling and Experiment, vol. 5, pp. 317–326 (2000)

Wallin, S., Johansson, A.: An Explicit Algebraic Reynolds Stress Model for Incompressible and Compressible Flows. J. Fluid Mech. 403, 89–132 (2000)

Wlicox, D.C.: Turbulence Modelling for CFD, 3rd edn. DCW Industries, LaCanada (2006)

Wollblad, C., Davidson, L., Eriksson, L.E.: Large Eddy Simulation of Transonic Flow with Shock Wave/Turbulent Boundary Layer Interaction. AIAA J. 44(10), 2340–2353 (2006)

Comparison of Hybrid RANS-LES Calculations within the Framework of Compressible and Incompressible Unstructured Solvers

K.A.Weinman and M. Valentino

Abstract. Unstructured methods are now more common in industry due to the ease with which grids for complex geometries can be generated. Massively separated flows are of interest to industry and traditional RANS models cannot provide adequate modelling for these cases. Hybrid RANS-LES methods have been shown to provide superior performance for massively separated flows and it is of interest to compare the performance of compressible and incompressible unstructured methods within a hybrid RANS-LES framework. . A modified JST scheme is proposed for the TAU code which reduces the non-physical damping of turbulence at higher wave numbers and allows full resolution of the resolved turbulent field on a hexahedral mesh. The influence of grid anisotropy is also examined. It is seen that the local grid anisotropy influences the dissipation at high wave numbers and suggests that isotropic mesh elements should be favoured within LES regions. A comparison is then made between TAU solutions obtained using DDES and LES for a cylinder at Re=3900. Differences between DES and LES solutions obtained on a hexahedral mesh (structured) and a tetrahedral mesh (unstructured) for TAU are compared against THETA LES solutions for the unstructured mesh.

1 Introduction

Bluff body flow is characterised by complicated phenomenon including separation and shear layer breakdown. Numerical estimation of these phenomena using statistical turbulence modelling has not been successful to date. Direct Numerical Simulation (DNS) requires significant computational resources and is not affordable at engineering flow Reynolds numbers. Fully resolved LES methods are marginally cheaper than DNS. Between LES and RANS lie hybrid methods, such as DES, which offer an advantage in that required boundary resolution corresponds to RANS requirements. This results in considerable economy over fully resolved or even wall-modelled LES. This is the motivation for industrial application of hybrid methods – improved solution physical content without the high-end cost of full DNS/LES methods. The object of this paper is to examine

K.A.Weinman · M. Valentino
Center for Computational Applications in Aerospace Science and Engineering ($C^2A^2S^2E$), Göttingen, Germany
keith.weinman@dlr.de

S.-H. Peng et al. (Eds.): Progress in Hybrid RANS-LES Modelling, NNFM 111, pp. 329–338.
springerlink.com

and compare the performance of two different solvers, the DLR TAU and TETA codes, for LES simulations about a cylinder at Re= 3900, using identical meshes The performance of the Spalart-Allmaras based DES hybrid method is then compared against the LES solutions for the TAU code.

2 Computational Framework

Two flow solvers are used in this work. They are based on unstructured dual grid metric (Oliker et al., 1996). Tetrahedral, prismatic, and hexahedral mesh elements are supported for both codes. The DLR THETA code uses a matrix-free finite volume formulation (Wiley et al., 1997) for the solution of the incompressible form of the Navier-Stokes (i.e. density variations are a function of the temperature field but independent of the pressure field) and the other governing equations. A pressure correction projection method (Goda, 1979, Chorin, 1968, Timmermans et al., 1996) provides velocity-pressure coupling, and multigrid algorithms (Hachbusch, 1983, Tottenberg et al, 2001) provide convergence acceleration for the solution of the pressure equation. The DLR TAU code uses a finite volume scheme to solve the conservation form of the compressible Navier-Stokes equations with associated additional transport equations for turbulence modelling and scalar transport. Integration in time is achieved using second and higher order Runge-Kutta schemes or an implicit LU-SGS implicit scheme (Dwight, 2004). Convergence acceleration is achieved through multigrid methods with residual smoothing. Both codes are second order in space and up to third order in time. Additional information can be found in Schwamborn et al, 2006.

3 Calibration of DES and LES Models

In order to ensure that LES or hybrid model returns a reasonable representation of turbulent flow it is necessary to validate against a suitable flow.

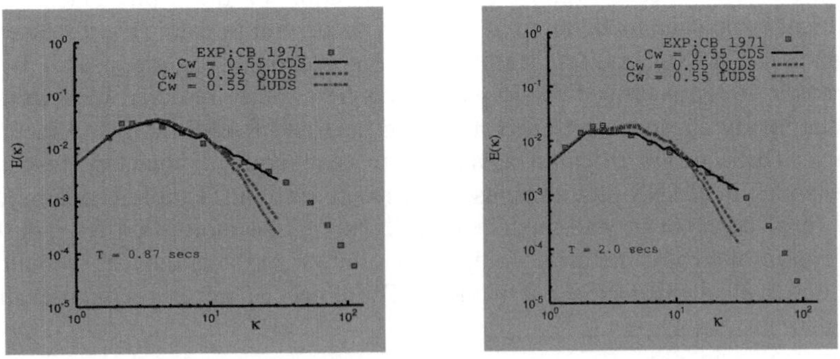

Fig. 1 Resolved energy Spectra obtained with Theta

It is conventional to calibrate sub-grid scale (SGS) model coefficients against decaying isotropic turbulence, see for example (Shur, 1999).

For the current work both codes are calibrated against the experiment of Comte-Bellot (Comte-Bellot et al., 1973). Figure 1(a) illustrates calibration results obtained for the THETA code using the WALE SGS model (Nicoud et al, 1999). Upwind schemes, LUDS and QUDS (Peric, 1985) provide excessive dissipation over the upper part of the resolvable wave number spectrum.. The central scheme returns correct levels of resolved turbulence over the entire wave number range. However, the central scheme is known to have a reduced stability radius on comparison with the upwind schemes and is prone to unphysical oscillatory behaviour (Breuss, 2005). Stability of central schemes for compressible solvers requires addition of artificial dissipation terms (Jameson at al.; 1981), referred to as the Jameson Schmidt Turkel (JST) scheme. The JST scheme was developed for Euler methods: Euler equations contain no dissipative mechanisms and artificial dissipation is required to damp solution oscillations. The JST scheme is written as an interface flux between nodes j and j+1 as

$$F_{j+1/2} = F\left(\frac{U_{j+1}+U_j}{2}\right) - d_{j+1/2}, \tag{1}$$

where the artificial dissipation term

$$d_{i+1/2} = \varepsilon^{(2)}_{j+1/2} D_2(U_{i+1/2}) - \varepsilon^{(4)}_{j+1/2} D_4\left(U_{i+1/2}\right), \tag{2}$$

provides additional dissipation. D_2 and D_4 are second and fourth difference operators respectively. The coefficients $\varepsilon^{(2)}$ and $\varepsilon^{(4)}$ are written as

$$\begin{aligned}
\varepsilon^{(2)}_{j+1/2} &= k^{(2)} R_{j+1/2} \psi_{j+1/2} \\
\varepsilon^{(4)}_{j+1/2} &= \mathrm{MAX}(0, k^{(4)} R_{j+1/2} - \varepsilon^{(2)}_{j+1/2}) \\
\psi_j &= \left| \frac{p_{j+1} - 2p_j + p_{j-1}}{p_{j+1} + 2p_j + p_{j-1}} \right| \\
\psi_{j+1/2} &= \mathrm{MAX}\left(\psi_j, \psi_{j+1}\right)
\end{aligned} \tag{3}$$

Here R is the spectral radius of the flux Jacobian at the respective cell face and the sensor ψ is designed to be of O(1) near discontinuities and O(0) in smooth regions.

The Navier-Stokes equations contain viscous dissipation effects which are insignificant in inviscid portions of the flow (Shalman et al., 1998). Artificial dissipation terms are required since viscous dissipation cannot usually stabilize the numerical scheme. The matrix form of the JST operator proposed by (Turkel, 1988) (MJST) reduces the component dissipation. Figures 2(a),(b) illustrates that the MJST scheme in TAU suffers from excessive dissipation over the upper range of the resolvable wave numbers. (Ducros, 1999) proposed a modified form for the JST scheme for compressible LES where the pressure sensor ψ is rescaled with the local function

$$\phi = \frac{(\nabla .u)^2}{\left((\nabla .u)^2 + \omega^2 + \varepsilon\right)} \tag{4}$$

varying from 1 near a shock to O(0) in weakly compressible regions. The parameter ε prevents numerical overflow. A slightly modified form of the Ducros expression is used in TAU where $\varepsilon^{(2)}$ is written as

$$\varepsilon^{(2)}_{j+1/2} = k^{(2)} R_{j+1/2} \psi_{j+1/2} Q(\phi_{j+1/2}) \tag{5}$$

where the function $Q(\phi) = \tanh(\alpha\phi^p)$ provides additional scaling. The scalar constant α is set to 4.53 which ensures that $Q(r) \le 1$ for $r > 0$. Here r can be any suitable sensor, for example the sensor proposed by Ducros or the ratio of local Mach number to a reference Mach number have both been tested. The parameter α varies Q(r) between zero and one. Q(r) is set to one in the RANS layer using the DDES grid protection switch (Spalart et al., 2006). Figure 2 shows that the modified MJST scheme (MMJST) resolves the full range of wave numbers and that its performance for the DIT problem is equivalent to low Mach number preconditioning (Choi et al, 1993). Mesh influence on the N=32 cubed grid are seen in figure 2(c). As the number of connected faces in the dual grid mesh increase (hexahedral → tetrahedral) the wave number at which the maximum dissipation occurs (κ_{md}) is reduced. The dissipation integral is computed by summing over the number of connected faces, thus grid topologies with a larger number of connected faces have an increased numerical dissipation, which can short-circuit the turbulent energy cascade between adjacent wave numbers.

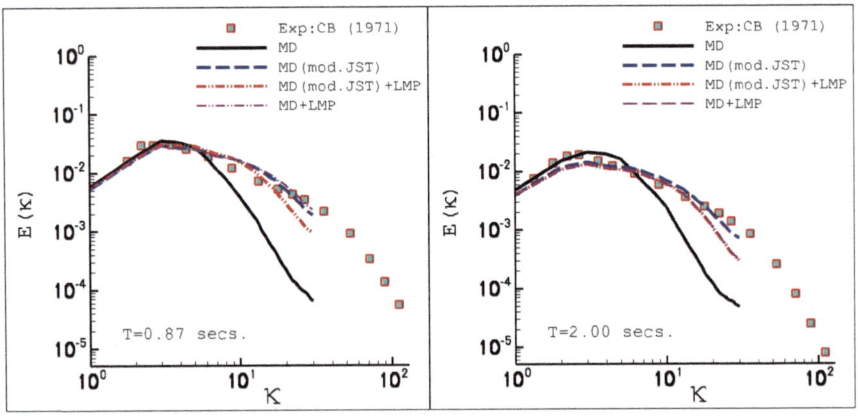

(a) t=0.087 secs, N=64 (b) t=2.0 secs, N=64

Fig. 2 Resolved energy spectrum returned with TAU. N(u) is the Navier-Stokes operator excluding time derivative term, so u. N(n) is the dissipation contribution from source terms. Peak dissipation wave number shift due to differing edge numbers on dual grids

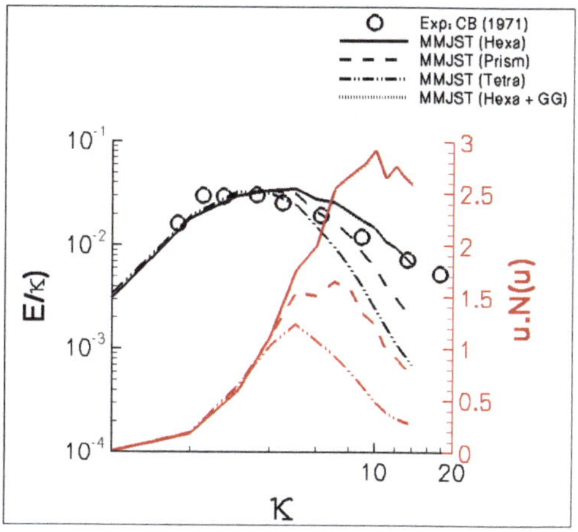

c) Mesh influence, N=32

Fig. 2 *(continued)*

4 Comparison of the Cylinder at $Re_D = 3900$

Cross flow about a cylinder remains a challenging problem due to the existence of a range of Reynolds number dependent flow regimes. Structured and unstructured solutions obtained with TAU for Spalart-Allmaras delayed DES (Spalart et al. 2006) [DDES] and LES using the WALE sub-grid model (Nicoud et al, 1999) are compared against THETA solutions using the WALE model on a fully unstructured mesh (see Figure 3 for mesh details). Table 1 gives the cylinder test cases that are presented in this paper. The range of tests undertaken at present allows for limited comparative testing on the influence of structured and unstructured meshes as well as comparisons between the TAU and THETA codes on the unstructured mesh.

Table 1 Nomenclature and description of the cylinder computations undertaken. Note that the number of convective time units (CTU) computed for the problem is given here

Test Case	TAU	THETA	Mesh	LES	DES	U_∞ (m/s)	CTU
TaLG1	X		G1	X		30	138
TaDG1	X		G1		X	30	130
TaLG2	X		G2	X		30	50
TaDG2	X		G2		X	30	50
ThLG2		X	G2	X		8	200

(a) Mesh G2: Full domain

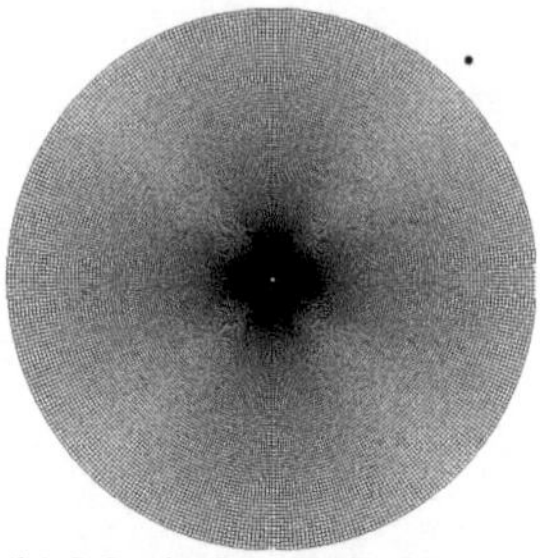

(b) Mesh G1. Full domain

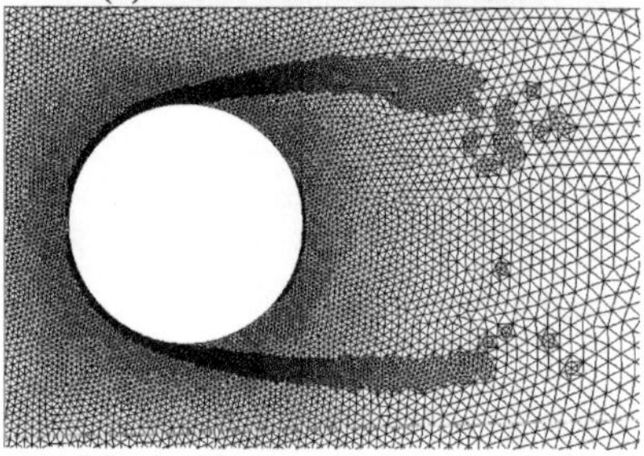

(c) Mesh G2: Interior near cylinder.

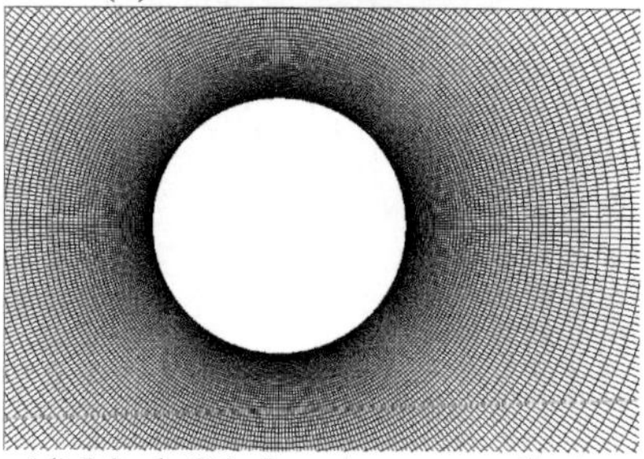

(d) Mesh G1: Interior near cylinder

Fig. 3 Comparison of the meshes used to compute the cylinder flow problem, Note that the DLR adaptation tool was used to refine mesh G2 in order to improve resolution of the shear layer break downstream

5 Averaged Flow Fields

Average force coefficients acting on the cylinder are significant for practical CFD applications. Table 2 presents a comparison of experimental data and computations for the mean drag (C_d), Strouhal number, St = fD/u, the recirculation length (Lr/D) and the minimum stream wise velocity along the wave centreline (Umin). All calculations, with the exception of TaDG1, return good estimations of the Strouhal number. LES results are superior (slightly) to the DDES. Recirculation lengths are under-predicted with exception of TaLG2 and TaDG2, which over predict the experiment. The reason for this behaviour on the unstructured mesh may be due to local numerical dissipation issues. LES calculations return better agreement with experiment in comparison to DDES which suffers from under sampling and further calculation is required. The maximum velocity deficit returned by TaLG1 and TaLG2 are close to the experiment, but lie outside of the error band. In all cases, with the exception of TaDG2, the base pressure coefficient is over predicted. All calculations overestimate bulk drag coefficient, with the best match against experiment being returned by TalG1. TaDG1 is the only computation which under predicts the Strouhal number. Averaged pressure coefficient is an important engineering parameter and figure 4 presents the pressure coefficient profiles returned by the calculations. Results obtained using the structured mesh G1 are both time and span averaged. Tests showed that for the averaging times used (see table 2) the differences between combined time/span averaged solutions and time averaged solutions only was not significant.

Table 2 Comparison of integral parameters returned by the computations

Data	Cd	-Cp	St	L_r/D	Umin
Exp.	0.99 ± 0.05	0.88 ± 0.05	0.215 ± 0.05	1.2 ± 0.1	-0.24 ± 0.1
TauLG1	1.08	1.13	0.213	1	-0.27
TauDG1	1.27	1.08	0.187	0.78	-0.22
TauLG2	3.1	0.75	0.218	1.47	-0.32
TauDG2	3.2	0.79	0.224	1.3	-0.31
ThetLG2	1.21	1.27	0.216	0.89	-0.27

Solutions obtained from mesh G2 were time-averaged only. Computational results differ from experiment in the pressure coefficient after separation and about the cylinder base. From figure 4 it can be seen that the influence of sub-grid model and grid combination is small at distances greater than three diameters from the cylinder. TaLG1, TaDG1 and ThLG2 underestimate the average reattachment length with the best representation being given by TaLG1. Comparisons of the stream-wise and cross-flow velocity components at two positions downstream from the cylinder inside the cylinder wake are given in figure 5. Computed velocity profiles underestimate the stream-wise velocity component, with the exception of TaDG2 at x/D = 1.06. However this is probably due to the under-predicted stream-wise momentum deficit along the wake centreline as seen in figure 4. At x/D=2.04 all computation under predict the stream wise momentum component. Maximum cross flow velocities are under predicted which implies that modelled cross-flow mixing will be under-predicted. Further examination of the resolved Reynolds stress components is required and this will be performed at a later date.

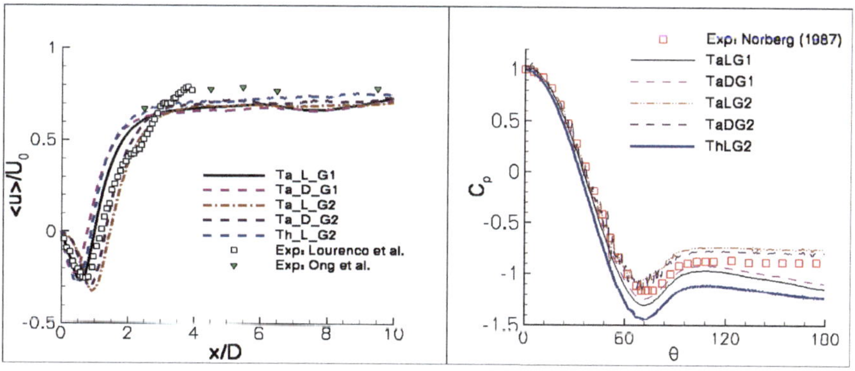

(a) Mean streamwise velocity (b) Surface pressure coefficient

Fig. 4 Mean stream wise velocity profiles along the centreline of wake behind the cylinder (a) and surface pressure coefficient (b)

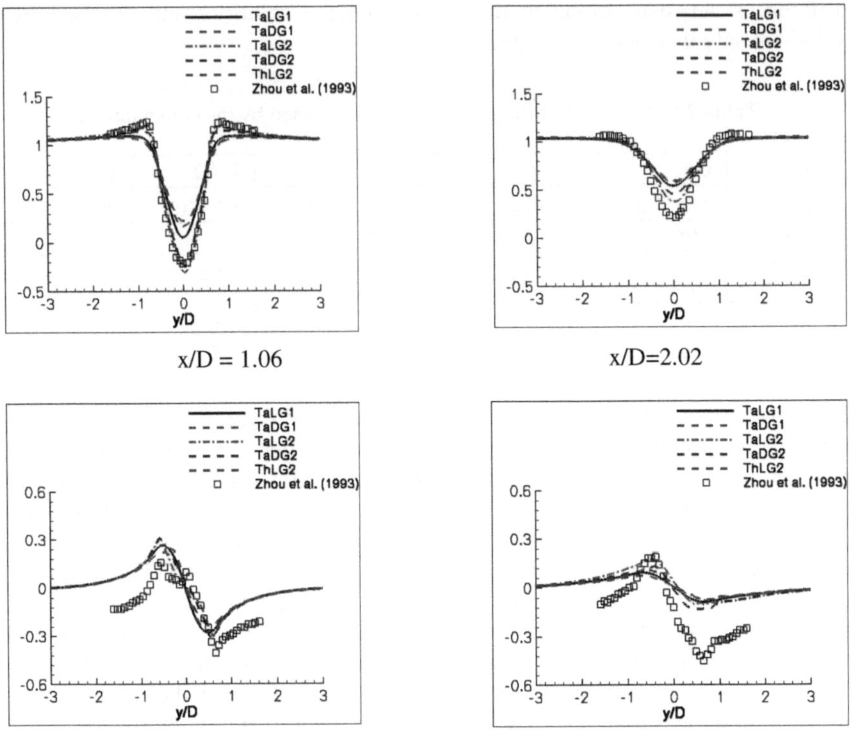

x/D = 1.06 x/D=2.02

x/D = 1.06 x/D = 2.02

Fig. 5 Comparison of stream-wise and cross-flow velocity components

6 Conclusions

The literature is well populated with data regarding the performance of LES/DES on structured meshes, however little is apparent for LES computations performed on unstructured or hybrid meshes. DIT computations using the TAU code show that the effective numerical dissipation increases as the number of connected faces on the dual grid mesh are increased. A good resolved energy spectrum can be obtained with hexahedral meshes (6 connected edges per point) however the high wave number resolution degrades as the number of edges is increased. This suggests that best results are be obtained using mesh elements with the minimum number of connected edges (hexahedra) in the LES regions of the flow. The compressible TAU solver was validated on a structured mesh after which comparisons against THETA using a fully unstructured mesh were made. The TAU results on the unstructured mesh differed substantially to those obtained on the structured mesh as well as the result obtained from THETA. LES and DES results were similar on identical grids for TAU suggesting that grid related numerical dissipation effects play a dominant role. The JST central scheme dissipation operator requires additional work for fully unstructured grids with lower-order numerical methods, however our results suggest that care in mesh design should be an absolute priority for any LES/DES endeavour. Algorithmic differences between the compressible and incompressible codes make direct comparisons non-trivial however the numerical dissipation effects (related to grid element topology and the numerics used) play a role in the flow development. A preliminary mesh adaptation study in the shear layer showed improvement in predicted centreline velocity for the unstructured mesh with the THETA code, and this suggests that some concerns in the initial grid design can be successfully handled via the implementation of an appropriate adaptation strategy. In the most general terms our results highlight the importance of the initial grid design as a tool to reduce the unwanted influence of numerical dissipation, particularly for lower-order numerical methods based on a dual grid metric. While additional treatment of the JST dissipation term can bring significant improvement, it is unlikely that a simple treatment along the lines suggested in the paper will provide a stabilising numerical dissipation that is independent of grid topology.

Acknowledgement. The authors are grateful to the helpful advice from Dr. R. Kessler in work relating to the DLR THETA code. Computations were funded from the DLR internal project RETTINA. Computational resources from the Center for Computational Applications in Aerospace Science and Engineering ($C^2A^2S^2E$), Braunschweig, Germany, were also used in this work.

References

Kröner, D.: Numerical Schemes for Conservation Laws. Wiley and Teubar (1997)

Goda, K.: A multistep technique with implicit differnce schemes for calculation two- or three-dimensional cavity flows. J. Comput. Physics 30, 76–95 (1979)

Chorin, A.J.: Numerical solution of the Navier-Stokes equations. Maath. Comp. 22, 745–762 (1968)

Timmermans, L.J.P., Minev, P.D., Van De Vosse, F.N.: An approximate projection scheme for incompressible flow using spectral elements. Int. J. Numer. Methods Fluids 22, 673–688 (1996)

Hackbusch, W.: Multigrid Methods and Applications. Springer, Heidelberg (1985)

Trottenburg, U., Oosterlee, A., Schüller, A.: Multigrid. Academic Press, London (2001)

Oliker, L., Biswas, R., Strawn, R.C.: Parallel implementations of an adaptive scheme for 3D unstructured grids on the SP2. In: Saad, Y., Yang, T., Ferreira, A., Rolim, J.D.P. (eds.) IRREGULAR 1996. LNCS, vol. 1117, pp. 35–47. Springer, Heidelberg (1996)

Jameson, A.: Time dependent calculations using Multigrid, with application to unsteady flows past airfoils and wings, AIAA 91-1595 (1991)

Dwight, R.: Time-Accurate Navier.Stokes calculations with Approximately Factored Implicit Schemes. In: Groth, C., Zwingg, D.W. (eds.) Computational Fluid Mechanics. Springer, Heidelberg (2004)

Smagorinsky, J.: General circulation experiments with the primitive equations: I. The basic equations. Mon. Weather Rev. 91, 99–164

Shur, M., Spalart, P.R., Strelets, M., Travin, A.: Detached-eddy simulation of an airfoil at high angle of attack! In: Rodi, W., Laurence, D. (eds.) Engineering Turbulence Modelling and Experiments-4. Elsevier Science Ltd, Amsterdam (1999)

Peric, M.: A finite volume method for the prediction of the three-dimensional fluid flow in complex ducts. PhD Thesis, Imperial College, London (1985)

Comte-Bellot, G., Corrsin, S.: Simple Eulerian time correlation of full- and narrow-band velocity signals in grid-generated, 'isotropic' turbulence. J. Fluid. Mech. 48, Part 2, 273–337 (1971)

Haase, W., Braza, M., Revell, A.: DESIDER – A European effort on Hybrid RANS-LES Modelling. Notes on Numerical Fluid Mechanics and Multidisciplinary Design 103 (2009)

Breuss, M.: An Analysis of the influence of data extrema on some first and second central approximations of hyperbolic conservation laws. ESAIM 39(5), 965–993 (2005)

Jameson, A., Schmidt, W., Turket, E.: Numerical solution of the Euler equations by finite volume methods using Runge-Kutta time stepping schemes, AIAA paper 81-1259

Shalman, E., Yakhot, A., Shalman, S., Igra, O., Yadlin, Y.: Attenuating Artifical Dissipation in the Computation of Navier-Stokes turbulent Boundary Layers. Journal of Scentific Computing 13(2) (1998)

Turkel, E.: Improving the accuracy of central difference schemes. In: 11th Intl. Conf. Numerical Methods in Fluid Dynamics. Lecture Notes in Physics, vol. 323, pp. 586–591. Springer, NewYork

Ducros, F., Ferrand, V., Nicoud, F., Weber, C., Darracq, D., Gacherieu, C., Poinsot, T.: Large-Eddy Simulation of the Shock Turbulence Interaction. Journal of Computational Physics 152, 517–549 (1999)

Spalart, P.R., Deck, S., Shur, M., Squires, K.D., Strelets, M.K., Travin, A.: A new version of detached.eddy simulation, resistant to ambiguous grid densities. Theor. Comput. Fluid. Dyn. 20, 181–195 (2006)

Broderson, O., Ronzheimer, A., Ziegler, R., Kunert, T., Wild, J., Hepperle, M.: Applications using MegaCads. In: Cross, M., Eiseman, P., Häuser, J., Soni, B.K., Jhompson, J.F. (eds.) Proc. 6th Int. Conference on Numerical Grid Generation in Computational Field Simulation, University of Greenwich, London, pp. 793–802 (1998)

Melber-Wilkending, S., Broderson, O., Kallinderis, Y., Wilhelm, R., Sutcliffe, M., Wild, J., Ronzheimer, A.: Hybrid Unstructured Grid Generation in MEGAFLOW. Notes in Numerical Fluid Mechanics and Multidisciplinary Design. Springer, Heidelberg (2006)

Kravenchenko, A.G., Moin, P.: Numerical studies of flow over a circular cylinder at Re_D=3900. Physics of Fluids 12(2), 403–417 (2000)

Mansy, H., Yang, P.M., Williams, D.R.: Quantitative measurements of three-dimensional structures in the wake of a circular cylinder. J. Fluid Mech. 270, 277–296 (1994)

Williamson, C.H.K., Wu, J., Sheridan, J.: Scaling of streamwise vortices in wakes. Physics of Fluids 7, 2307 (1995)

Breuer, M.: Numerical and modelling influences on large eddy simulations for the flow past a circular cylinder. Int. Journal of Heat and Fluid Flow 19, 512–521 (1998)

Nicoud, F., Ducros, F.: Sub-grid stress modelling based on the square of the velocity gradient tensor. Flow. Turb. Combust. 62, 183–200 (1999)

Choi, D., Merkle, C.L.: The application of preconditioning to viscous flows. Comput. Phys. 105, 207–223 (1993)

Schwamborn, D., Gerhold, T., Heinrich, R.: The DLR TAU code: Recent applications in research and industry. In: European Conference in Computational Fluid Mechanics, Egmond aan Zee, ECOMASS CFD 2006 (2006)

Description of Test Section and Measurement in a Nozzle for Shock Upstream Mach Number M=1.45

P. Doerffer

Abstract. The test case described stems from the European UFAST project – see next chapter 30 and Doerffer (2009). Firstly, it was intended to being used as a conference test case, but because of late provision of data, the conference organizer decided to postpone this case and to use it for a likely follow-on conference, the fourth symposium for Hybrid RANS-LES methods then – as it was requested already by several participants of the present third symposium.

1 Introduction

For the test case chosen, the following measurements were carried out, forming a comprehensive data set for validation on the basis of hybrid methods:

- Static pressure along the wall at centerline
- Schlieren photography
- Oil visualization at rectilinear wall
- Boundary layer measurements by pneumatic probes
- CTA Measurements
- Shock oscillations measurements

The experimental investigations for the flat wall nozzle were carried out in the test section, which is presented in Fig. 1. The origin coordinate, X=0, is located at 17.5 mm downstream of the nozzle throat as shown in Fig. 1. Based on the flow parameters used, the shock is located in a constant velocity region in the nozzle. The location of the beginning of the pressure jump upstream of the shock is at coordinate X=237 mm. The location of three traverses of boundary layer measurements and CTA (constant temperature anemometry, a hot wire anemometry) measurements is shown in Fig. 1, too. The traverse upstream of the shock is located at X=212 mm. The traverses downstream of the shock are located at X=267 mm and X=292 mm, respectively.

Piotr Doerffer
Institute of Fluid Flow Machinery, Polish Academy of Sciences, Fiszera 14, 80-952 Gdansk, Poland
doerffer@imp.gda.pl

S.-H. Peng et al. (Eds.): Progress in Hybrid RANS-LES Modelling, NNFM 111, pp. 339–344.
springerlink.com

Table 1 presents the height of CTA and shock oscillations measurements, accordingly.

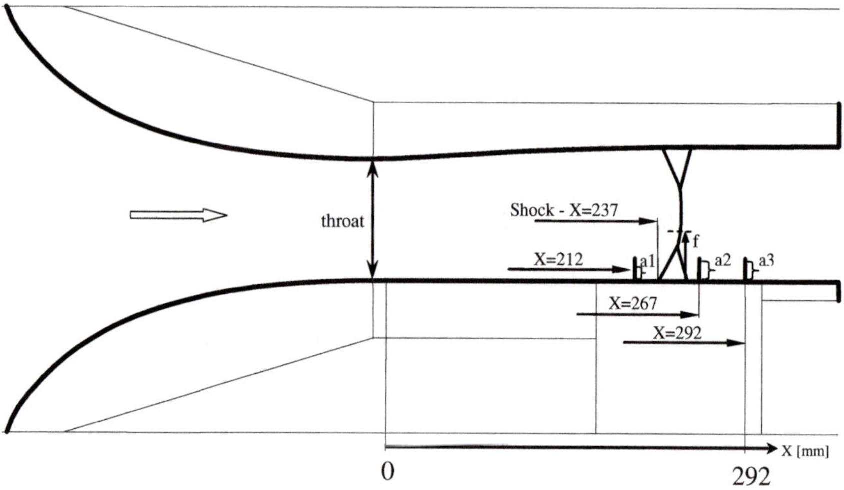

Fig. 1 Test section layout

The CTA measurements were carried out in the middle of the channel, while shock oscillation is measured in a horizontal section at f = 26 mm above the rectilinear wall, compare Figure 6, below.

Table 1 Height of CTA and shock oscillations measurements

M	a1 [mm]	a2 [mm]	a3 [mm]
1.45	2.5	5	5

Geometrical data to be used for mesh generation are provided for upper wall and lower wall geometry, with a channel depth of 100 mm.

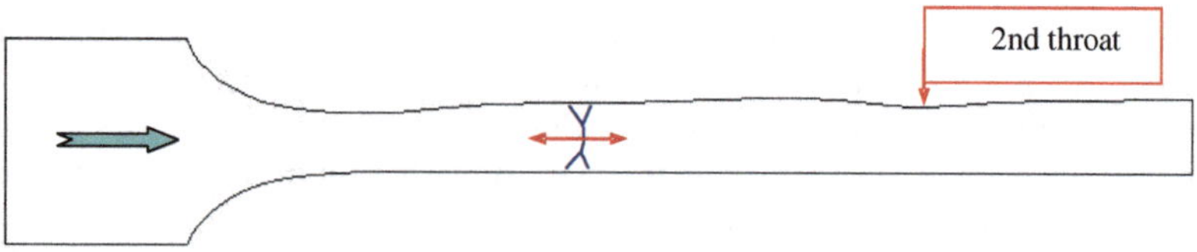

Fig. 2 View on flat wall nozzle geometry

The location of the shock wave is "adjusted" by the second throat size as it is sketched in Figure 2. Although is it somewhat "inconvenient" from the CFD point of view, in order to obtain natural unsteady effects, the downstream part of the test section must (!) be taken into account. An adjustment of the shock location of about ± 5 mm is sufficient because the Mach number is constant in that area.

2 Flow Conditions

2.1 Inlet Conditions

Inlet conditions are ambient flow conditions, with

- Total pressure = 101000 Pa
- Total temperature = 283 K
- Turbulence intensity = 1%
- Turbulent viscosity ratio = 10

2.2 Outlet Condition

The shock location is fixed by the opening of the 2nd throat (Fig.2). In the 2nd throat, a Mach number equal to 1 should be obtained.

3 Experimental Results

In the following, experimental data are provided, to be taken as a guideline for numerical treatment – they can be offered in electronic form on request. Please contact Pawel Flaszynski, pflaszyn@pg.gda.pl.

Figure 3 describes the mean value of the static pressure along the centre line, while Figure 4 exhibits the isentropic Mach number along the centerline, as determined from stagnation pressure at inlet and local static pressure.

Flow features can be detected from Figures 5 and 6, showing a Schlieren picture and oil flow pattern on the lower channel surface, respectively. Note the wide oil accumulation zone upstream of separation. Oil dots, visible at reattachment indicate unsteadiness.

Figure 7 exhibits a plot of typical shock oscillations at the position f=26 mm, measured from the lower surface.

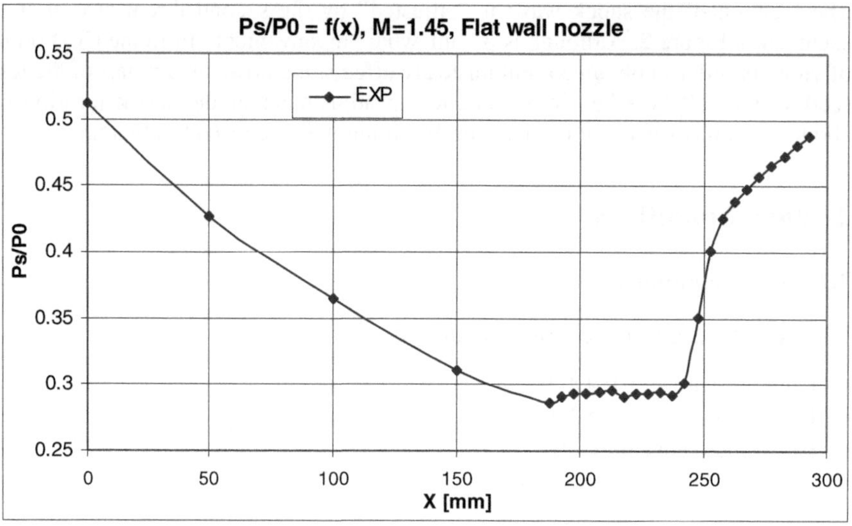

Fig. 3 Static pressure along the centre line, mean value

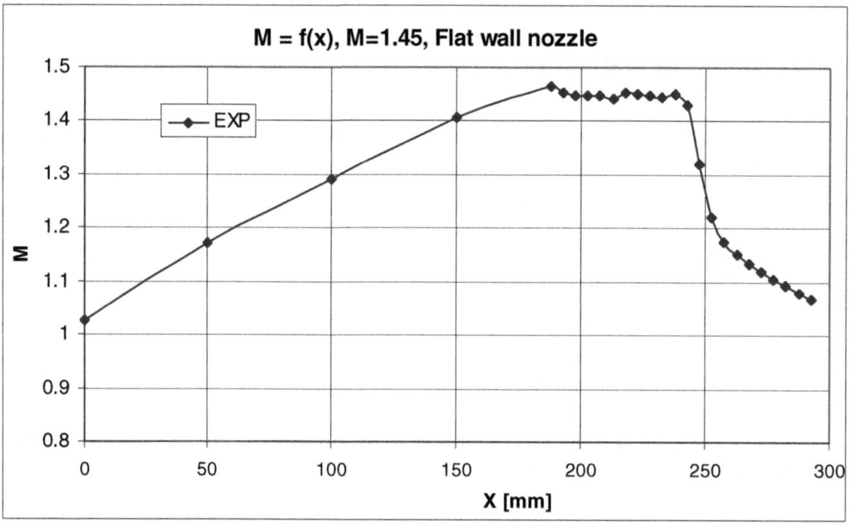

Fig. 4 "Isentropic Mach number" along the centerline; determined from stagnation pressure at inlet and local static pressure

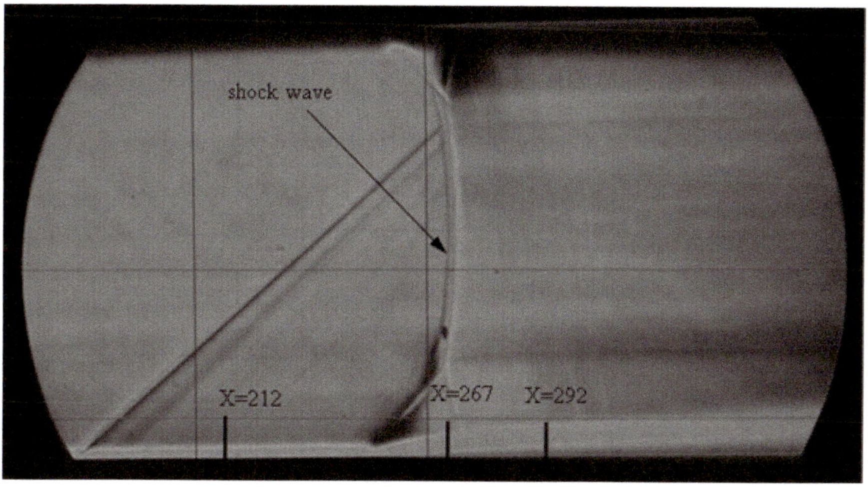

Fig. 5 Schlieren photo

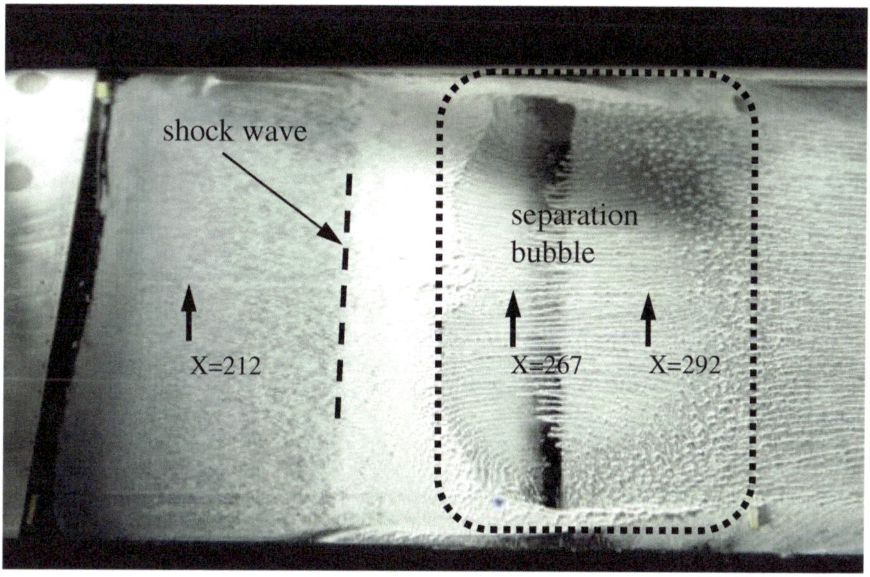

Fig. 6 Surface oil visualisation

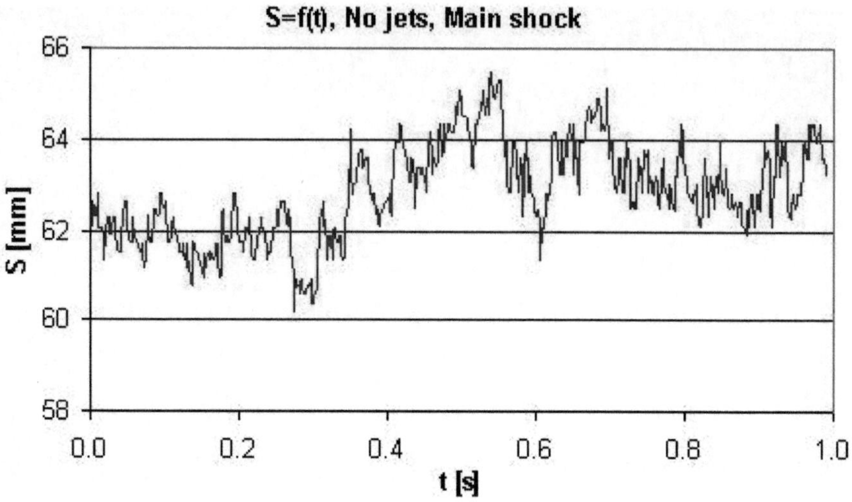

Fig. 7 Typical shock fluctuation at f = 26 mm above the wall

Reference

Doerffer, P.: UFAST Experiments Data Bank, Unsteady effects of shock wave included separation. Wydawnictwo IMP PAN, Gdansk, Poland (2009), ISNB 978-83-88237-46-1

UFAST Project – Unsteady Effects in Shock Wave Induced Separation

P. Doerffer

1 Introduction

Aligned with the needs of the aeronautics industry the general aim of the UFAST project is to foster experimental and theoretical work in the highly non-linear area of unsteady shock wave boundary layer interaction (SWBLI). Although in the past several EU projects were aiming at transonic/supersonic flows, the area of unsteady shock wave boundary layer interaction has not yet been treated. Moreover, experimental methods as well as numerical approaches have been improved considerably.

The main study cases, shock waves on wings/profiles, nozzle flows and inlet flows, will serve as a sound basis for open questions posed by the aeronautics industry and can easily be exploited to enable more complex applications to be tackled. In addition to basic flow configurations control methods (synthetic jets, electro-hydrodynamic actuators, stream-wise vortex generators and transpiration flow) will be investigated for controlling both interaction and inherent flow unsteadiness.

Emphasis is placed on closely coupled experiments and numerical investigations to allow for feeding back numerical results to the experiments and vice versa for the sake of identifying and overcoming weaknesses in both approaches.

Using RANS/URANS and hybrid RANS-LES methods, UFAST will assess new methods for turbulence modelling in particular for unsteady, shock dominated flow. Moreover, LES methods will be applied to resolve the large coherent structures that govern SWBLI. This way, UFAST will investigate the "range of applicability" between RANS/URANS and LES.

2 Unsteady Effects in Shock Wave Induced Separation

The causes of unsteadiness depend on flow geometry, therefore many configurations of interaction should be studied. Three Tasks have been identified:

- transonic interaction with local supersonic domain
- interaction in nozzles where a shock is dividing supersonic and subsonic part of the flow
- oblique shock wave reflection from a flat wall.

Piotr Doerffer
Institute of Fluid Flow Machinery, IMP PAN
Fiszera 14, 80-952 Gdansk, Poland
doerffer@imp.gda.pl

S.-H. Peng et al. (Eds.): Progress in Hybrid RANS-LES Modelling, NNFM 111, pp. 345–347.
springerlink.com

These are typical shock interaction cases which can be encountered in many industrial applications. Our belief is that this wide scope of interaction flow fields will make it possible to extract the main physical phenomena independent of a particular flow configuration. Basic interaction phenomena are studied in a Work Package 2.

The goal of WP3 is to determine the potential of several novel concepts for the control of shock wave/boundary layer interactions. In particular the ability of control methods to suppress/delay shock induced separations and the effect of control on SWBLI unsteadiness will be examined. Apart from novel control methods several older techniques will also be investigated to enable a comparison with previous research. The investigation of SWBLI control for separated unsteady flows is not only important because flow control is a crucial technology for next generations of aircraft, but also because it increases our knowledge of the interaction phenomenon. Apart from the natural unsteadiness of SWBLIs this work package also investigates the effect of control under forced unsteady shock conditions. Such forced excitations cannot be ignored because they may dominate the characteristics of the interaction behaviour.

The results of WP2 will form the baseline for an evaluation of the effectiveness and physical influence of each control technique. In particular, the following methods will be investigated:

- periodic profile pitch and/or oscillating aileron deflection
- electro-hydrodynamic actuators (EHD)
- synthetic jets (SJ)
- traditional vortex generators (VG)
- sub-boundary-layer vortex generators (SBVG)
- air jet vortex generators (AJVG)
- perforated wall suction.

A complete review and critical assessment of all control methods will be undertaken, resulting in recommendations for their application to various interaction situations. Following a comparison with numerical simulations, an assessment will also be made about the level of confidence in the use of CFD for practical investigations. The final report will contain all key results as well as recommendations concerning the potential integration of control devices into future aircraft.

One of the main outcomes expected from UFAST is the improvement of simulation techniques which should increase the confidence in predicting these complicated flows using traditional numerical approaches, based on RANS and URANS approximations associated with various turbulence models WP4. To support this evaluation of (U)RANS based CFD methods, it is essential to validate the simulations against the new experimental measurements obtained in WP2 and WP3. The resulting improved understanding of the involved phenomena may help to reconsider current turbulence modelling issues. Successful improvement of the CFD modelling carried out in WP4 will constitute the main achievement of the project.

Nowadays, in industrial CFD practice statistical turbulence modelling is used almost exclusively, via the Reynolds-averaged Navier-Stokes equations (URANS).

Yet, URANS is not suitable for flows which are instationary over a wider range of scales. Also URANS cannot predict critical flow phenomena which appear only in a statistically insignificant number of realizations. These problems can be addressed, however, by Large-Eddy simulation (LES). LES does not represent a statistical flow description but a single flow realization. Computational savings are achieved by reducing the range of resolved spatial scales, at the expense of a subgrid-scale model.

The objective of WP 5 is to bridge the gap between the current RANS and URANS modelling strategies for turbulent flows in industrial CFD applications and future LES for physically complex supersonic flows. For many practical situations it is unnecessary to apply LES over the entire flow domain. In particular for external flows physical complexity is confined to regions near a body surface and in its wake, whereas the remainder of the flow domain can be well treated by RANS approaches. Also, RANS approaches are considered for the near-wall region in high-Reynolds-number turbulent boundary layers in order to reduce the computational cost of wall-resolving LES. These latter approaches are so-called hybrid-RANS-LES models which are also considered in this Work Package. Hybridization can be done by interfacing RANS and LES regions, where RANS and LES models can be of entirely different type, or by using modifications of RANS models which assume the character of an LES model for sufficiently fine grids. The latter is frequently called Detached-Eddy Simulation (DES). DES has been developed to handle massively separated flows where separation is most frequently fixed by the geometry (e.g. base flow).

The UFAST project will deliver an extensive data bank on unsteady effects of shock induced separation. In the same time it will present the ability of modern CFD to handle such complicated flow phenomena.

From FLOMANIA via DESider to ATAAC

W. Haase and D. Schwamborn

Beginning in the late 80's of the last century, European efforts on validation of CFD codes were initiated by the European Commission programme Brite-EURAM. Since then a number of collaborative programmes have been carried out indicating the importance of flow-physics modelling in aerodynamic predictions.

To highlight the progress achieved as well as the still existing deficiencies of turbulence modelling in an industrial framework, this presentation deals with three consecutive projects: FLOMANIA, DESider and ATAAC, the last one just started in April 2009. We will present the major objectives of these projects and the outcome existing so far, not to depict a few colourful results but to reflect and guideline future needs:

As in aircraft design, efficiency is determined by the ability to accurately and reliably predict and model turbulent flows, the development of models and their validation became the main objectives of the FLOMANIA project, starting in 2002 (until 2004). Thus the main goal in the area of RANS computations was to increase predictive accuracy for complex geometries and complex flows taking care of a necessary change in paradigm in CFD which means that prediction of delta-value between two consecutive solutions is replaced by the need to provide absolute and more accurate values, e.g. in drag. This demand on improved designs, better flight performance and enlarged flight envelopes leads to further three objectives in enhancing flow-physics modelling. In FLOMANIA these were firstly related to improving current industrially used turbulence models, secondly to employing sophisticated differential Reynolds stress models which represent physics in the whole flow domain much better and thirdly to detecting the limits in RANS computations by using DES for massively separated flows.

As a direct consequence of the first positive attempts with DES in the FLOMANIA project, i.e. with hybrid RANS-LES methods, the DESider project (launched in 2004 and finalized June 2007) was motivated by the increasing industrial demand on turbulent unsteady aerodynamic flows. Thus the major aim was to overcome known weaknesses in the different hybrid approaches - in order to support the European aeronautics industry with methods with increased

W. Haase
formerly EADS Military Air Systems
office@haa.se

D. Schwamborn
Center of Computer Applications in Aerospace Science and Engineering, German Aerospace Center, Bunsenstrasse 10, D37073 Göttingen
Dieter.schwamborn@dlr.de

S.-H. Peng et al. (Eds.): Progress in Hybrid RANS-LES Modelling, NNFM 111, pp. 349–351.
springerlink.com

predictive accuracy for flows exhibiting (massive and hence unsteady) separation. Emphasis was placed on the enhancement of both the modelling capability of RANS-LES methods and the alleviation of computational costs. In that respect, DESider was trying close the gap between RANS and LES but was not treating LES as a stand-alone method – as it is known that full LES for practical use will (likely?) not be available in the next decade. An additional effort was placed on an investigation on how far RANS-LES methods can be well applied to multi-disciplinary topics as there are aero-acoustics (noise reduction) and aero-elastics (reduced A/C weight, unsteady loads, fatigue issues, improved A/C safety), improving by this a cost-effective design.

Despite the very promising outcome of the DESider project, it has become clear that a concentration on hybrid RANS-LES approaches alone is not sufficient to fulfil the industrial needs. Even with the further increase in the computing powers available to industry less demanding methods are needed allowing large numbers of computations in relatively short times. Thus the new ATAAC project revisits also Differential Reynolds Stress Models (DRSM), which bear the prospect of improved modelling at marginally higher cost that e.g. two-equation models, as was demonstrated already in FLOMANIA.

As LES will not be affordable for high Reynolds numbers real-life flows in the next decades, ATAAC focuses on approaches below the LES level, namely Differential Reynolds Stress Models (DRSM), advanced Unsteady RANS models (URANS) including Scale-Adaptive Simulation (SAS), and different hybrid RANS-LES schemes, including the latest versions of DES, as well as Wall-Modelled LES and Embedded LES.

The work in the project is following "roadmaps", which link challenges in aircraft performance prediction rooted in turbulence modelling issues to a chain of more basic stepping stone configurations which allow an easier identification of underlying modelling issues and assessment of improvements.

The final goals of ATAAC are:

- to recommend one or at most two "best" DRSM for conventional RANS and URANS;
- to provide a small set of hybrid RANS-LES and SAS methods that can be used as "reference" turbulence-
- to resolve approaches in future CFD design tools;
- to formulate clear indications of areas of applicability and uncertainty of the proposed approaches for aerodynamic applications in industrial CFD.

With all contributions towards reliable industrial CFD tools, ATAAC will have a direct impact on the predictive capabilities in design and optimisation, and directly contribute to the development of Greener Aircraft.

References

Haase, W., Aupoix, B., Bunge, U., Schwamborn, D. (eds.): FLOMANIA - A European initiative on flow physics modelling. Notes on Numerical Fluid Mechanics and Multidisciplinary Design, vol. 94. Springer, Heidelberg (2006)

Peng, S.-H., Haase, W. (eds.): Advances in Hybrid RANS-LES Modelling. Notes on Num. Fluid Mech. And Multidisciplinary Design, vol. 97. Springer, Heidelberg (2007)

Haase, W., Braza, M., Revell, A. (eds.): DESider – A European Effort on Hybrid RANS-LES Modelling. Notes on Numerical Fluid Mechanics and Multidisciplinary Design, vol. 103. Springer, Heidelberg (2009)

Scrutinizing a Seamless Hybrid LES/RANS Approach for Turbomachinery Applications

D. Borello, G. Delibra, K. Hanjalic, and F. Rispoli

In recent years, the need to improve the design procedures for turbomachinery forces towards the development of more accurate numerical tools for the predictions of turbulent phenomena. Particular relevance assumes the prediction of losses driven by turbulent fluctuations causing noise and vibrations. LES is particularly appropriate for this class of flows but its use is strongly limited by its computational demand due to the very high Reynolds numbers and the complexity of the geometries. In hybrid RANS/LES, LES is applied in regions away from the walls, and an advanced U-RANS is used in the near wall region. It is then possible to use hybrid models for resolving (part of) the large scale turbulence spectrum requiring much lower computational costs with respect to conventional LES.

However, the coupling of LES and U-RANS regions remains an open problem and many approaches are at present available and none of them demonstrated a clear predominance (Frolich and Von Terzi, 2008). In a previous work (Borello et al., 2009) we proposed a seamless hybridisation of the ζ-f model with LES by sensitizing the U-RANS model to recover a portion of the turbulence spectrum.

In this coupling the dissipation rate in the k-equation (1) is multiplied by a grid-detection function in term of the ratio of the RANS and LES length scales (2).

$$\frac{Dk}{Dt} = \frac{\partial}{\partial x_j}\left((\nu + \nu_t) \frac{\partial k}{\partial x_j} \right) + P - \alpha\varepsilon \qquad (1)$$

$$\alpha = \max\left(1, c\frac{L_{RANS}}{L_{LES}}\right); \qquad L_{RANS} = \frac{k_{\text{mod}}^{3/2}}{\varepsilon}; \qquad L_{LES} = (\Delta x \Delta y \Delta z)^{1/3} \qquad (2)$$

It is important to note that at the beginning of the simulation, when the field of turbulent variables has not developed yet, a simplified version of L_{RANS}, is adopted assuming that the length scale is proportional to the distance from the nearest wall. The hybrid model imposes a further constraint based on the criterion:

$$\nu_t = \max(\nu_t^{RANS}, \nu_t^{LES}) \qquad (3)$$

where $\nu_t^{RANS} = c_\mu \zeta k^2 / \varepsilon$ and ν_t^{LES} is obtained by the dynamic Smagorinsky model.

D. Borello
Dip. di Meccanica e Aeronautica, Sapienza Università di Roma, Italy
borello@dma.ing.uniroma1.it

S.-H. Peng et al. (Eds.): Progress in Hybrid RANS-LES Modelling, NNFM 111, pp. 353–355.
springerlink.com

In the full paper the potential of this formulation will be exploited with reference to a flow in a pin fin array reproducing the internal cooling process near the trailing edge of a turbine blade, previously investigated by using URANS in Delibra et al. (2009) and a flow in a 3D compressor cascade with tip leakage.

For a preliminary assessment of the model we analysed the fully developed channel flow at Re_τ=2000 of Hoyas and Jimenez (2006). In the DNS 632 cells are placed in the direction normal to the wall while in our computations we used a URANS type discretisation (64 cells). In Fig.1 a comparison of the *k* profiles is shown. It is possible to note that in hybrid computations the total *k* (modelled *plus* resolved) is in acceptable agreement with DNS also if a slight under-prediction of *k* is noted in the region where URANS and LES overlap up to y^+ equal to 500.

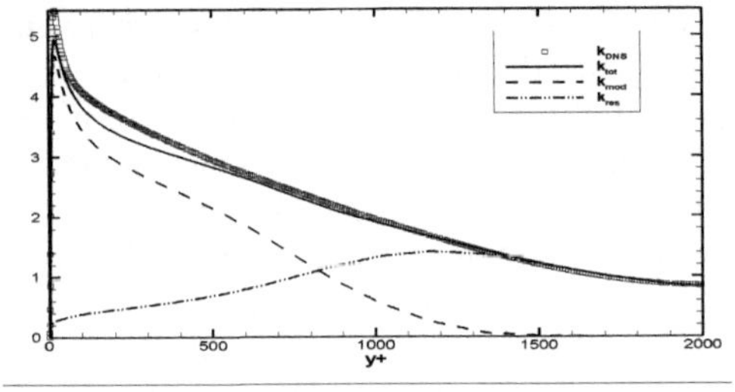

Fig. 1 Turbulent kinetic energy profile at Re_τ=2000

Table 1 Unsteadiness in the first pin

	Exp	LES	Hybrid	URANS
Strouhal	0.234	0.240	0.230	0.244
Fluctuations (max value)	-	2.6 m/s	0.6 m/s	0.004 m/s

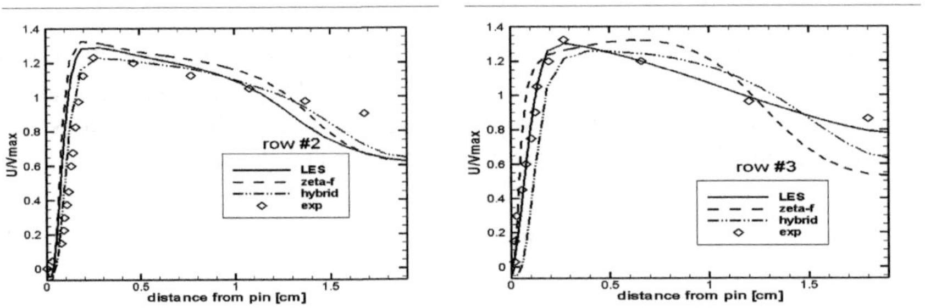

Fig. 2 Pin fin array: velocity profile normal to the wall in the midplane between two pins in spanwise direction

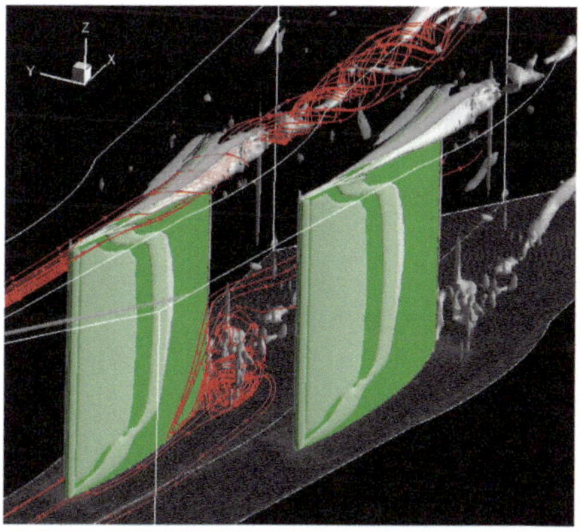

Fig. 3 Flow in a 3D compressor cascade – details of wake region: hybrid LES/RANS

Now some preliminary results relative to the pin-fin array will be discusssed. The URANS (zeta-f in the legend) and hybrid results have been obtained on the same grid of 1.3 M cells, while the reference LES results refer to a 2 M cells grid. In Fig. 2 predictions of the velocity profile in a section placed in the mid between two consecutive pins placed in spanwise direction are shown. The superior performance of the hybrid model w.r.t. to the URANS model is evident. Hybrid model always demonstrates a better agreement with LES and experiments w.r.t. URANS solved on the same grid probably due to its capability to recover the influence of flow unsteadiness. This is demonstrated by the data in Tab.1 relative to the unsteadiness in a point located in the aft region of the first pin subjected to the vortex shedding. For all the simulations, the Strouhal numbers show a good agreement with the experimental results. On the other hand, the max excursion of the fluctuations shows that the URANS is almost steady in this location while LES and hybrid simulations are able to predict a strong fluctuating field. In Fig. 3 we show the turbulent structures in a 3D compressor cascade. Such structures mainly arise from tip leakage, hub corner and wake. Simulations obtained by using hybrid model show the relevance of the resolved structures in the LES region.

Acknowledgement. The work of K. Hanjalić has been made possible by the award of an EU Marie Curie Chair at DMA, 'Sapienza' University of Rome.

References

Borello, D., Delibra, G., Hanjalic, K., Rispoli, F.: URANS and Hybrid LES/RANS computations of tip-leakage and secondary flows in axial compressor cascade. In: 8th European Turbomachinery Conference, Graz (March 2009)

Delibra, G., Borello, D., Hanjalic, K., Rispoli, F.: U-RANS of Flow and Endwall Heat Transfer in a Pinned Passage Relevant to Gas-Turbine Blade Cooling. International Journal for Heat and Fluid Flows (2009) (in press)

Frolich, J., Von Terzi, D.: Hybrid LES/RANS methods for the simulation of turbulent flows. Progress of Aerospace Sciences 44, 349–377 (2008)

Turbulent Flow Modeling via Galerkin Method and Finite Time Thermodynamics

M. Morzyński, B.R. Noack, M. Schlegel, and G. Tadmor

The Navier-Stokes equations are the trusted description for many flows of practical relevance. Ample physical and engineering expertize is necessary to accommodate this description to an industrial setting. Presently and in predictable future there will be no CFD method to simulate the flow ranging from large coherent structures to the Kolmogorov scale. The flow modeling process can be considered as the sufficiently accurate reduction of the system. The key challenge in turbulence modeling is the reduction of the dynamical system to a coarser discretization. For spectral methods, or more generally: Galerkin models, discretization corresponds to the adequate base of expansion modes.

Here, a recent development of Finite Time Thermodynamics (FTT) (Noack et al. 2008 JNET) approach to turbulent flow modeling is presented. FTT shows that the turbulent flow can be described by dynamical system in which each phase space direction can be considered a thermodynamic degree of freedom (DOF). The use of thermodynamic DOFs was postulated earlier by L. Osager (1949) for point vortices in 2D flow and R.H Kraichnan (1955,1989). for an absolute equilibrium ensemble of Fourier modes. The proposed FTT theory treats the turbulent flow modeling as an thermodynamic energy problem. The modes, being the thermodynamical DOFs, are parametrized by energy levels.

For the Galerkin system built on this base, the FTT represents a statistical closure. The Galerkin system contains constant, linear and quadratic terms. The constant term can be removed by a translation to the fixed point. The linear term can be brought in normal form based on stability eigenmodes. The role of the quadratic nonlinear term on the attractor presents the major difficulty. It is shown that quadratic terms of Galerkin system are energy preserving. The FTT formalism elucidates how linear and nonlinear processes balance between the extremes of linear instability and thermal equilibrium by nonlinearity. External interactions of modes are determined by energy levels of involved modes. Internal interactions via triadic interactions are derived from first principles and yield a generalized Fick's law. In contrast to Kraichnan's DIA, our triadic interaction term fulfils a

M. Morzyński
Poznan University of Technology, Poland
morzynski@stanton.ice.put.poznan.pl

B.R. Noack · M. Schlegel
Berlin Institute of Technology MB1, Germany

G. Tadmor
Northeastern University, Boston, MA, USA

S.-H. Peng et al. (Eds.): Progress in Hybrid RANS-LES Modelling, NNFM 111, pp. 357–358.
springerlink.com

vital realizability condition. In this way, FTT yields approximate versions of the modal Reynolds and TKE equations.

The FTT model is successfully applied to shear flows with simple to complex dynamics. It is demonstrated for the vortex shedding behind a cylinder and 3D homogeneous shear turbulence. The FTT formalism opens the first path to a fully nonlinear, infinite-horizon control theory, a unified description of the normal and inverse turbulence cascade, and represents an alternative to the eddy viscosity concept.

References

Noack, B.R., Schlegel, M., Ahlborn, B., Mutschke, G., Morzyński, M., Comte, P., Tadmor, G.: A finite-time thermodynamics of unsteady flows. J. Non-Equilibr. Thermodyn. 33(2), 103–148 (2008)

Kraichnan, R.H., Chen, S.: Is there a statistical mechanics of turbulence? Phys. D 37, 160–172 (1989)

Author Index

Notes on Numerical Fluid Mechanics and Multidisciplinary Design

Available Volumes

Volume 111: Shia-Hui Peng, Piotr Doerffer, and Werner Haase (eds.): Progress in Hybrid RANS-LES Modelling – Papers Contributed to the 3rd Symposium on Hybrid RANS-LES Methods, Gdansk, Poland, June 2009. ISBN 978-3-642-14167-6

Volume 110: Michel Deville, Thien-Hiep Lê, and Pierre Sagaut (eds.): Turbulence and Interactions – Proceedings the TI 2009 Conference. ISBN 978-3-642-14138-6

Volume 109: Wolfgang Schröder (ed.): Summary of Flow Modulation and Fluid-Structure Interaction Findings – Results of the Collaborative Research Center SFB 401 at the RWTH Aachen University, Aachen, Germany, 1997-2008. ISBN 978-3-642-04087-0

Volume 108: Rudibert King (ed.): Active Flow Control II – Papers Contributed to the Conference "Active Flow Control II 2010", Berlin, Germany, May 26–28, 2010. ISBN 978-3-642-11734-3

Volume 107: Norbert Kroll, Dieter Schwamborn, Klaus Becker, Herbert Rieger, Frank Thiele (eds.): MEGADESIGN and MegaOpt – German Initiatives for Aerodynamic Simulation and Optimization in Aircraft Design. ISBN 978-3-642-04092-4

Volume 106: Wolfgang Nitsche, Christoph Dobriloff (eds.): Imaging Measurement Methods for Flow Analysis - Results of the DFG Priority Programme 1147 "Imaging Measurement Methods for Flow Analysis" 2003–2009. ISBN 978-3-642-01105-4

Volume 105: Michel Deville, Thien-Hiep Lê, Pierre Sagaut (eds.): Turbulence and Interactions - Keynote Lectures of the TI 2006 Conference. ISBN 978-3-642-00261-8

Volume 104: Christophe Brun, Daniel Juvé, Michael Manhart, Claus-Dieter Munz: Numerical Simulation of Turbulent Flows and Noise Generation - Results of the DFG/CNRS Research Groups FOR 507 and FOR 508. ISBN 978-3-540-89955-6

Volume 103: Werner Haase, Marianna Braza, Alistair Revell (eds.): DESider – A European Effort on Hybrid RANS-LES Modelling - Results of the European-Union Funded Project, 2004–2007. ISBN 978-3-540-92772-3

Volume 102: Rolf Radespiel, Cord-Christian Rossow, Benjamin Winfried Brinkmann (eds.): Hermann Schlichting – 100 Years - Scientific Colloquium Celebrating the Anniversary of His Birthday, Braunschweig, Germany 2007. ISBN 978-3-540-95997-7

Volume 101: Egon Krause, Yurii I. Shokin, Michael Resch, Nina Shokina (eds.): Computational Science and High Performance Computing III - The 3rd Russian-German Advanced Research Workshop, Novosibirsk, Russia, 23–27 July 2007. ISBN 978-3-540-69008-5

Volume 100: Ernst Heinrich Hirschel, Egon Krause (eds.): 100 Volumes of 'Notes on Numerical Fluid Mechanics' - 40 Years of Numerical Fluid Mechanics and Aerodynamics in Retrospect. ISBN 978-3-540-70804-9

Volume 99: Burkhard Schulte-Werning, David Thompson, Pierre-Etienne Gautier, Carl Hanson, Brian Hemsworth, James Nelson, Tatsuo Maeda, Paul de Vos (eds.): Noise and Vibration Mitigation for Rail Transportation Systems - Proceedings of the 9th International Workshop on Railway Noise, Munich, Germany, 4–8 September 2007. ISBN 978-3-540-74892-2

Volume 98: Ali Gülhan (ed.): RESPACE – Key Technologies for Reusable Space Systems - Results of a Virtual Institute Programme of the German Helmholtz-Association, 2003–2007. ISBN 978-3-540-77818-9

Volume 97: Shia-Hui Peng, Werner Haase (eds.): Advances in Hybrid RANS-LES Modelling - Papers contributed to the 2007 Symposium of Hybrid RANS-LES Methods, Corfu, Greece, 17–18 June 2007. ISBN 978-3-540-77813-4

Volume 96: C. Tropea, S. Jakirlic, H.-J. Heinemann, R. Henke, H. Hönlinger (eds.): New Results in Numerical and Experimental Fluid Mechanics VI - Contributions to the 15th STAB/DGLR Symposium Darmstadt, Germany, 2006. ISBN 978-3-540-74458-0

Volume 95: R. King (ed.): Active Flow Control - Papers contributed to the Conference "Active Flow Control 2006", Berlin, Germany, September 27 to 29, 2006. ISBN 978-3-540-71438-5

Volume 94: W. Haase, B. Aupoix, U. Bunge, D. Schwamborn (eds.): FLOMANIA - A European Initiative on Flow Physics Modelling - Results of the European-Union funded project 2002 - 2004. ISBN 978-3-540-28786-5

Volume 93: Yu. Shokin, M. Resch, N. Danaev, M. Orunkhanov, N. Shokina (eds.): Advances in High Performance Computing and Computational Sciences - The Ith Khazakh-German Advanced Research Workshop, Almaty, Kazakhstan, September 25 to October 1, 2005. ISBN 978-3-540-33864-2

Volume 92: H.J. Rath, C. Holze, H.-J. Heinemann, R. Henke, H. Hönlinger (eds.): New Results in Numerical and Experimental Fluid Mechanics V - Contributions to the 14th STAB/DGLR Symposium Bremen, Germany 2004. ISBN 978-3-540-33286-2

Volume 91: E. Krause, Yu. Shokin, M. Resch, N. Shokina (eds.): Computational Science and High Performance Computing II - The 2nd Russian-German Advanced Research Workshop, Stuttgart, Germany, March 14 to 16, 2005. ISBN 978-3-540-31767-8

Volume 87: Ch. Breitsamter, B. Laschka, H.-J. Heinemann, R. Hilbig (eds.): New Results in Numerical and Experimental Fluid Mechanics IV. ISBN 978-3-540-20258-5

Volume 86: S. Wagner, M. Kloker, U. Rist (eds.): Recent Results in Laminar-Turbulent Transition - Selected numerical and experimental contributions from the DFG priority programme 'Transition' in Germany. ISBN 978-3-540-40490-3

Volume 85: N.G. Barton, J. Periaux (eds.): Coupling of Fluids, Structures and Waves in Aeronautics - Proceedings of a French-Australian Workshop in Melbourne, Australia 3-6 December 2001. ISBN 978-3-540-40222-0

Volume 83: L. Davidson, D. Cokljat, J. Fröhlich, M.A. Leschziner, C. Mellen, W. Rodi (eds.): LESFOIL: Large Eddy Simulation of Flow around a High Lift Airfoil - Results of the Project LESFOIL supported by the European Union 1998 - 2001. ISBN 978-3-540-00533-9

Volume 82: E.H. Hirschel (ed.): Numerical Flow Simulation III - CNRS-DFG Collaborative Research Programme, Results 2000-2002. ISBN 978-3-540-44130-4

Volume 81: W. Haase, V. Selmin, B. Winzell (eds.): Progress in Computational Flow Structure Interaction - Results of the Project UNSI, supported by the European Union 1998-2000. ISBN 978-3-540-43902-8

Volume 80: E. Stanewsky, J. Delery, J. Fulker, P. de Matteis (eds.): Drag Reduction by Shock and Boundary Layer Control - Results of the Project EUROSHOCK II, supported by the European Union 1996-1999. ISBN 978-3-540-43317-0

Volume 79: B. Schulte-Werning, R. Gregoire, A. Malfatti, G. Matschke (eds.): TRANSAERO - A European Initiative on Transient Aerodynamics for Railway System Optimisation. ISBN 978-3-540-43316-3

Volume 78: M. Hafez, K. Morinishi, J. Periaux (eds.): Computational Fluid Dynamics for the 21st Century. Proceedings of a Symposium Honoring Prof. Satofuka on the Occasion of his 60th Birthday, Kyoto, Japan, 15-17 July 2000. ISBN 978-3-540-42053-8

Volume 77: S. Wagner, U. Rist, H.-J. Heinemann, R. Hilbig (eds.): New Results in Numerical and Experimental Fluid Mechanics III. Contributions to the 12th STAB/DGLR Symposium, Stuttgart, Germany 2000. ISBN 978-3-540-42696-7

GPSR Compliance

The European Union's (EU) General Product Safety Regulation (GPSR) is a set of rules that requires consumer products to be safe and our obligations to ensure this.

If you have any concerns about our products, you can contact us on ProductSafety@springernature.com

In case Publisher is established outside the EU, the EU authorized representative is:

Springer Nature Customer Service Center GmbH
Europaplatz 3
69115 Heidelberg, Germany

Batch number: 10755329

Printed by Printforce, the Netherlands